Die Maintenance Handbook

Die Maintenance Handbook

David A. Smith

Society of
Manufacturing Engineers
Dearborn, Michigan

Association for
Forming & Fabricating
Technologies/SME

987654321

Library of Congress Catalog Card Number: 2001094396
International Standard Book Number: 0-87263-528-7

Additional copies may be obtained by contacting:
Society of Manufacturing Engineers
Customer Service
One SME Drive, P.O. Box 930
Dearborn, Michigan 48121
1-800-733-4763
www.sme.org

SME staff who participated in producing this book:
Cheryl Zupan, Staff Editor
Rosemary Csizmadia, Production Supervisor
Kathye Quirk, Graphic Designer/Cover Design
Frances Kania, Administrative Coordinator
Jon Newberg, Production Editor

Printed in the United States of America

Dedication

I would like to dedicate this book to the many people that made it possible, especially my wife Marlyn Smith who helped with the formatting and proofreading of the original training manuals that were used to produce this work. I would also like to thank members of the production and editorial departments at SME who went the extra mile to give this work polish and continuity. Those individuals include Cheryl Zupan, Bob King, Rosemary Csizmadia, Kathye Quirk, Jon Newberg, and Frances Kania.

About SME

The Society of Manufacturing Engineers (SME), headquartered in Dearborn, Mich., is the world's leading professional society serving the manufacturing industries. Through its publications, expositions, professional development resources, and member programs, SME influences more than 500,000 manufacturing executives, managers, and engineers. Founded in 1932, SME has members in 70 countries and supports a network of hundreds of chapters worldwide.

ABOUT AFFT/SME

The Association for Forming and Fabricating Technologies of the Society of Manufacturing Engineers (AFFT/SME) provides SME members and the manufacturing community with accessible technical knowledge, professional development opportunities, and industry trends. AFFT focuses on the technologies and processes that efficiently make products from metal sheet, coil, plate, tube, or pipe stock. Typical industries served are automotive, off-highway, aerospace, defense, appliance, furniture, and consumer electronics products. Core processes include general pressworking—stamping, drawing, forming, bending, and shearing, for example—as well as the fabricating technologies of punching, cutting, sawing, welding, and others.

Table of Contents

Preface

Developing a popular book on the topic of die repair has been a challenging task. The goal of every shop should be to minimize die maintenance expenditures through good die design. To do this, each of the metal forming, pressworking processes need to be understood from a simple, yet analytical, point of view. Thus, this reference work on die maintenance deals with metal-forming processes in an elementary way, before looking at the bigger picture of cost containment using damage avoidance and robust die designs.

CHOOSING THE CONTENT OF THIS BOOK

A goal of this reference work is to produce a handbook that will make solving stamping-process problems understandable by newcomers to the trade. At the same time, it is intended to serve the experienced journeyman with reference information for solving difficult problems. An essential theme is that die maintenance is a cost that needs to be minimized. Too often, die maintenance is treated as unavoidable and unacceptable, with its very real costs lumped in with other overhead. Success in any critical mechanical repair or defect correction necessitates that required tasks occur correctly and in the proper sequence. This takes both skill and the ability to document results for future reference.

Those who have spent most of a lifetime in the tool and die trade hope for the day when educators will quit teaching that hourly employees in industry have chosen a second-rate alternative to completing a college degree. In reality, many journeymen also hold two-year, technical degrees. Some go on to complete advanced degrees in engineering and business management. Yes, these skilled workers will continue to hold engineering and top management positions in the future.

Most community colleges and trade schools do a good job of teaching the basic principles of computer numerical control (CNC) machining, industrial welding, hydraulics, pneumatics, and electronic technology. However, schools lack hands-on training in the basic theories of die making and metal-forming processes. Hands-on experience needs to occur in a die building and/or repair facility under the guidance of skilled workers.

Apprentices or trainees are not a source of low-cost labor. Die builders and the manufacturers who depend on pressworking tooling in their processes have a duty and an obvious, personal stake to provide the best training to all in the trade, without regard to experience level. Anything less deprives both the craftsperson and employer of the critical skills that are essential to remain innovative and competitive in a world market.

BOOK ORGANIZATION

An essential die-repair skill is the ability to modify existing tooling to correct design mistakes and problems that can affect part quality and proper tool operation. A good understanding of metal-forming processes in pressworking tooling is necessary to maintain and correct die problems.

This book begins with an overview of basic die operations and continues with the anatomy of how these processes work. Troubleshooting, or using insight into the process, is an essential factor in problem solving. Formability theory and practice begin with having basic knowledge of bending and stretching metals through complex, forming states. Essential, analytical methods that are needed to solve problems are also covered. Here, the information is in a simple format, yet it is comprehensive enough to be technically correct from a tool-engineering perspective.

The appropriate use of electronic instrumentation for process protection and optimization is essential in modern pressworking. This ties in with pressworking safety—both of the process and especially of the pressroom personnel.

Troubleshooting of large dies, including casting repair, is based on the experience of co-workers and the author with large, sheet-metal-forming dies. The reader will find that much of the information here is either hard to find, or that it has never been put in print. Topics include solving dirt problems in automotive, outer-skin-panel, forming operations, as well as repair methods for broken cast-iron die shoes. This unique text is meant to aid those faced with these repair and troubleshooting problems. Tool engineering, including problem solving, is not a black art. It is a discipline based mainly on mechanical engineering and material science.

The material science behind die materials, heat treatment, and tool coatings is objective and based on expert review. Objective information is a problem area because of large amounts of conflicting information and negative comparisons prevalent today in trade advertising.

This book offers detailed procedures for establishing a die maintenance program. Problem tracking and cost control to start or improve a system are explained. Examples of die standards are generic and recast subject to applicable law and good engineering practice.

It is hoped that the individuals, organizations, and firms who have contributed to this work will recognize the benefits gained from cooperation and mutual assistance that extend far beyond any financial recompense. By donating time, talent, and design examples, contributors have provided a wealth of practical design and troubleshooting information. Their help has made this work possible.

Contributors

The author wishes to acknowledge the following individuals for their contributions to this book:

Taylan Altan, Ph.D.
Professor and Director
ERC/NSM
Ohio State University
Columbus, Ohio

Gary Amin
Supervisor
Die Design and Standards
Ford Motor Company
Dearborn, Mich.

Jim Austin
Director of Continuous Improvement
Michigan Seat Company
Jackson, Mich.

James Barrett, Jr., Ph.D.
President
Link Systems
Nashville, Tenn.

John Bates
President and CEO
Heidtman Steel Products, Inc.
Toledo, Ohio

Lisa Bisson
Editor
Eagle Publications
Novi, Mich.

Rollin Bondar
President
MPD Welding, Inc.
Orion, Mich.

Arthur Brown
Publisher
Eagle Publications
Novi, Mich.

Douglas Burke
Forms Consultant
Superior Business Solutions
Kalamazoo, Mich.

Bill Christiansen
Senior Process Engineer
Jostens, Inc.
Princeton, Ill.

Daniel N. Falcone
Vice President
Toledo Transducers, Inc.
Holland, Ohio

Kevin M. Fessler
Electrical Maintenance Consultant
Centerville, Ind.

Chuck Fitch
Director of Fluorescent Operations
Cooper Industries
Americus, Ga.

Jeff Fredline
President
Fredline & Associates
Okemos, Mich.

Henry H. Gehlmann
Blanking Area Manager (retired)
Ford Motor Company
Woodhaven, Mich.

Leo Goepfrich
Tool and Die Manager (retired)
Ford Motor Company
Chicago Heights, Ill.

Bob Green
Engineering & Facilities Manager
Cooper Industries
Americus, Ga.

Roger P. Harrison
Director of Training
Rockford Systems, Inc.
Rockford, Ill.

Bill Hays
District Maintenance Supervisor
Governor's Office of Technology
Frankfort, Ky.

John Heffelfinger
Manufacturing Manager
Norlen, Inc.
Schofield, Wisc.

James Hicks
Professor of Chemistry
Cumberland College
Williamsburg, Ky.

Robert Hurst
President
Phoenix Specialty Manufacturing Company, Inc.
Bamberg, S.C.

Casimir S. Janiszewski
President and CEO
Superior Die Set Corporation
Oak Creek, Wisc.

Stuart P. Keeler, Ph.D.
President
Keeltech Technologies
Grosse Ile, Mich.

Karl A. Keyes
President
Feinblanking Ltd.
Fairfield, Ohio

Gregory A. Kreps
Tooling Engineer
Capital Die, Tool and Machine Company
Columbus, Ohio

James L. Lehner
Applications Manager
HMS Products Company
Troy, Mich.

Ernie Levine
President and CEO
Admiral Tool and Manufacturing
Chicago, Ill.

Cecil Lewis
Corporate Vice President
Midway Products Group, Inc.
Monroe, Mich.

Albert A. Manduzzi
Supervisor (retired)
Die Design and Standards
Ford Motor Company
Dearborn, Mich.

John McCurdy
President (retired)
W. C. McCurdy Company
Oxford, Mich.

Bradley K. Mettert
Field Engineering Manager
Toledo Transducers, Inc.
Holland, Ohio

Carl Meyer
President
Progressive Tool Company
Waterloo, Iowa

Harry J. Micka
Manufacturing Engineer
LA-Z-BOY Chair Company
Dayton, Tenn.

Richard Micka
Vice President of Administration
LA-Z-BOY Chair Company
Monroe, Mich.

Arnold Miedema
President
Green Meadows Engineering
Hudsonville, Mich.

Ernest E. Morgan
Consulting Structural Engineer
Henderson, Ky.

Eugene J. Narbut
Die Room Unit Leader
Auto Alliance International
Flat Rock, Mich.

Al Nichols
Area Leader
Stamping Engineering
Auto Alliance International
Flat Rock, Mich.

Mark Nichols
Corporate Safety Director
LA-Z-BOY Chair Co.
Maplewood, Minn.

Michael F. Ollie, Jr.
President
Henry & Wright Corporation
Fairforest, S.C.

Pat Ontrop
Applications Engineer
Minster Machine Company
Minster, Ohio

Dick Peoples
Metallurgist
Heidtman Steel Products, Inc.
Toledo, Ohio

Jim Pienta
Production Manager
Jostens, Inc.
Princeton, Ill.

Gerald A. Pool
Consultant
Royal Oak, Mich.

Ronald E. Prater
President, Koppy Corporation
Orion, Mich.

Frank Randall
Engineering Manager (retired)
Zippo Manufacturing Company
Bradford, Pa.

Anthony Rante
Director of Engineering
Fitzpatrick Company
Elmhurst, Ill.

Jim Ricketts
Production Manager
LA-Z-BOY Chair Co.
Dayton, Tenn.

Mike Riley
Mechanical Engineer, P.E.
Consultant & Writer
Pinson, Ala.

Brad Sauer
Industrial Engineer
LA-Z-BOY Chair Co.
Monroe, Mich.

Daniel A. Schoch
Mechanical Engineer, P.E.
Minster Machine Company
Minster, Ohio

Mike Searson
Machine Shop Supervisor
Phoenix Specialty Manufacturing Company, Inc.
Bamberg, S.C.

Aniese Seed
President (retired)
Toledo Transducers, Inc.
Holland, Ohio

Kenneth L. Smedberg
President
Smedberg Machine Corporation
Chicago, Ill.

Andrew D. Smith
Mechanical Engineer, T. R. W
Livonia, Mich.

Gary D. Smotherman
International Representative (retired)
U. A.W.
Detroit, Mich.

Fredric Spurck
President
Webster Industries, Inc.
Tiffin, Ohio

Robert Stadler
Tool Engineer
Ranger Tool & Die
Grand Rapids, Mich.

Allan Steinbock
Vice President
Superbolt, Inc.
Carnegie, Pa.

Robert Storer
President
Toledo Transducers, Inc.
Holland, Ohio

Leslie Suthers
Manager
Manufacturing Services
Norlen, Inc.
Schofield, Wisc.

Thomas A. Suthers
President
Norlen, Inc.
Schofield, Wisc.

Nick Tarkany
Director
Research & Technical Education
Dayton Progress Corp.
Dayton, Ohio

Ed Tarney
Director
CPM Development
Crucible Materials Corporation
Camillus, N.Y.

Jack Thompson
President
Penda Corporation
Portage, Wisc.

Ed Tremblay
Vice President, Engineering and Manufacturing
P/A Industries
Bloomfield, Conn.

Robert Vance
Consulting Marine Engineer
Glasgow, Ky.

Bob Wahls
Chief Engineer
Michigan Seat Company
Jackson, Mich.

Keith Weaver
Press Engineer
Capital Die, Tool and Machine Company
Columbus, Ohio

Joe Wise
Maintenance Supervisor
H. J. Heinz Company
Freemont, Ohio

Wayne Wittaker
Vice President of Operations
Unisorb Machinery Installation Systems
Jackson, Mich.

Greg Yirga
Metal Department Manager
LA-Z-BOY Chair Co.
Dayton, Tenn.

1

An **Overview** of Basic Die Operations

Basic die operations are divided into several categories such as cutting, bending, forming, drawing, and squeezing. All of these operations require that portions of the workpiece undergo plastic deformation. The yield point of the material must be exceeded to accomplish this.

Understanding how these processes are appropriately applied is key to designing and maintaining dies for successful pressworking. Knowledge of material and die interactions is also required for cost-effective design, repair, and troubleshooting.

In a successful stamping enterprise, the goal is to be the lowest-cost producer, with the best quality. Every effort must be made to use the best processes, appropriate tooling designs, and efficient die maintenance.

CUTTING, PUNCHING, AND PERFORATING

Cutting operations are essentially controlled processes of *plastic deformation*, or yielding of the material leading to fracture. Both tensile and compressive strains are involved. Bending or stretching of the scrap portion, and occasionally the work, may also occur. Figure 1-1 shows a sectional view of a simple cutting die for punching a round hole.

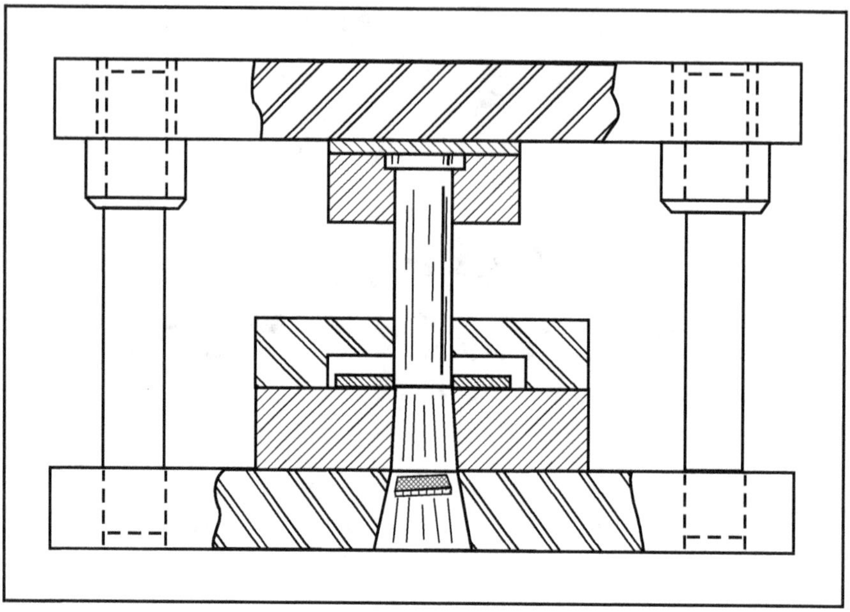

Figure 1-1. Sectional view of a simple cutting die for punching a round hole.

BENDING AND FORMING

Simple *air bending* involves both tensile and compressive straining of the material in the bend zone. Upon release from the die, residual stresses result in springback of the material. *Springback* is a result of the partial release of residual stresses in the material.

To lessen the effect of residual stresses, the metal in the bend may be squeezed or *coined*. In this process, the entire thickness of the material in the bend is subjected to compressive forces above the yield point of the material.

Another method of compensating for springback is to overbend the material, so that when it springs back, the correct amount of bend is achieved. This is done easily with simple air bending tooling.

Figure 1-2 illustrates an example of a part made by bending in a die or with press brake tooling. Another process used to form such shapes is *roll forming*, where a flat strip is passed through a series of forming rolls to produce the desired shape.

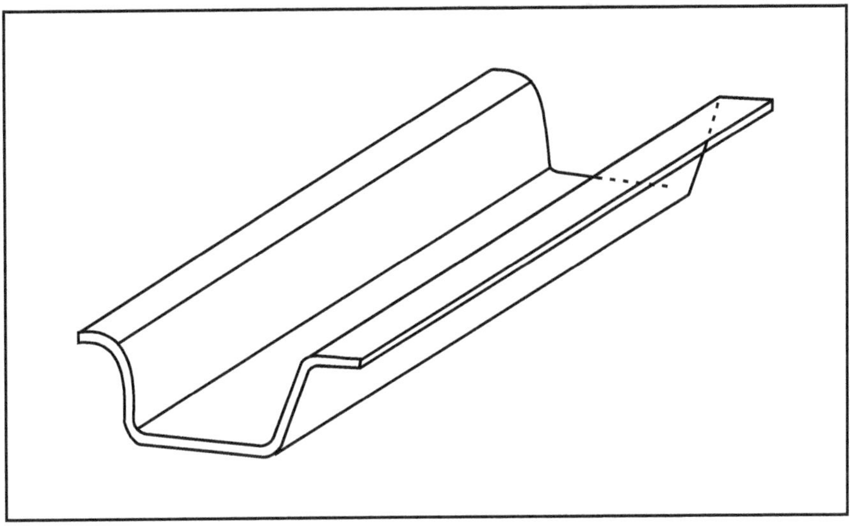

Figure 1-2. A simple part produced by a bending or forming process. The part can be produced easily in small quantities on a press brake. High-volume production can prove more economical when it is done by die or roll forming.

Roll forming is normally a continuous process. The desired length of the part is cut off with a die in a shearing device that travels on a carriage at the same linear velocity as the forming process. This device is termed a *flying shear*.

BENDING AND STRAIGHTENING

A simple type of die employed for both bending and straightening is used to make U-shaped cross sections with right-angle flanges, as shown in Figure 1-3. Parts of this type are used as stiffeners attached to flat panels in many applications. Examples are automotive body frame rails, and cross members assembled by welding.

Bending and straightening operations, in conjunction with other forming processes, are often used to produce complex stampings. A simple example of bending and straightening occurs in the sidewalls of formed, box-shaped, rectangular shells.

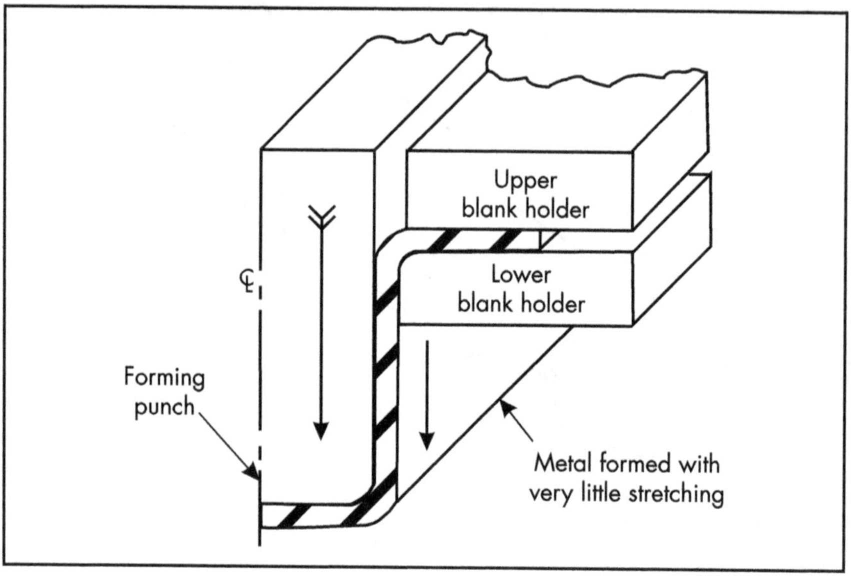

Figure 1-3. A bend-and-straighten operation in which the metal is bent and straightened in the die, permitting very large deformations with little or no thickness change.

STRETCHING

Another process used to form metal is stretching. Forming processes involving cold working, such as stretching, often improve the mechanical properties of the material. Familiar products made by stretching include many automotive and aircraft skin panels.

If the metal is deformed in only one direction, it is called *plane strain* deformation. This process permits less deformation before a fracture occurs than stretching the metal in two directions.

Figure 1-4 illustrates an example of plane-strain stretching, where the metal is clamped tightly by a lock bead in the blank holder. Forming metal by plane-strain stretching rapidly thins the material. If the thinning is severe, a fracture may occur.

Biaxial Stretch Forming

If the blank is tightly restrained by a blank holder, as shown in Figure 1-5, and stretching of the width and length of the part

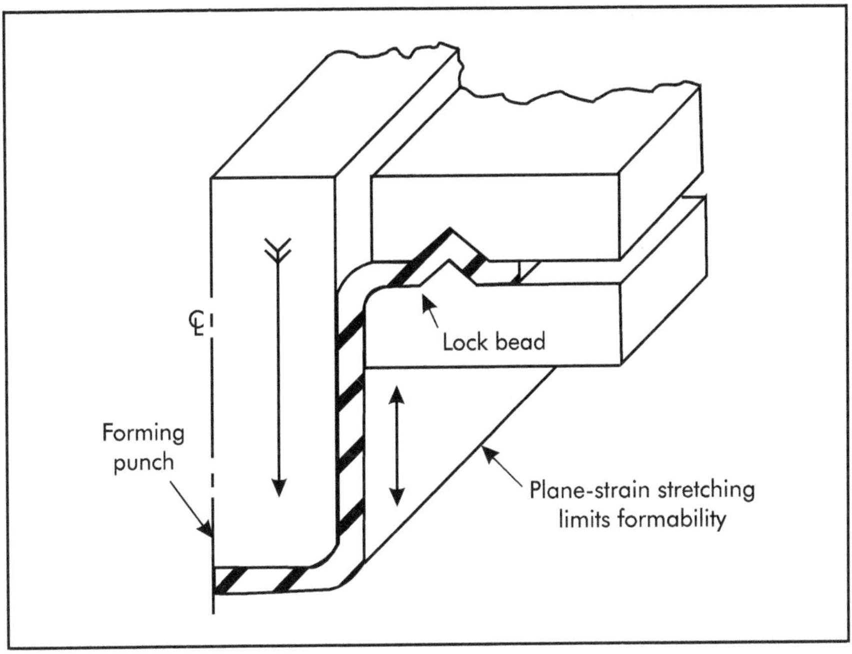

Figure 1-4. Plane-strain stretching or deformation.

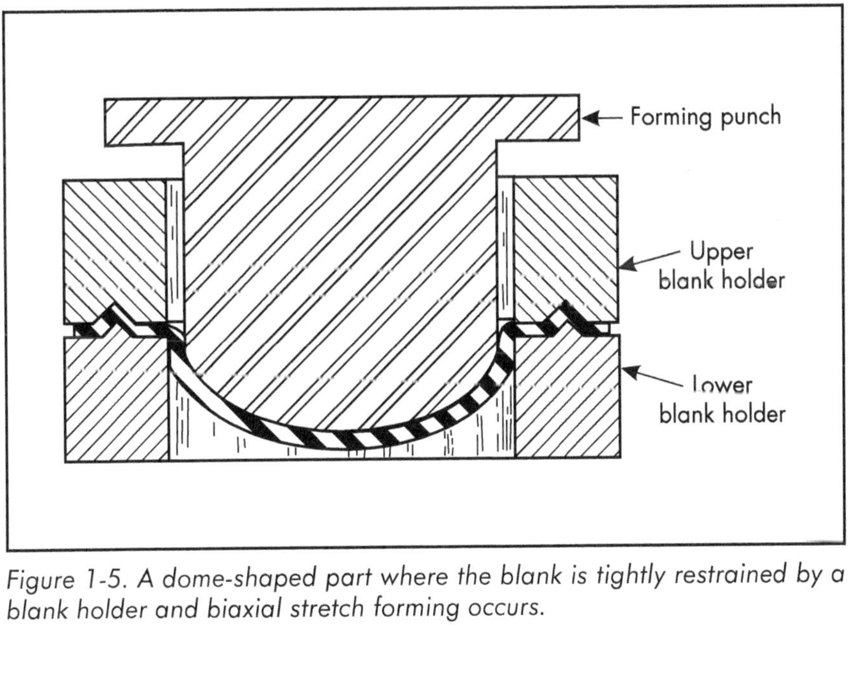

Figure 1-5. A dome-shaped part where the blank is tightly restrained by a blank holder and biaxial stretch forming occurs.

occurs, the process is called *biaxial stretch forming*. Biaxial stretch forming permits greater deformation than plane-strain stretching.

DRAWING

Popular shop terminology often applies the term *draw die* to any die where metal is stretched or drawn into the die cavity. Many such operations form the part by stretching or bending, rather than drawing. *Round cup drawing*, where the metal is restrained by a blank holder as it is drawn into the die cavity by a draw punch, is an example of a simple true drawing operation (Smith 1990).

A large body of empirical data describes the known practical limits of drawing when using conventional tooling. Large deformations may involve multiple redrawing operations.

True Cup Drawing

In cup drawing (see Figure 1-6), the metal thickens or is subject to circumferential compression while restrained by the blank holder as it is drawn into the die cavity. As the metal is drawn over the die radius, it is thinned. Fracture of the metal due to thinning limits the severity of this process.

SQUEEZING OR COINING OPERATIONS

Coining is a process for cold-forming metals that is similar to cold forging. In *coining*, the metal is shaped between two dies. It is shaped in a manner such that relief depressions in the die surfaces are filled by displacing the metal in a plastic state. Coining involves cold-working deformation of the metal well below its recrystallization temperature. This results in strain hardening.

Squeezing operations, where the thickness dimension of the metal is subjected to pressures that exceed the material-yield point to change its thickness or shape, are often called *coining*. This term is based on the most familiar example: the minting of coins (see Figure 1-7).

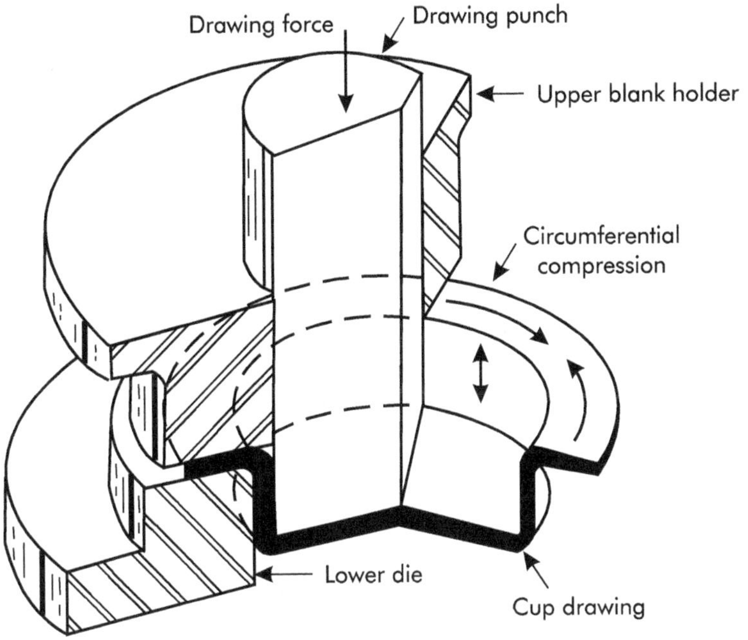

Figure 1-6. A simple cup drawing die.

Squeezing operations may also be called cold forging, depending on the purpose and severity of the operation. Based on the material and the operation, in-die pressures can exceed the yield strength of the material by a factor of two or more. It is important to use parameters such as hardness and area of the metal at the completion of coining when designing dies and determining press requirements.

If metal is not constrained in a die cavity, lateral movement can occur, increasing the area being cold worked. Lateral stock movement in coining dies with fine detail or relief subjects any projections to shearing forces and metal fatigue problems. This can lead to crack propagation. This tool failure mode occurs frequently in coining dies.

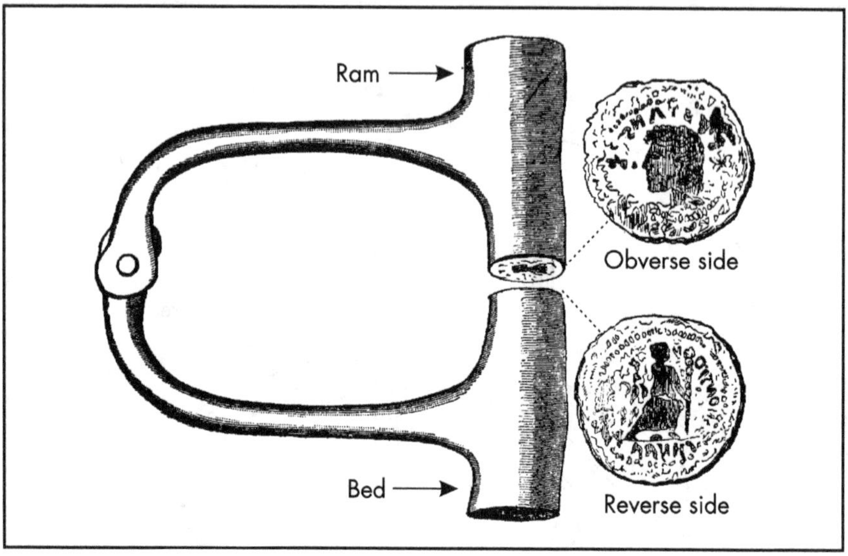

Figure 1-7. An ancient hinged coining press for minting coins (Smith 1896).

ANALYTICAL TOOLS

An objective understanding of the behavior of metal undergoing cutting, shaping, and forming processes is key to modern analytical techniques. Today, we have the ability to design easily manufactured metal products and solve many problems.

The analytical tools include circle grid analysis (CGA) and the forming limit diagram (FLD). Computer-aided formability analysis is also used to determine if stampings can be successfully manufactured.

The large amount of data based on proven die designs and metal formability has taken troubleshooting pressworking operations from a trial-and-error process to an engineering science.

REFERENCES

Smith, D. 1990. *Die Design Handbook, Third Edition.* Section 10, Displacement of Metal in Drawing. Section 11, Product Development for Deep Drawing. Dearborn, MI: Society of Manufacturing Engineers.

Smith, Oberlin. 1896. *Press Working of Metals*. New York: John Wiley & Sons, Inc.

BIBLIOGRAPHY

Keeler, S. 1986. *Circle Grid Analysis*. Livonia, MI: National Steel Corporation Product Application Center.

Sheet Steel Formability. August 1, 1984. Washington, DC: The Committee of Sheet Steel Producers, American Iron and Steel Institute.

Tharrett, M. August 25-27, 1987. *Computer-Aided Formability Analysis*. Dearborn, MI: SME Die and Pressworking Tooling Clinic.

2

Troubleshooting Punching and Cutting Operations

Cutting, which includes shearing, is the most common press-working operation. A single-formed stamping, such as a sieve or an automobile inner door, has many holes that are produced by cutting. Perforating occurs in the case of a sieve, where the holes are in a regular pattern. The use of the same cutting process applies to producing a single hole in either the sieve or the inner door of the automobile.

Repair and troubleshooting of cutting dies are both very important skills. The work varies from maintaining simple hole-punching dies, to maintaining complex, progressive and transfer dies that perform many metal-forming operations, in addition to cutting and shearing.

Effective die maintenance starts with the designer. If designed correctly, a die will only require infrequent, simple repairs. This chapter starts with the basics that all experienced die makers, designers, and engineers may already understand. The material included here is intended to serve as a reference tool for all workers, in addition to those just starting in the trade.

UNDERSTANDING THE SEQUENCE OF OPERATIONS

Figure 2-1a illustrates a sectional view of a simple die for punching a hole in a part. Such dies may have several punches. In addition, if the die is fed with strip or coil stock, a cut-off shearing

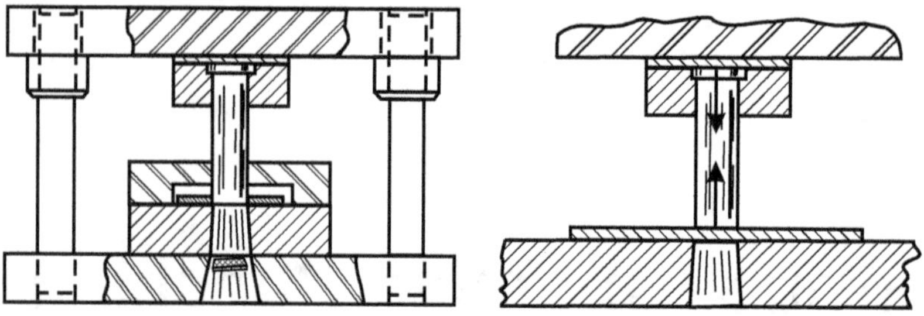

Figure 2-1. (a) Sectional view of a cutting die for producing a round hole is shown; (b) the punch is compressed after making initial contact with the stock.

operation may be included. The punch is fastened to the upper die shoe with a retainer that has a hardened backing plate. A slug falling through the lower die shoe can be seen. On the upstroke of the press, the stock is stripped from the punch by a simple, fixed, or tunnel stripper.

As the press and die close, the punch or punches first make contact with the stock. Exactly what occurs after the moment of the punch's initial contact on the stock varies. This variation is based on the following factors:

- the speed and mass of the press ram or slide;
- the thickness and hardness of the stock; and
- the force required to initiate a yielding of the stock.

Figure 2-1b illustrates a punch making initial contact with the stock. As the die closes, a compressive strain stores energy within the punch body. This is like the compression in a spring.

Cutting operations are a controlled process of plastic deformation, or a yielding of the material leading to fracture. Both tensile and compressive strains are involved. Bending, or distortion of scrap metal, is trimmed away, and slugs may occur.

Figure 2-2a shows compressive strain developing in the punch as the die closes. Once the force developed by the closure of the press is transmitted through the punch, shown in Figure 2-2b, and equals the yield point of the stock, plastic deformation starts occurring as seen in Figure 2-2c.

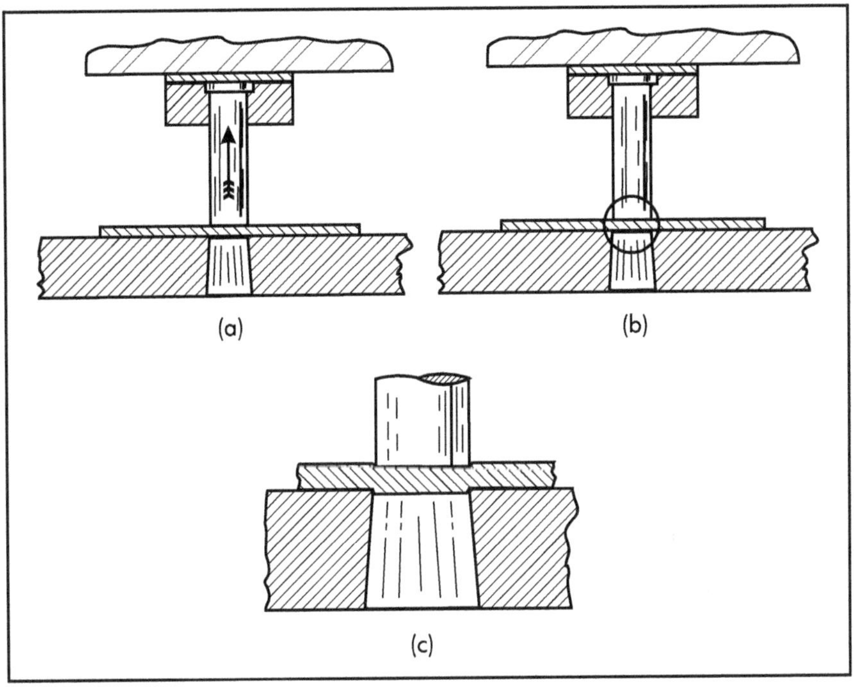

Figure 2-2. (a) Compressive strain in the punch develops as the die closes; (b) the force caused by the press closure is transmitted through the punch; (c) once the force equals the yield point of the stock, plastic deformation starts occurring.

In a simplified sequence of operations, as the press closes, the punch remains in contact with the stock until the force transmitted through the punch is great enough to exceed the shear strength of the material. The material continues to yield (see Figure 2-3a), until complete fracture occurs. Next, the compressive strain in the punch releases the slug, and it is completely pushed into the die opening (see Figure 2-3b) at the bottom of the press stroke.

Most slugs fall into the die opening by gravity as shown in Figure 2-4a. If this does not occur, there are a number of ways for the die designer and die maker to assure proper slug discharge. The final step is the withdrawal of the punch from the stock on the press upstroke (see Figure 2-4b).

Figure 2-1a shows a tunnel or fixed stripper. Tunnel strippers have the advantage of simplicity and are adequate for many applications.

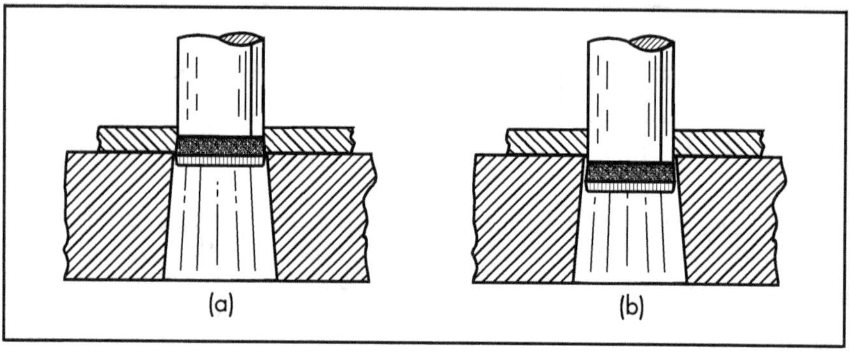

Figure 2-3. (a) The stock continues to yield until a complete fracture occurs; (b) the slug is pushed into the die opening at the bottom of the press stroke.

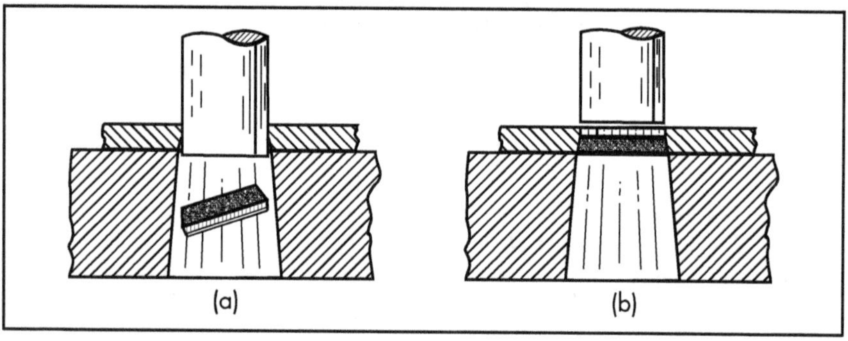

Figure 2-4. (a) The slug falls into the die opening at the bottom of a stroke; (b) the punch is withdrawn.

However, spring-loaded strippers are often required for precise stock control throughout the cutting process.

Roughness of Fracture

Plastic deformation of the material occurs throughout the cutting process. Cutting involves a controlled failure of the material. The fractured portion of the edge will be somewhat rough, due to the tearing action that occurs. In thick materials, this roughness may be quite pronounced. Applying the term "fracture" to metal-cutting operations can be easily misunderstood. The fracture is seldom a sudden parting of the material.

For example, when cutting low-carbon steels, such as SAE-AISI 1005-1008, fracture occurs from plastic deformation of the material wherein the stock is torn apart. Cutting involves a controlled, tensile failure of the material.

EFFECT OF DIE CLEARANCE

There are no absolute rules governing the amount of clearance between the punch and die. This clearance is expressed as a percentage of stock thickness, per side. For mild steel, the clearance per side varies between 5–12% of the stock thickness. In general, tight clearances will result in holes having a high ratio of shear or burnish to fracture. This will also result in less taper at the expense of accelerated tooling wear.

Figure 2-5a illustrates a sectional view of a punch, die stock, and fractured slug where normal clearance is being used. Normal clearance typically results in one-third burnish and two-thirds fracture.

Excessive clearance (see Figure 2-5b) may result in die roll at the point of punch entry and a large burr on the underside of the part. Here, there is very little shear or burnish. The fracture is rough and it has a large taper on the cut edge.

The fracture, which starts from each side, may not meet evenly. One or more sharp projections may result. Depending on the material being cut, optimizing the amount of die clearance may require some experimentation to minimize an uneven, fractured-edge condition. By increasing the die clearance within reasonable limits, cutting pressure is lowered and tool life can be extended. A limiting factor is the amount of taper permitted in the hole and the allowable burr height.

Insufficient Clearance

Double breakage, shown in Figure 2-5c, occurs when too little clearance is used. The fracture, which starts from each side, may not meet evenly, and one or more sharp projections may result. Optimizing the amount of die clearance to suit the material cut may be required to minimize an uneven, fractured-edge condition.

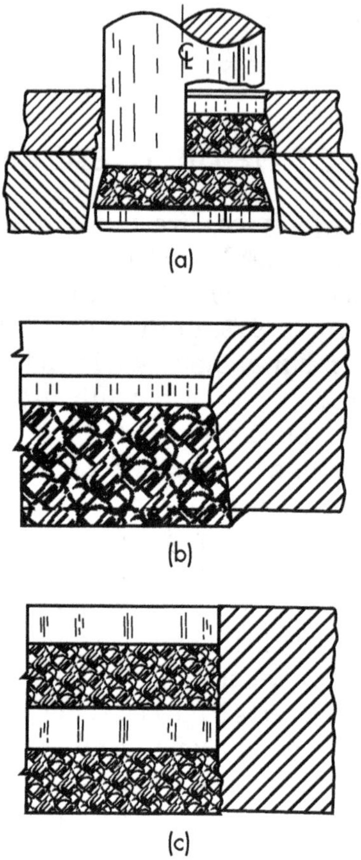

Figure 2-5. (a) Normal clearance results in approximately one-third shear; (b) very large clearance results in excessive die-roll and burr height, and a large, tapered fracture; (c) insufficient clearance may result in double breakage.

Figures 2-6a and 2-6b show the progression of a fracture that does not meet due to insufficient clearance. Figure 2-7b shows a cut edge with double breakage.

Essential indicators for the die maker to look for are secondary, shiny areas on the inside of the hole and/or slug, and a rough, torn fracture. The die may burnish the torn peaks of the fracture.

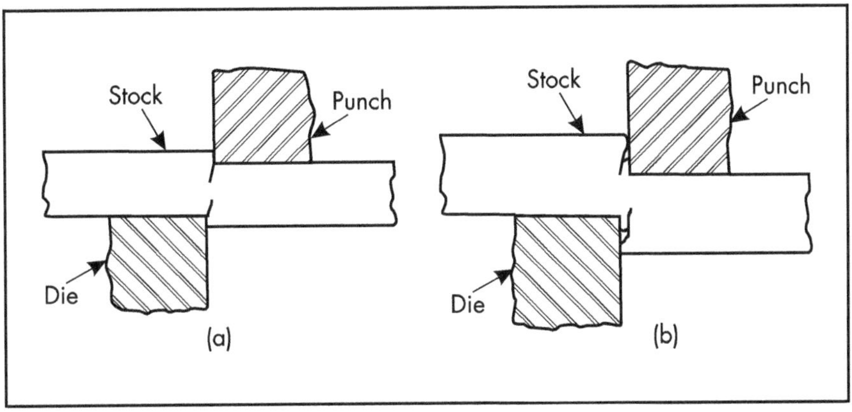

Figure 2-6. (a) Cutting is illustrated with insufficient clearance; (b) as fracture continues, rough tearing occurs and the fracture paths do not meet.

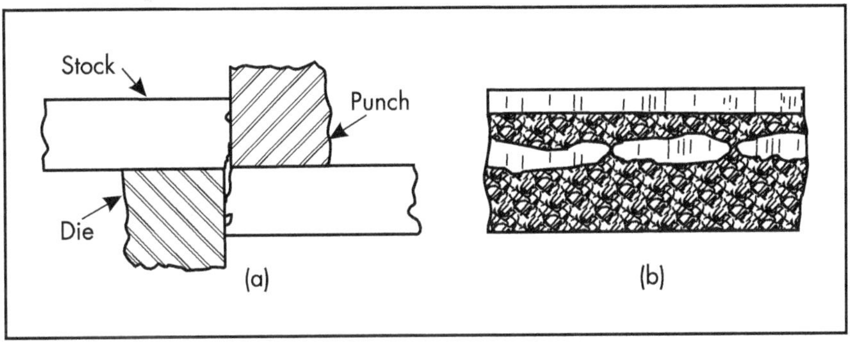

Figure 2-7. (a) Insufficient clearance results in a double fracture or breakage condition; (b) the appearance of a cut edge shows a double-breakage condition.

Double breakage can be resolved by increasing the punch-to-die clearance as shown in Figures 2-8a and 2-8b. Making the punch smaller or the die opening larger is frequently the solution. This will change the size of the hole and/or the size of the part. There will be an increase in the taper of the fracture.

Within reasonable limits, increasing the amount of clearance between the punch and die will decrease the required cutting force as shown in Figure 2-9a. Lowering the cutting force will usually increase the number of parts produced before the tool requires sharpening.

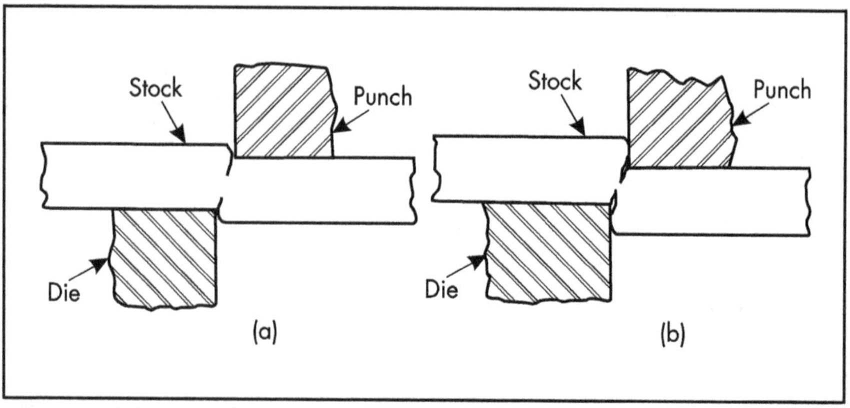

Figure 2-8. (a) Punch penetration and fracture starts with a relatively large, punch-to-die clearance; (b) the fracture paths meet evenly.

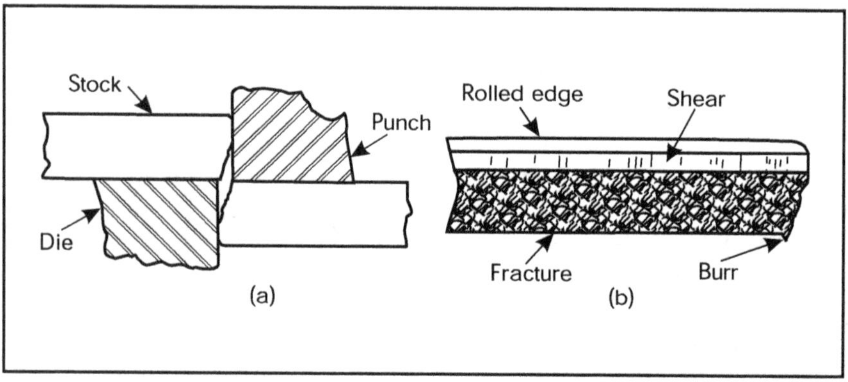

Figure 2-9. (a) This illustrates the completion of punch penetration and fracture with a generous punch-to-die clearance; (b) shown is a view of a cut edge with large punch-to-die clearance.

Clearances of 12–15%, per-side, may be required to eliminate double breakage problems in soft steels. The cut edges may have a pronounced taper in their fractured portions. Die roll and burr height also may be more pronounced. These edge conditions, shown in Figure 2-9b, may be acceptable for many applications.

If the clearance is very large, higher than normal forces may be required. The cut edges may have a very pronounced die-roll, taper, and burr height. Very large clearances also result in high lateral forces on the punch and die that can shorten tool life.

Correct Clearance

There are no absolute rules governing correct die clearances. Generally, tight clearances in the 3–5% per side range result in:

- parts with less taper on the cut edge;
- less tendency for the slug to be pulled from the die opening;
- higher cutting forces; and
- a tendency to have double-breakage problems, especially in thick materials.

As larger clearances in the 7–14% ranges are used, the results are often:

- a longer punch and die life between resharpening;
- a need to prevent slug pulling;
- lower cutting forces;
- an avoidance of double breakage; and
- greater edge taper and burr height.

Case Study Example

This case study looks at how the company, Western Electric, optimized die clearance to produce coil cores for telephone relays. It looks at a high-volume operation requiring many identical, progressive dies punching thick, silicon-steel. The hard, abrasive nature of this material resulted in short runs, even with progressive die design/construction. The punches and die sections for these tools used a combination of air-hard tool steel, tungsten high-speed tool steel, and cemented tungsten carbide.

The normal clearance for a new tool is 5% per side, with a 0.25°-die taper, per side. The die openings have a straight land of approximately 0.040 in. (1 mm) to increase die life.

Production requirements were very high. Several identical dies were in use at all times to meet production demands. The tool was pulled and replaced frequently with a sharpened tool, thus avoiding part defects. Using this method, a minimum amount of material was needed to be ground away to sharpen the tool.

It should be noted that careful inspection of the die edges, conducted under high magnification with a toolmaker's microscope, measured the amount of die height to grind off. Workers entered the die-height amount to sharpen the tool on the work order. New dies typically produce 60,000–75,000 parts before being pulled for sharpening.

Strategy to Maximize Tool Life

The goal at Western Electric was to provide trouble-free service at the lowest possible cost. The die changeover time routinely required less than 10 minutes. This fast changeover was comparable to the norm established several decades before with the single-minute exchange of dies in Japan. *Single-minute exchange of dies* refers to changeover expressed in a single digit or in less than 10 minutes. This required maintaining constant shut heights, providing positive die locators, and having rapid die-fastening devices.

The key to minimizing die cost per part was to sharpen the tool before excessive burr height became a problem. Attempting to get several thousand more parts out of a dull tool usually would mean the difference between grinding 0.005 in. (0.13 mm) versus 0.015 in. (0.38 mm) of tool steel to obtain a sharp edge.

Tool Life Results

Excessively tight clearances result in high cutting forces. Another effect is extreme pressure on the cutting edges, which shortens tool life. It was found that as the Western Electric relay dies developed greater clearance, due to grinding into the 0.25° per side die taper, the cutting forces decreased, and the allowable runs became greater.

As a set of die sections approached the end of their useful life, the clearance would be more than 8%. At this clearance, the allowable runs typically increased from 60,000–75,000 parts, to over 130,000. When a part's dimensional tolerances and burr heights dictated rebuilding the die, runs were often as high as 150,000 pieces.

CUTTING FORCES

The pressure required for cutting through stock increases with the ultimate tensile strength of a material. Die cutting requires less energy than parting metal by tensile failure. There is no absolute relationship between tensile strength and shear strength. Generally, shear strength is between 60–80% of the ultimate tensile strength.

Generally, using the ultimate tensile strength to calculate the cutting forces provides a substantial safety factor. For example, AISI-SAE 1010 cold-rolled steel has an approximate ultimate tensile strength of 56,000 psi (386 MPa) and shear strength of 42,000 psi (290 MPa).

The shear strength of the material increases due to the fast strain rates encountered in high-speed pressworking. The ultimate tensile strength used to calculate the required force allows for a margin of safety.

Calculating Cutting Force Requirements

Systems of Length, Area, and Force Measurement

In North America, shops still carry out engineering calculations for stampings using measurements for length and thickness based on the inch. Shear or yield strengths are based on pounds per square inch (psi). Usually, the press force is in short tons, based on 2,000 pounds (lb).

The metric system is in standard use throughout most of the world. Metric linear and area measurements are the meter, centimeter, and millimeter. Pressworking forces in metric tons, based on 1,000 kilograms (kg), are common in Asia. Most of the metric world uses the kilo-Newton (kN) or mega-Newton (MN). The preferred metric unit for material strength is the kilo-Pascal (kPa) or mega-pascal (MPa).

The die designer or engineer *must* calculate force requirements to determine the required size and type of press. An assumption that the pressroom can somehow fit a new job into an existing press can be a foolish blunder.

Length of cut, material thickness, and shear strength are calculation entry items. Some computer aided design (CAD) programs can automatically calculate tonnage.

Determining Length of Cut

To calculate the force required for cutting or shearing materials, the actual measurement of the total-cut length is required. The part-print dimensions provide a starting point. For progressive die work, all pilot holes and all work done to cut the carrier strip must be included.

Determining Theoretical Peak Cutting Force

To determine the theoretical peak cutting force, multiply the total length of cut by the stock thickness to obtain the area of material cut. Then, multiply the total area of cut by the shear strength of the material. Equation 2-1 is useful for calculating cutting forces. Making no allowance for shear angles or timing of entry provides a safety factor. The formula is:

$$F_s = L \times t \times S_s \tag{2-1}$$

where:

F_s = force required to shear (lbf [kN])
L = length of cut (in. [mm])
t = thickness of material (in. [mm])
S_s = shear strength of the material as defined by ASTM tests (psi [MPa])

Determining Cutting Energy Requirements

The energy required to cut through metal is often surprisingly small, especially when compared to processes such as sawing, oxy-fuel gas, and laser cutting. This is because the actual shearing of the stock occurs when the punch or cutting steel penetrates the material at approximately one-third of the stock thickness. The following equation is used to determine cutting energy requirements:

$$E = F \times D \tag{2-2}$$

where:

E = energy (ft-lb [J])
F = cutting force (lbf [kN])
D = distance sheared (ft [cm])

Working in inch-tons may be more convenient. An inch-ton equals 166.67 ft-lb. For example, adding a safety factor, the shear strength of mild steel is approximately 50,000 psi or 25 short tons per square inch (345 MPa). Cutting a 12-in. (304.8-mm) diameter round blank 0.1875-in. (4.763-mm) thick requires the punch to penetrate only one-third of the material's thickness or 0.0625 in. (1.588 mm) before fracture occurs.

The total area sheared is 7.0686 in.^{2} (45.606 cm^2), requiring a force of 176.72 tons (1,572 kN). However, this force only acts through a distance of approximately 0.0625 in. (1.588 mm). The required energy is 11.04 in.-tons or 1,840.8 ft-lb (2,495 joules [J] or watt-seconds). Operating the press at 60 strokes per minute (SPM) produces one blank per second. Ignoring frictional losses, a motor output of only 3.347 horsepower (hp) (2.497 kW) is required to restore energy to the flywheel.

Effect of Cutting Speed on Force Requirements

In high-speed pressworking, the force required might have higher measured values than expected, based on the shear strength of the material. This is because the punches shear the stock more rapidly and the strain rate of the material increases. In high-speed perforating operations, the measured cutting forces may approach the ultimate material strength.

Side-thrust or Lateral Forces

Figure 2-10a illustrates the shear, tensile, and compressive forces that occur during the cutting process. The amount of lateral force varies with the cutting clearance and material.

For round and symmetrical holes, the lateral forces balance out. However, the die must be sufficiently strong to withstand high spreading forces. For notching, shearing, and other unbalanced operations, the alignment system of the die must not allow excessive deflections to occur.

Side-thrust or lateral force can result in excessive deflections of die components, such as punches, heel blocks, and guide pins. As lateral deflection occurs, clearances increase.

The lateral pressure can exceed the press force by a factor of three or more, due to a wedge-like, mechanical advantage. If not carefully controlled, the resulting misalignment can damage the tooling and produce scrap parts.

Equation 2-3 gives an approximation of the side thrust or lateral force generated when cutting or shearing. When applying this equation, adjustments for the type of material and die conditions need to be made. Figure 2-10a shows that cutting operations involve some bending. The graph in Figure 2-10b is based on Equation 2-3 (Smith 1990).

$$\frac{C}{T-P}=\frac{F_H}{F_V} \tag{2-3}$$

where:

C = clearance (in. [mm])
F_H = side thrust (lbf [kN])
T = material thickness (in. [mm])
P = penetration, typically $0.33 \times T$
F_V = cutting force (lbf [kN])

Effect of Die Clearance

Excessive clearances and dull die steels can result in very large side thrusts. In extreme cases, the side thrust or lateral force may be so great that the die may shatter due to extreme pressures and interference of die components.

It is the function of die alignment system components, such as guide pins and heel blocks, to limit deflections due to side thrusts to acceptable values. Figure 2-10c illustrates how side thrusts can cause punch deflection. The punch deflection increases the die clearance, thus leading to greater lateral forces.

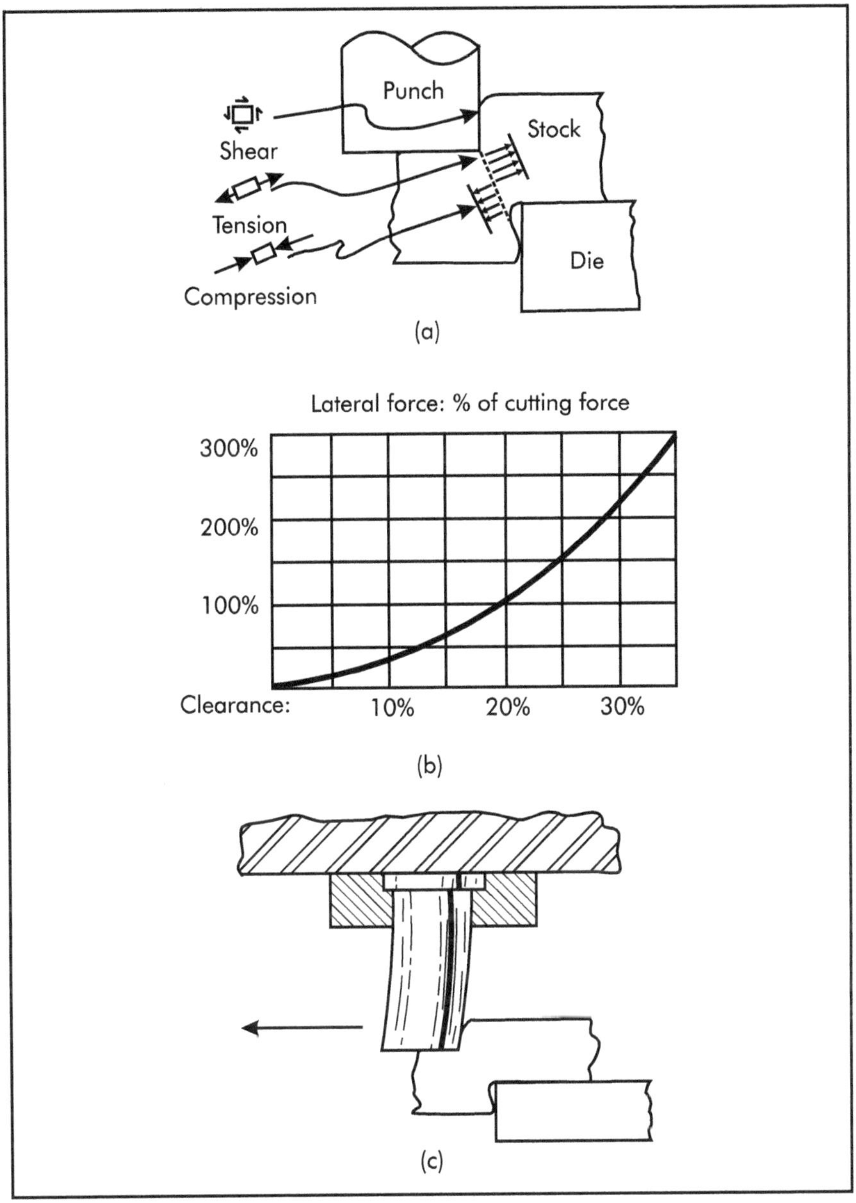

Figure 2-10. (a) Shear, tensile, and compressive forces occur during the cutting process; (b) as punch-to-die clearance increases, the lateral force, or the side thrust, rapidly increases; (c) as side thrust increases, the cutting clearance may increase, leading to greater side thrust or lateral forces, and potential tool failure.

REFERENCE

Smith, D. 1990. *Die Design Handbook*. Section 3, Die Engineering—Planning and Design. Dearborn, MI: Society of Manufacturing Engineers.

3

Designing Large EDM Die Sections for Repairability

Conventional die-sinking electrical discharge machining (EDM) developed in the 1940s. During the next two decades, it became a well-established, die-making machine tool. EDM technology reduced the cost of die-molding cavities, as well as the cost of many stamping-die components. In addition, it introduced a new capability to drill and tap accurate holes in solid tungsten carbide and hardened tool steel.

APPLICATION OF EDM TECHNOLOGY

Wire EDM has revolutionized cutting die construction. The precision cutting of die sections can now be completed with sidewall, taper, or draft angles. Previously, complicated dies were made of many individual, hardened sections that were produced by difficult surface-grinder setups using compound angles. Often, complex forms were dressed onto the grinding wheels. It is now possible to make large die sections as a single-piece detail. Figure 3-1a shows an example of a blank through a die opening made of four sections without wire EDM. Note that each section has two dowels to provide location accuracy and enough screws to prevent shifting.

Wire EDM eliminated most hand fitting by filing, stoning, and lapping. The main impact of wire EDM has been the reduction of time and cost to produce dies. Wire EDM has also resulted in a reduction of the number of die sections required for precision dies.

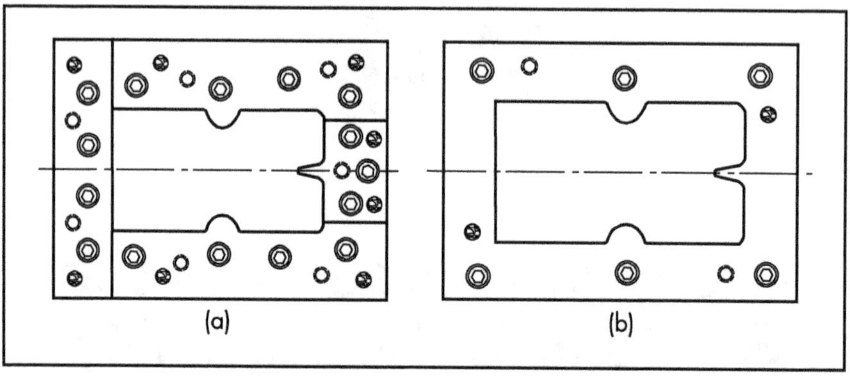

Figure 3-1. (a) An example of a blank-through-die opening made of four sections; (b) an example of a one-piece, blank-through-die opening made by the wire-EDM process.

Thus, die-sinking EDM has made tasks such as tapping holes in solid carbide possible, while reducing the cost of die cavity and opening production. Wire EDM has simplified die construction that otherwise would require difficult surface grinding and laborious hand fitting. It has also lowered the cost and need for some skills once required to make complex die sections.

THROWAWAY VERSUS REPAIRABLE DIE SECTIONS

Wire EDM can be used to make a large, one-piece, blank-through die from a single piece of hardened tool steel. An example of this construction is shown in Figure 3-1b. The only machining normally done before hardening is creating the counter-bored screw and pullout or jackscrew holes. The dowel holes can be made directly by the EDM process, although drilling starter holes reduces machine time. Making this die detail easy to repair would require breaking the detail down into four or more sections as shown in Figure 3-1a. One-piece construction is economical, but it makes repair of damage difficult. Although the cutting forces are balanced, the die-section size must provide sufficient strength to keep lateral deflection low. This type of construction is advisable if it is more cost effective to discard a damaged section rather than to repair a sectionalized design.

DIE SECTION LOCATION AND RETENTION

The use of dowel pins is the most widely used method to accurately position components in die assemblies. Preventing a shifting of the punch and die sections is accomplished by static friction. During the cutting process, both compressive and lateral forces are developed. Compressive forces add to the screw-clamping action. These forces prevent shifting, while lateral forces tend to cause shifting. It is wise to use enough screws to provide sufficient friction and prevent any shifting. The proper size and quantity of screws in the steel can be calculated from the screw manufacturer's data. A general rule is that the clamping force of the screws must be 6.7 times the side thrust, based upon a coefficient of friction of 0.15.

There is a misconception that dowels and keys prevent shifting. Since dowels and keys are elastic steel components, they will deflect under any load and allow some shift. Therefore, these details will only restrict or limit shifting to a minimum value, provided the loads do not result in plastic deformation. The correct use of dowels is to align components in the no-load situation to aid in die assembly and maintenance.

The die designer should follow the torque-versus-tension graphs to determine the screws to be used in die designs. Many screw manufacturers provide data for this purpose. Because of the wide variety of threaded fasteners used in die making and die setting, the individual manufacturer's data should be followed with regard to tensile strength, fatigue life, and tightening requirements.

Limitation of Shifting with Keys

For heavy work, keys may be needed to prevent die-section shifting. The prevention of shifting with screw tension alone may not be economical, because the required screw sizes would have to be very large. This is especially necessary in flanging dies where the wedge action can create a side thrust greater than the vertical force.

The modulus of elasticity under a shear stress solves the side thrust problem. This is sometimes called the "shear modulus of rigidity." For either plain carbon or alloy steel in the soft or hardened state, this modulus is about 12,000,000 psi (83 GPa). Engineering handbooks offer detailed procedures for the mathematical determination of key requirements.

The most frequent use of keys in die construction is to prevent lateral shifting. This is done by placing them in milled pockets in the die shoe in contact with die sections that are subject to shifting. The key may be a separate detail or an integral part of the section. Another application of keying to prevent shifting is shown in Figure 3-2a. Here, four key projections are designed into the top and bottom sections and mate with pockets in the end sections.

PUNCH AND DIE-STEEL DESIGN

Make Fragile Sections Replaceable

The projecting small section is dovetailed into a larger section in Figure 3-2b. This feature makes the section easily replaceable in case of excessive wear or breakage. Another advantage is that the section can be made of the most appropriate tool steel for the application. For example, grade D2 or M4 wear well in most applications, but may be subject to chipping. Here, grade S7 could be used to reduce a chipping problem, although it would wear more

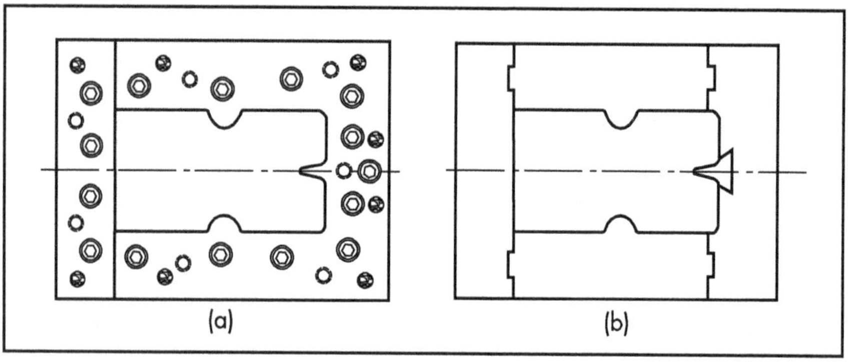

Figure 3-2. (a) Example of one-piece, blank-through-die opening having additional screw, dowel, and jackscrew holes. This section can be cut apart with wire EDM as needed to insert repair sections or to facilitate welding repairs. (b) How a blank-through-die opening can be designed with interlocking sections. This five-section design is held to the shoe with screws and dowels (not shown).

rapidly than D2 or M4. The small detail can be replaced as needed during periodic die maintenance.

Clearance

Punch and die steels must be designed for cutting to allow for the clearance increases caused by deflection (see Figure 2-10a). Figure 2-10b showed an example of how lateral shifting force increases exponentially as a function of cutting die clearance. Clearance increases on trimming or cutting steels can be caused by any or all of the following:

- shifting;
- tipping; and/or
- bending.

Screw Placement

Properly placed screws that are the correct size and tightness will prevent tipping or revolving of the back, lower edge of the die steel. Two very important rules apply to screw placement. First, screws should be placed as near as possible to the working edge of a section. This provides consistently strong cutting edges. These screws have a greater moment arm and are, therefore, better for restricting tipping.

Second, tipping is not as likely to be a problem if the height-to-width ratio of the steel is less than or equal to one. The vertical forces will then tend to counteract side thrusts without the aid of screws. Tipping is more of a problem in form dies, where the side thrusts exceed the vertical force.

Die-block thickness is governed by the strength necessary to resist the cutting forces and by the amount of material removed by sharpening. The type and thickness of the cut material is another factor. For very thin materials, 0.5-in. (12.7-mm) thickness may be sufficient. However, finished thickness is seldom less than 0.875 in. (22.23 mm), which allows for blind-screw holes and sharpening. The one-piece die block (see Figure 3-2a) has extra screw and dowel holes to permit cutting it into several sections to facilitate repair. Each section should have sufficient screws and dowels to provide proper location and prevent shifting.

CONCLUSION

EDM is an essential machining process upon which most modern die construction depends. It makes dies that are low in cost and easy to repair. EDM construction offers tradeoffs between initial die cost, service life, and repairability. The die designer should consider the anticipated die life and the operation's severity and design accordingly.

4

Die-shaving Operations

The edges of a conventionally punched hole are not straight. Figure 4-1 shows a punched hole with approximately one-third shear or burnish and two-thirds fracture. The fractured portion has an angular taper. For many applications, punched holes do not have critical requirements for straightness. If a hole with straight sides is required, an operation known as *shaving* may be used.

PURPOSE

Round-shaved holes are often required for applications where the hole must act as a sleeve bearing for a round shaft. Other applications include round and irregularly shaped holes that have close dimensional-tolerance requirements.

Figure 4-1. A typical hole-punching operation that produces a tapered hole.

Shaving is not limited to holes. Shaving part-edge features produces parts with tolerances that normally require machining. Any stamping application where close tolerances are required is a candidate for shaving. Usually critical dimensions of the stamping are shaved.

PUNCH-TO-DIE CLEARANCE

The clearance for the punch-to-die is normally much less than for conventional hole-punching operations. Typical clearances are about 0.001 in. (0.03 mm) per side. Another rule of thumb is to use 1.5% of stock thickness per side. The purpose of the small clearance is to assure as little burr as possible, while avoiding shearing the punch. This fit is as close to metal-to-metal contact as the die alignment system and punch deflection will permit.

Tables in die design books can be helpful (Smith 1990). These tables give a starting point for a shave-die clearance. However, often rework is needed to obtain the desired results. Ideally, the shaved edge will have nearly a 100% clean-shaved edge with little or no burr.

SIZE OF SHAVE PUNCH VERSUS PART HOLE SIZE

When specifying the actual size of the shave punch, it is important to remember that the finished shaved-hole size will be slightly smaller than the punch. This is due to the stretching and elastic recovery of the material during and after shaving.

MULTIPLE SHAVING OPERATIONS

Figure 4-2 illustrates the first step in the sequence of a shaving operation—initial contact of the shaving punch.

For close-tolerance stampings or when shaving thick materials with a pronounced taper, more than one shaving operation may be required. It is important to allow enough material for a second and third shaving operation to get beneath the work-hardened surface left by the previous shaving operation.

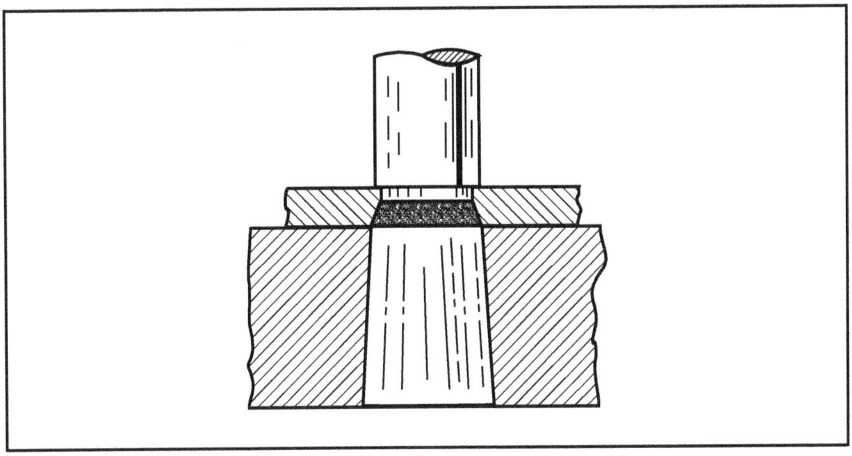

Figure 4-2. Step one—initial contact of the shaving punch.

In Figure 4-3, as shaving progresses, a chip or shaving results. The shaving drives ahead of the punch.

As the shaving operation is completed, the chip or shaving falls free at the end of travel (see Figure 4-4).

Depending upon the condition of the initial punched hole, two or more shaving operations may be required.

Work hardening of the surface and mill scale can shorten punch life. A pressworking lubricant may aid the process.

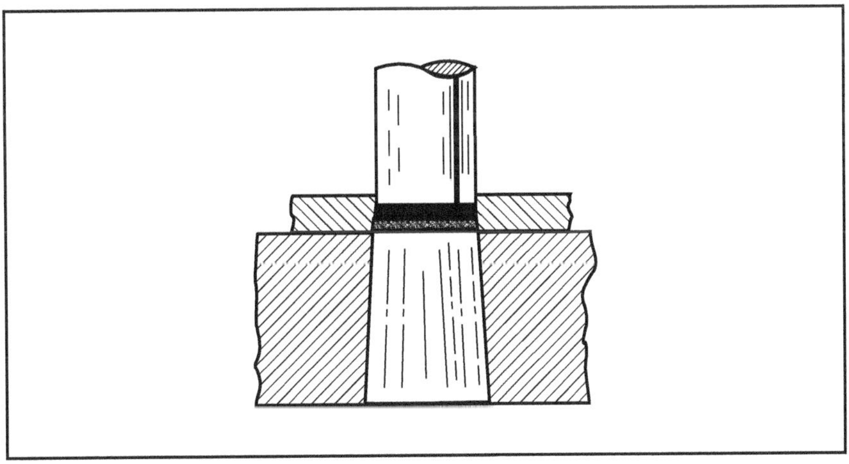

Figure 4-3. Step two—shaving in progress.

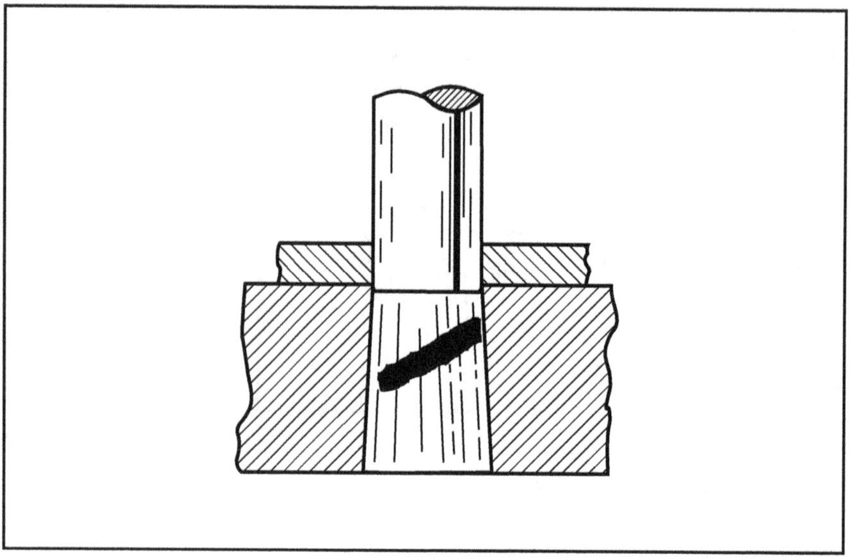

Figure 4-4. Shaving at the bottom of stroke.

Each material has an optimum rake angle. Some aluminum alloys may require as much as 45° of back rake. Some brass and bronze alloys require a zero or positive rake angle.

APPLICATION OF THE HOLLOW-GROUND CONCEPT

Figure 4-5 illustrates a hollow ground-shaving punch with a back-rake angle. This provides cutting-tool-like action, which produces a superior surface finish and extends punch life. Figure 4-6 illustrates a close-up view of the concept. An obvious application for the concept is the shaving of aluminum, a material that often causes metal pickup on the tooling. A typical back-rake angle of up to 45°, together with proper lubrication, is required for good tool performance. The use where possible of a light lubricant, such as vanishing oil, helps to avoid metal pickup or galling problems.

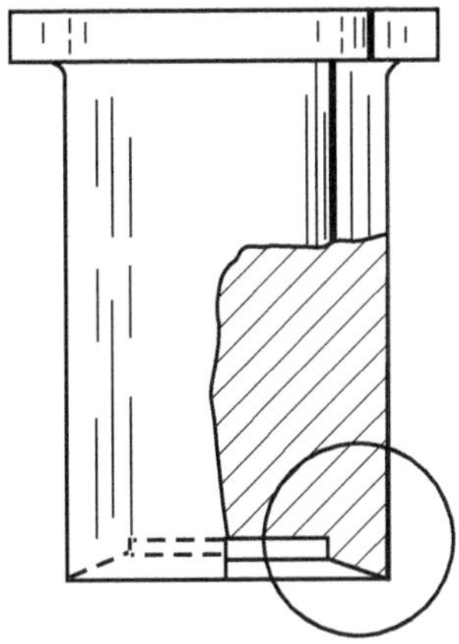

Figure 4-5. A section through a hollow ground-shaving punch showing back-rake angle. The circled area refers to the close-up view shown in Figure 4-6.

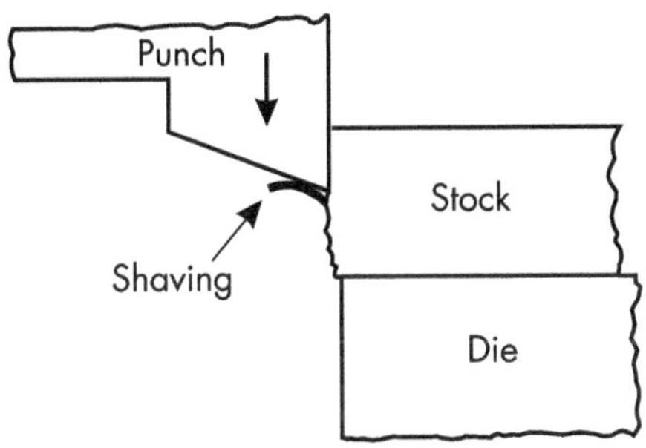

Figure 4-6. Close-up view of shaving punch.

REFERENCE

Smith, Dave. 1990. Die Design Handbook, Third edition. Dearborn, MI: Society of Manufacturing Engineers.

5

Solving Punch Head Breakage Problems

Punch head breakage can be a serious problem, especially in heavy punching and high-speed work. Punch compression occurs during cutting operations (see Figure 2-1b). Rapid strain rates that occur in high-speed stamping may increase the punch-cutting load from the normal shear strength to a value approaching the ultimate tensile strength.

PUNCH HEAD BREAKAGE

Contributing Factors

High stripping loads often contribute to head breakage. However, extremely high stripping loads alone cannot pull the head off the punch. The amount of stripping force depends on several factors. These include:

- type and thickness of stock;
- lubrication used;
- any galling or metal pickup on the punch; and
- sharpness of the punch and die.

Depending on the speed and severity of the work, another head failure mode is breakage due to extreme compressive forces and load reversal. This is especially a problem in high-speed operations and when cutting heavy stock. Figure 5-1 illustrates the part of the punch where the failure starts with a view of a punch head in a retainer having a hardened tool-steel backing plate. Typically, the backing plate is heat-treated for toughness to 38–45 R_C. See Figure 5-2 for a close-up view. In good severe work designs, the punch and retainer should have a small radius or relief to accommodate an unavoidable radius in the corner where the punch head joins the shank body.

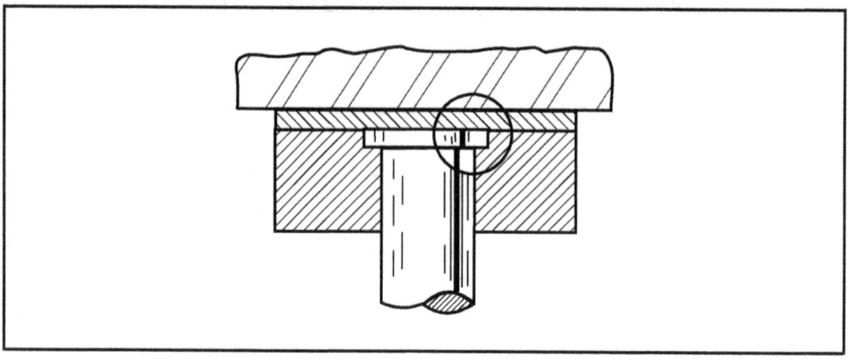

Figure 5-1. Head breakage problems usually start at the step where the head and punch body join.

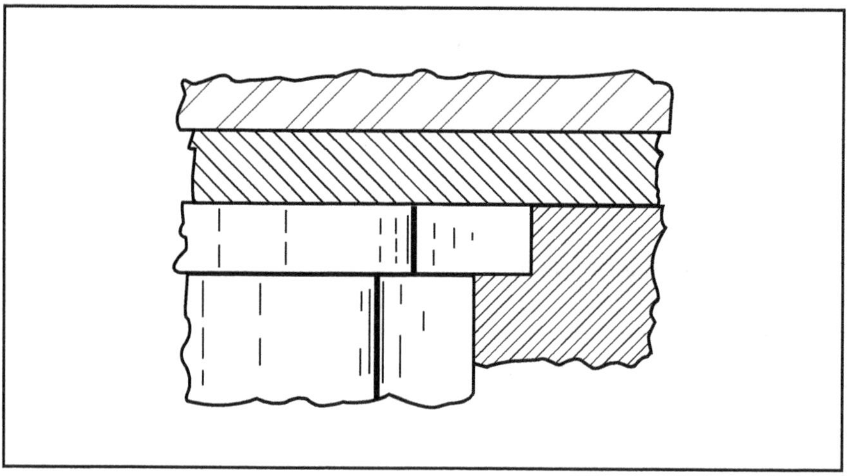

Figure 5-2. A close-up view of Figure 5-1 shows where the head and punch body join.

Punch Head Flexure and Crack Formation

As the punch cuts through heavy metal, compressive strain occurs in the punch body, backing plate, and die shoe. This results in deflection or bending of the punch head as shown in Figure 5-3.

At break through, there is a recoil action resulting in a tensile strain concentrating on the punch head as shown in Figure 5-4. Repeated flexure of the punch head can result in crack formation at the juncture of the punch head and body.

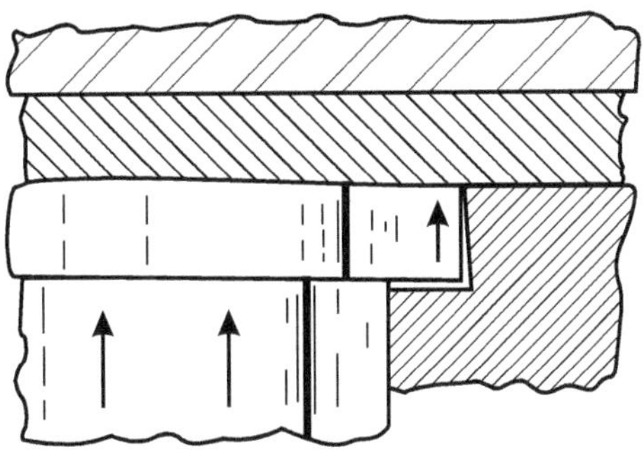

Figure 5-3. Deflected punch head is deformed and compressed into the backing plate.

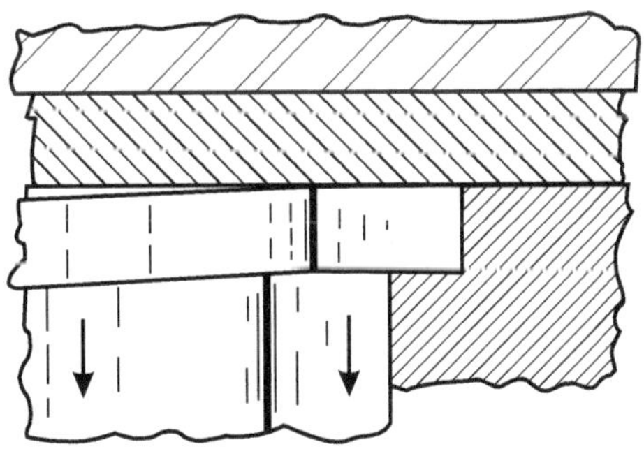

Figure 5-4. Compressive strains release as the punch cuts through the stock and stripping occurs, resulting in tensile strain on the punch head.

Once a crack forms, continued operation under high cyclical loading is almost certain to cause the crack to grow larger (see Figures 5-5 and 5-6). The ultimate result often is complete separation of the punch head from the body.

Addressing the Root Cause

Designs are available that resist head breakage. Good designs are usually effective in eliminating the head breakage problem. The punch head should have a slight interference fit in both the

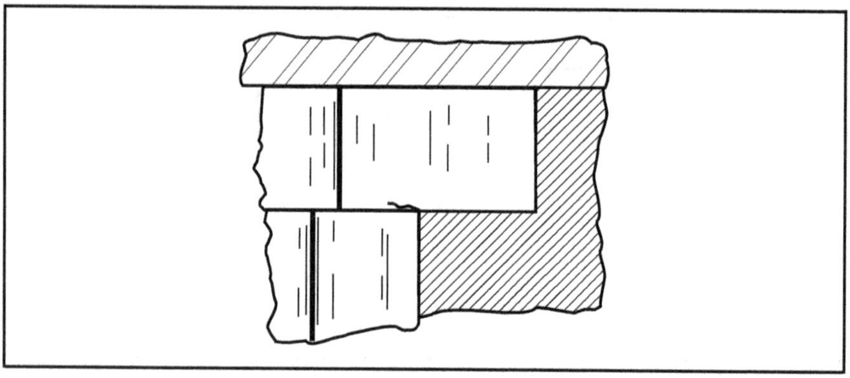

Figure 5-5. Stress concentration at the juncture of the punch head and body under heavy cyclical loading can result in crack formation.

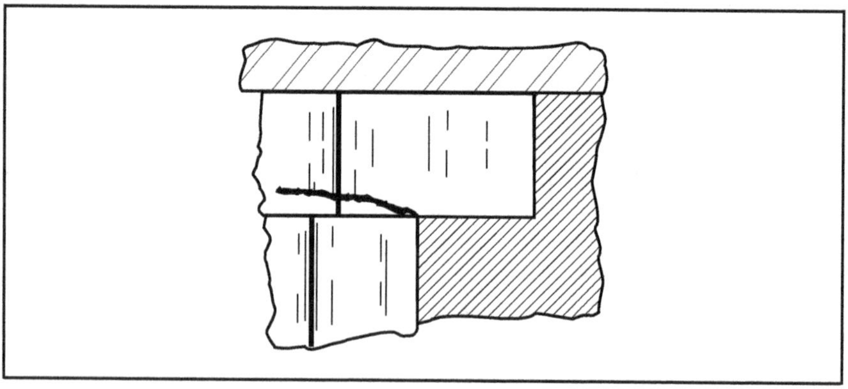

Figure 5-6. Stress concentration at the juncture of the punch head and body has caused crack formation. Continued operation results in crack propagation.

retainer hole and counterbore. The head thickness also should be slightly thicker than the counterbore depth. Keeping these parts tight will prevent breathing and movement.

Figure 5-7 illustrates two simple solutions that will eliminate all but the most serious head breakage problems. The first modification is to grind a slight back angle on the head to permit the punch to compress into the backing plate without flexing the punch head.

The second improvement is to provide a generous radius at the head-to-body juncture. This will reduce the stress concentration in this area when punch recoil occurs at break through.

Elastomer Washers

Figure 5-8 shows an elastomer washer placed under or around the punch head as a means to reduce shock. This method has been tried a number of times with very limited success. The elastomer material tends to break down and extrude around the punch shank.

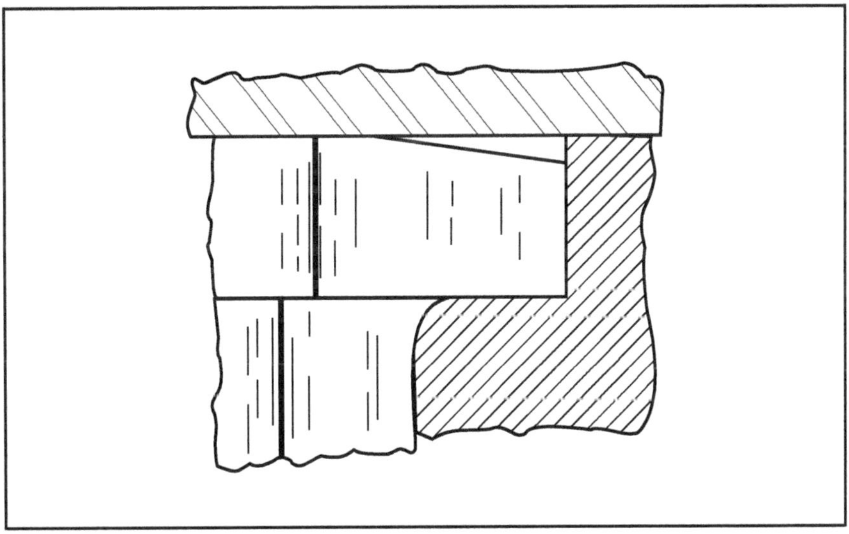

Figure 5-7. A back angle on the punch head and a large radius at the juncture of the punch head and body reduce stresses. This solution will solve most head-breakage problems.

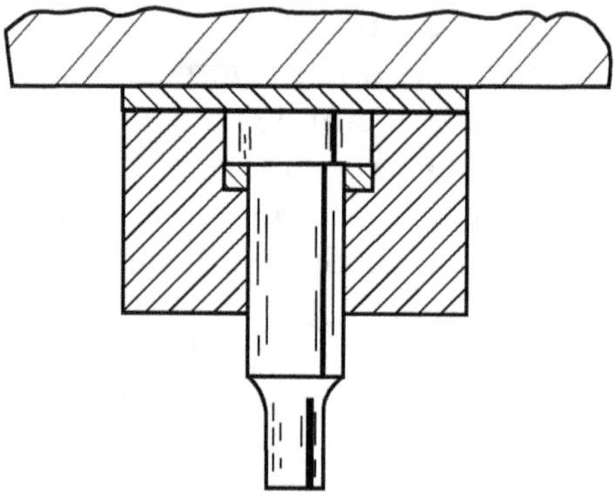

Figure 5-8. An elastomer washer is placed under the punch head as a means to reduce shock.

Head Modifications

If head separation as shown in Figures 5-5, 5-6 and 5-7 is not a frequent occurrence, grinding an angular taper around the head as illustrated in Figure 5-7 is simple to accomplish for round shank punches. The angular relief recommended is approximately 3–6°. It leaves a flat surface in the middle the same size as the shank diameter. Grinding the angle is a simple procedure. The methods include:

- the use of a cylindrical grinder;
- using a lathe with a tool-post grinding attachment;
- grinding the tapered land with a simple "whirligig" hand-actuated spinning fixture on a surface grinder; and
- the use of universal cutter grinders that can hold punch shanks and grind the taper easily and quickly.

Applying a Stress-reducing Head Radius

A new punch must be made if the radius shown in Figure 5-7 is required. The retainer must have clearance for the radius. If the punch retainer is heat-treated to no more than 48 R_C, it should be possible to machine the radius in the retainer with a high-speed steel or counterbore.

A radius is easy for a skilled cutter grinder or toolmaker to machine from drills, milling cutters, and existing counterbores. Carbide counterbores are excellent for this work. However, a jig grinder, inside diameter (ID) grinder, or conventional electrical discharge machine are other options.

Increasing the Ratio of the Head to the Point

Another solution to the head-breakage problem is to make the body and head diameter much larger than the point. This effective solution usually will work if other methods fail. However, there are two drawbacks to this solution. A large-diameter, expensive tool-steel blank is required. Secondly, the large head diameter limits how closely together punches can be spaced (see Figure 5-9).

SAFETY OF PERSONNEL

If the head separates from the punch body, it can fall out of the retainer while the press is being cycled. Figure 5-10 illustrates complete separation of the punch body from the head. If this occurs, severe die damage can result. In addition to the pressworking equipment damage, there is a great possibility of serious injury to

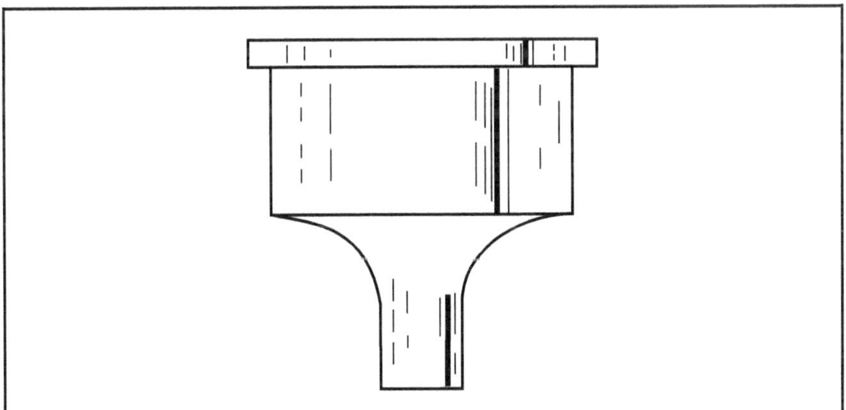

Figure 5-9. A large body-to-point diameter ratio is an effective, but expensive, solution. It also limits how closely punches can be spaced together.

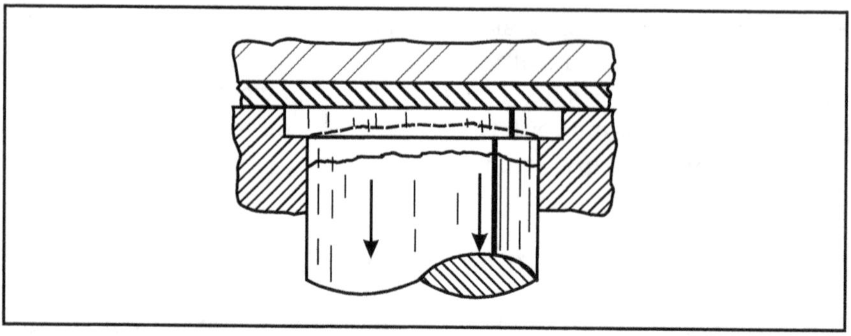

Figure 5-10. Complete separation of the punch head can result in the punch falling out of the holder. This can result in die damage and injury to personnel.

pressroom personnel. The punch and other tool-steel die components may shatter, causing sharp fragments to become airborne at high velocity. Serious or fatal injury to pressroom personnel has occurred in this way.

The main emphasis of pressworking safety regulations involves keeping body parts out of the point of operation during pressworking operations. Safety distance, guard-opening sizes, and other safeguarding measures address the safety of personnel.

All pressroom injuries are avoidable by proper process planning, process control, employee training, and good pressworking practices. Good safety practice in the case of dies that can shatter and eject objects requires a physical barrier guard. This is clearly stated in all government and insurance safety regulations.

If there is any danger of a die shattering in a pressworking operation, a physical barrier guard is necessary. Clear plastic of sufficient thickness and strength to arrest any airborne shattered die components is a solution that may work in some cases. If this is insufficient, solid steel guards of adequate strength may be required.

FUNCTION OF THE PUNCH BACKING PLATE

The purpose of the backing plate is to provide a hard surface to back-up the punch. A frequently used backing plate material is oil-hardening gage stock. The usual practice is to heat-treat the backing plate to approximately 38–45 R_C to provide for toughness and wear resistance. This hardness is in the spring-temper range giving long flexure life. Some experts consider this hardness range to provide the best vibration damping capacity. The damping capacity of steel is low in both the hard and soft state. In any case, the backing plate, properly heat-treated, serves to spread the load over the underlying plate without cracking.

As shown in Figure 5-3, the punch head will compress into the backing plate. The backing plate in turn will compress into the die shoe. This compression occurs each time the punch penetrates the stock. Recoil as shown in Figure 5-4 occurs as the punch breaks through the stock.

IRONWORKER- AND BULLDOZER-STYLE PUNCHES

The ironworker-style punch, shown in Figure 5-11, has been in use for many years in portable equipment used to punch rivet and bolt holes in heavy plate and structural members. A modern ironworker is a special type of steel fabricating machine capable of cutting steel bars, punching holes, cutting steel angles, and bend-

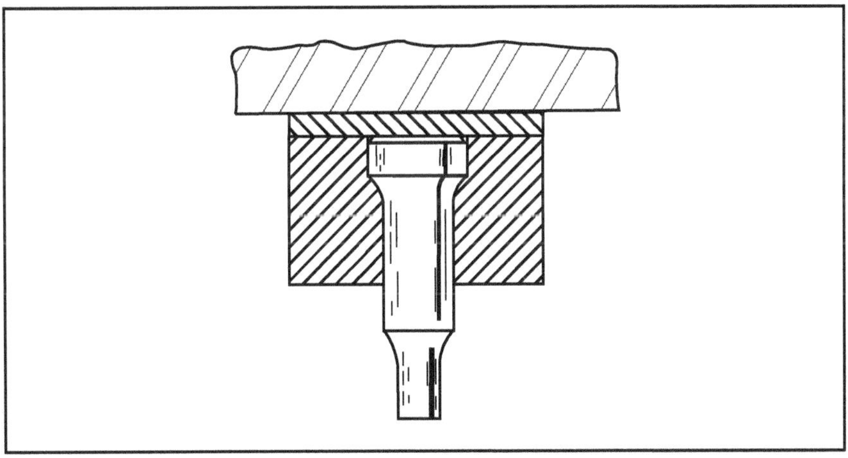

Figure 5-11. An "ironworker"-style punch is an old design that avoids stress concentrations in any area of the head.

ing operations. Mechanical bulldozers have also made use of this type of punch.

The mechanical bulldozer originated in the middle of the 19th Century as essentially a powerful C-frame press placed on its back. The early mechanically driven machines were typically double- or triple-back-geared and, hence capable of exerting great forces. The design found widespread application in all sorts of heavy-plate fabrication such as punching rivet holes and bending operations. Metalworking bulldozers were in use before the development of the familiar earth-moving bulldozer.

A derivation of the 19th Century design was adopted for punch head types used for boiler shell and other fabricated wrought iron-work. The design probably evolved at the hands of clever black-smiths who produced simple tooling for this type of work. The advantage of this style of punch is the highest possible reduction of stress concentration in the head configuration.

The design is both economical to produce and seldom fails under heavy punching loads. A disadvantage when applying the design to pressworking tooling is the need to use a punch retainer that may not be readily available.

Many good punch designs for modern ironworker punches simply use a generous angle rather than a sweeping radius. These have proven to avoid the punch-head breakage problem in heavy-steel fabrication work.

6

The Cost of Punch and Die Maintenance

RECORD CAUSAL FACTORS

Punch and die maintenance is a very costly budget item for most pressrooms. Tooling repair issues and causal factors must be identified before the true cost of die maintenance can be determined. Repair cost includes the diemaker's time spent in the press to adjust for stock conditions. All time and money expenditures should be identified, categorized, and tracked.

Careful record keeping of repair-cause codes will reveal the true cost of maintenance. It is advisable to use a computerized maintenance-management database. There are hundreds of companies who supply maintenance-management software. It is important to invest the time to determine the best one for your purposes.

Cause Codes

In any instance where conditions are expected to improve continuously, it is important to use facts as a management tool. Before improvement can occur, a means must be in place to measure that improvement. Finger pointing, without improvement, is simply a way of accepting smashed dies as a normal condition.

In metal stamping, the analysis of die damage and work-cause codes is vital to reducing die-repair costs. Typical cause codes include:

- sharpening dies before burr height becomes a problem;
- scheduled inspection of springs and nitrogen cylinders for weakness or lack of charge;
- die washing, inspection, and lubrication; and
- replacement of perishable details that are subject to predictable wear.

So far, the examples of cause codes cover bench maintenance that is necessary and scheduled. In a perfect shop, all needed maintenance is predictable. In addition, there is no time charged to die damage. There are also no adjustments required for dealing with material variations and no press-maintenance conditions. In reality, however, problems do arise in pressrooms. These problems can cause die damage, ranging from minor to catastrophic. While it is true that the goal must be to prevent problems from ever occurring, from time to time, die damage will occur in the pressroom. The reasons range from lack of training and human carelessness, to unforeseen press- and auxiliary-equipment problems, to overlooked items in the equipment maintenance program.

The following are die-damage descriptions or cause codes that are external to the die itself. Carefully recording all downtime items, including the time, cost of correction, and the long-term-avoidance plan are very important to minimize the occurrence of such problems. Examples include:

- hard stock or stock defects that damage the die;
- slug buildup due to die problem;
- slug buildup due to scrap-removal problem;
- stock mislocated due to misfeed;
- problems achieving feed-progression uniformity and/or repeatability;
- die damage due to part-ejection failure;
- slug pulling;
- broken springs;
- low nitrogen pressure; and
- operator error.

Make Use of the Data

Gathering data is useless, unless it is intended for managing the business. Cause codes need to be reported accurately, audited for correctness, and charted at least weekly, or monthly, if weekly is not possible.

THE REAL COST OF DIE DAMAGE

Conservative estimates by stamping managers indicate that at least half of all die-repair work is due to avoidable damage in the pressroom. The true amount of die repair is probably higher. Avoidable damage is observed daily in the toolroom. Examples include sheared cutting edges, bent dowels, and distorted dowel holes. This needed repair work can be an indicator of damaged dies, misfeeds, and other common pressroom damage.

It is essential that the data be carefully reviewed, with appropriate notes made by the tooling manager. To effect corrections, copies of notes should be sent to all department heads and action plans should be agreed on. This data also should be posted conspicuously with an explanation of the needless impact caused by die damage. Part of the solution may lie in proper procedures training.

THE VALUE OF GOOD WORKMANSHIP

A skilled observer can tell a great deal about stamping shops and die-construction facilities through simple observation. For example, it can be assumed that a production shop that routinely shims die details with large stacks of poorly fitting shims will need to perform breakdown maintenance on the press.

Likewise, die-building shops that are dirty and poorly organized are not good sources for quality tooling. This author does short-term contract consulting to companies on production jobs that are experiencing problems. Years ago there was a job that involved a critical, deep-drawn, automotive component supplied by a large, top-tier supplier. Upon examining the poor workmanship of the transfer-press tooling, it was determined that substantial revisions were necessary to make the dies worthy of production. The client at the shop wanted to have the work done at no additional cost by the same shop that had designed and built the original problematic tooling. Upon inspection, the shop was found to be poorly managed and a disorganized mess. "Upon using their wash room, I asked how to flush the commode," said author/consultant, David A. Smith. "I was told to draw a bucket of water from a large nearby sink and flush it that way. They explained that there was no profit in maintaining the toilet." The recommendation to this company, which was subsequently followed by the firm, was to cut its losses by shipping the tooling to a shop known for good design capabilities and workmanship. It made sense that any shop that could not maintain a simple device such as a flush toilet probably could not build good dies either.

BIBLIOGRAPHY

Smith, D. 1991. *Quick Die Change*. Chapter 27, Selecting Maintenance Software. Dearborn, MI: Society of Manufacturing Engineers.

7

Good Maintenance Practice

In addition to the maintenance of cutting dies, pressworking tooling with punches also includes cast iron, alloy-drawing punches for automotive outer-skin panels, and form punches for frame rails used in heavy trucks. This text is limited to the discussion of cutting punches and dies.

GOOD GRINDING PRACTICE

The routine maintenance of cutting tools nearly always involves grinding. In the event of severe damage, welded repairs followed by grinding may be required. It is essential to use the best procedures when grinding tool-steel die parts. Poor procedures compromise the life and dependability of the die. Figure 7-1 shows the worn end of a punch being ground to sharpen it. Figure 7-2 depicts a close-up view.

Wheel Selection

Surface grinders are machine tools and grinding is a machining process. The cutting tools are the projecting, sharp abrasive grains on the wheel. Abrasive grains used for sharpening tool steels are typically aluminum oxide and, less frequently, silicon carbide. Cubic boron nitride (CBN) and diamond are also used for toolroom work. Grains are held together by a bonding material. Vitrified bonds are most popular for toolroom work because they are stable in

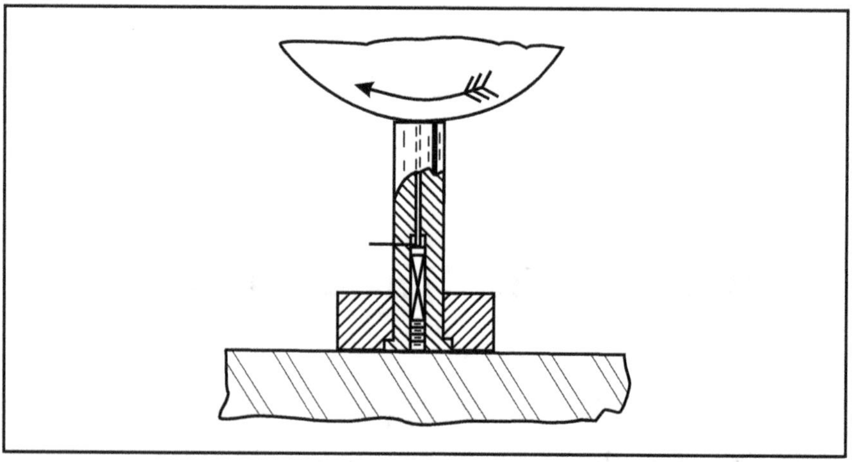

Figure 7-1. The worn end of a punch is ground to sharpen it. A wire inserted through a breather hole retracts the spring-loaded, slug-shedder pin. (Courtesy Dayton Progress Corp.)

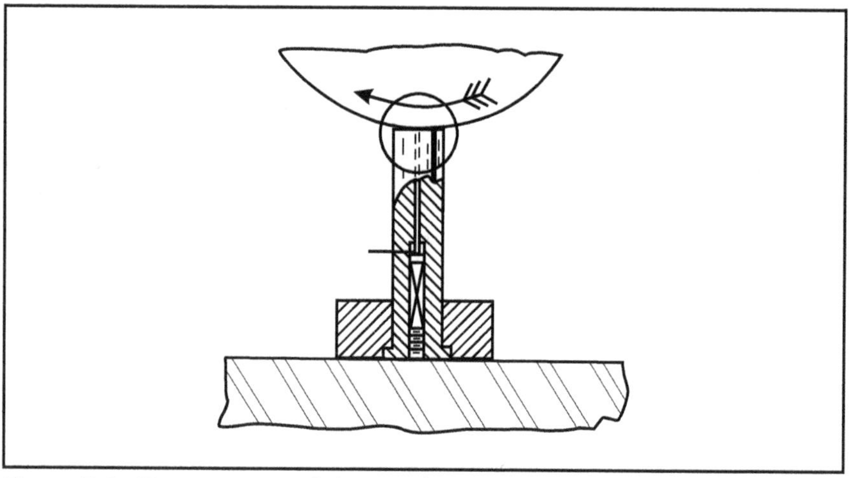

Figure 7-2. Close-up view of the grinding wheel in Figure 7-1 that is sharpening a punch.

wet and dry work. The bond can be compounded with a wide range of densities and strengths for job-specific hardness and self-sharpening breakdown characteristics.

If the correct wheel is selected, extremely hard, abrasive grains will cut the hardened tool steel until the wheel becomes dull. Like any dull tool, more cutting pressure is developed than if a sharp tool is used. Increased pressure will cause the dull grain to break away from the bond material. Thus, grain size and type is chosen for the best cutting action and finish on the tool steel being ground. Bond structure and type determine how strongly the abrasive grains are attached to the wheel. Abrasive grains should break away as they become dull, exposing sharp grains as the wheel wears.

Wheel Nomenclature

Toolroom grinding wheels used in the United States are usually marked with the ANSI Standard B74.13-1990 identification code. Figure 7-3 shows the ANSI standard marking system for aluminum oxide, silicon carbide, and aluminum zirconium grinding wheels. The code is a series of short numbers and single letters printed in the paper washer in the center of the wheel. The series starts with a prefix number followed with a letter.

The most important information is shown in bold print. The abrasive type A is aluminum oxide, the most common type for toolroom surface-grinder work. The next most common abrasive is silicon carbide, designated C. Silicon carbide is useful in the toolroom for heavy off-hand grinding wheels, as well as for surface grinding of cast iron and hard-bronze alloys.

The grain size is given in the screen-mesh size that will pass the grains given in inch dimensions. Commercial grinding-wheel-grain sizes run from approximately 8, a very course grit, through 220, which is fine. Sharpening flat-die sections on a manual-surface grinder requires gentle-stock removal and good surface finishes. A suggested range of A or aluminum oxide grit sizes is grade 36 through 60 mesh. The relative bond hardness is designated alphabetically from A through Z and ranges from soft to hard. The bond should be soft enough to allow the wheel to break down, exposing new grains as the grains become dull. Bond-hardness G and H are among those suitable for flat-surface, grinder-die-sharpening work. If a sharp corner must be ground square, a fine hard wheel, such as a 220-grain J-bond, may be required to square up the corner. A perfectly square corner cannot be ground or cut with EDM processes. Two tool-steel sections are normally butted together to form a perfectly square corner.

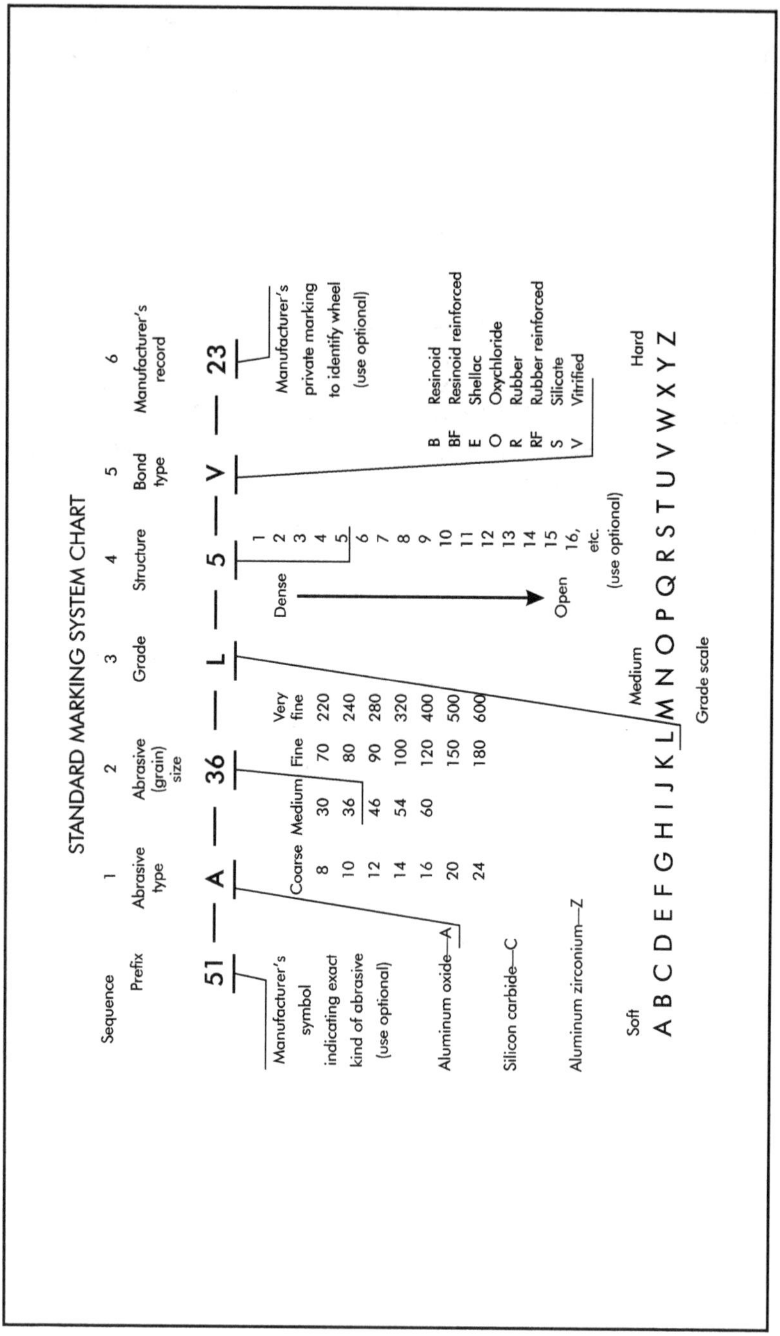

Figure 7-3. Standard marking system chart for aluminum oxide, silicon carbide, and aluminum zirconium grinding wheels (ANSI 1990).

Wheel Usage

Grinding wheels for sharpening tool-steel punches and dies should have an open structure with a moderately soft bond. Fine-grained wheels with hard bonds tend to dull quickly and create excessive heat. A soft, open-bond wheel will break down in use and remain sharp. A grinding wheel functions by cutting, not by frictional melting of the metal being ground. As the wheel wears down to a smaller diameter, the lower cutting velocity will result in the wheel acting softer than a full-size wheel.

Good storage practice for used wheels is to place them on wall-mounted pegs near the surface grinder. When selecting a used wheel, choosing a smaller-diameter, worn wheel will result in the wheel acting softer by a grade letter or more.

Wheel Dressing

Most tool-steel sharpening with a surface grinder requires an aluminum-oxide wheel having a soft bond structure. Wheel dressing requires a single point or diamond cluster set in a holder to dress the wheel. When using a single-point diamond to dress the wheel, turn it in its holder from time to time to expose a sharp corner. Diamonds wear and become blunt.

Make enough passes across the diamond to true the wheel. The crosswise motion of the diamond should be fast enough to continually dislodge old grains, without remaining in contact long enough to dull the newly exposed abrasive. Make the last pass rapidly. In no instance should the diamond take a final pass without removing wheel material. This will avoid blunting the exposed abrasive edges by the much harder dressing diamond.

Wheel Safety

Toolroom work using machine and off-hand grinding tools must be carried out safely. Always use proper guards. Never alter wheel-mounting methods or exceed the manufacturer's rated speed. Examine all wheels before use. New wheels could be damaged in shipping and handling. When gently tapped, a sound wheel will have a clear bell-like tone. This is called *ring-testing* the wheel. The wheel is held by the center hole and gently tapped with a soft object, such as a wooden hammer handle or screwdriver handle—never a metal object. Cracked wheels will emit a dull sound—these should be destroyed.

Grinding Burrs and Sharp Edges

When sharpening punches and dies by grinding, a very sharp edge or burr may occur. Grinding is a machining process and the sharp projections on the wheel-abrasive grains act as cutting tools. A small burr normally results. This burr is illustrated in Figure 7-4.

To insure long tool life, the burr must be removed. Also, sharp edges should be blunted slightly. A dead, sharp edge is subject to extreme pressures during the cutting process. Since the cutting process is really a controlled plastic deformation, a dead, sharp edge is often undesirable because it will chip and/or otherwise dull quickly.

Tool-steel edges should be blunted slightly with an Arkansas or fine India stone. Cemented tungsten-carbide tools may be blunted with a very fine silicon carbide or boron nitride polishing slip. A copper lap charged with diamond dust may be used for critical work.

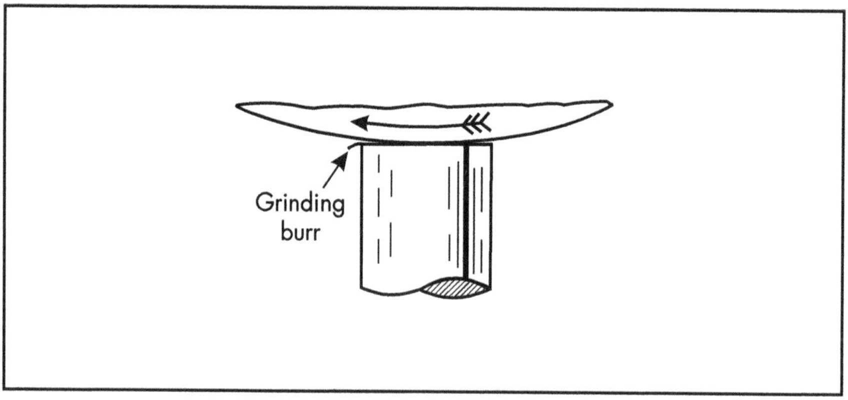

Figure 7-4. Close-up view of grinding wheel while sharpening a punch. Note the grinding burr created on the punch edge.

Correct Edge-stoning Procedure

Exactly the manner and amount that a cutting edge is stoned will vary from one type of job to another. Since the purpose of stoning the grinding burr is to avoid edge chipping, thereby increasing the part quality and tool life, any additional blunting of the edge should be done in a way that is correct for each individual job. Cutting thin and/or soft materials requires a sharp edge and tight cutting clearances. Here, the correct procedure is to hold the stone against the die opening in line with the punch, and remove only the burr to leave a sharp edge.

Gentle versus Aggressive and Abusive Grinding

The craftsperson sharpening the die must use great care not to damage the tool-steel sections by abusive grinding. Die sections are expensive production tools and extra care is justified to obtain the best possible tool life. The wheel is selected with great care for maximum metal removal consistent with the desired results. Attempting to hurry the sharpening process through excessive metal removal rates can overheat and damage die sections and punches. Recirculating flood coolant is useful to remove heat.

Figures 7-5 and 7-6 illustrate examples of mistakes a craftsperson might make if not properly instructed. Figure 7-7 shows the position of a sectional view through the punch depicted in Figure 7-5. The surface grinder is being used much like a milling machine or shaper. The punch being sharpened is badly overheated. The hardness will be drawn back as a result and, potentially the punch face may be cracked.

Effect and meaning of discoloration. Figure 7-6 shows an end view of a punch that overheated during sharpening. The color or hue of the discolored area is an indicator of how hot the punch became during an abusive grinding process. The oxides of iron take on a characteristic color in this instance when heated in air, depending on the peak temperature attained. The colors that indicate temperature are based on plain carbon steel. Tool-steel alloys will not discolor in exactly the same way. Closer views of this area are shown in Figures 7-8 and 7-9.

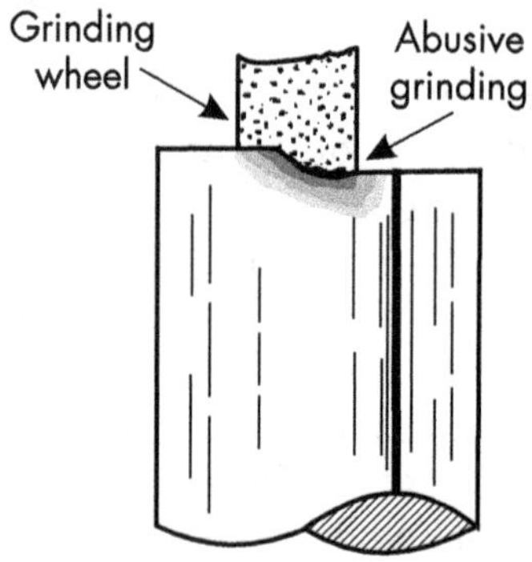

Figure 7-5. An extreme example of an abusive grinding technique that has damaged the punch. Light passes should be taken to minimize heat buildup.

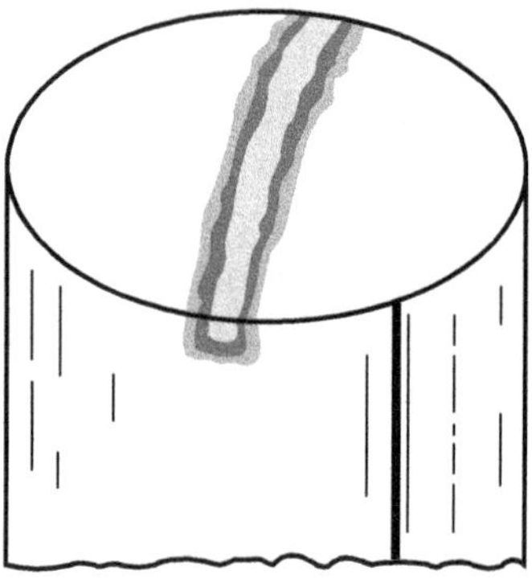

Figure 7-6. An end view of a punch that has overheated during sharpening.

Figure 7-9 shows Rockwell-C hardness levels after the punch is overheated. The base metal before sharpening was 62 R_C. An area drawn to 48 R_C exhibits a blue coloration. If the area becomes hot enough to turn whitish gray, it is nearing red heat—hot enough to anneal carbon steel.

Crack formation. When extreme localized overheating of hardened tool steel occurs, the hot area expands and some metal is upset or forced outward. Once the heat source is removed, the hot area is rapidly cooled by the surrounding cold mass of the punch.

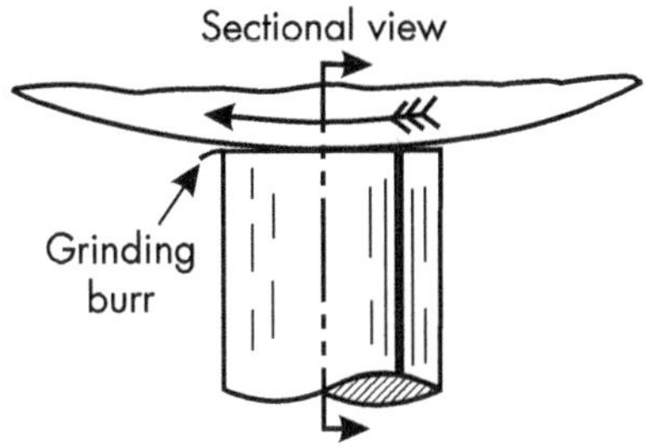

Figure 7-7. Position of a sectional view through the punch depicted in Figure 7-5.

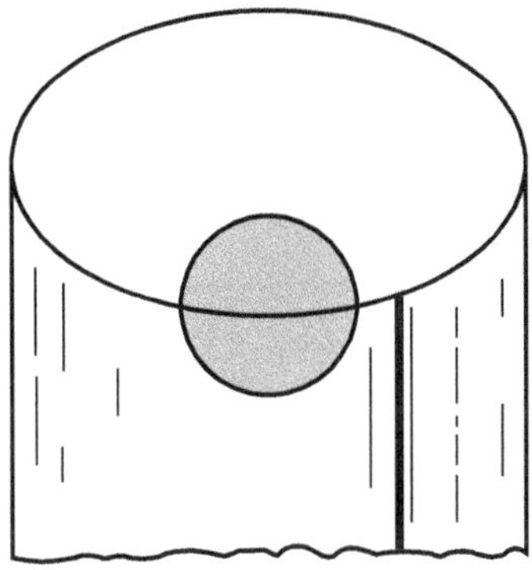

Figure 7-8. Area highlighted for close-up view in Figure 7-9.

If the tensile strength of the hot area is exceeded as cooling occurs, cracks will form at the surface as shown in Figure 7-10. Any cracks will weaken the punch by providing a local point for stress concentration.

Don't hide the problem. Unless carefully instructed, craftspersons may have a tendency to make mistakes and then attempt to find an easy solution. After noticing discoloration, it might make sense to dress the wheel and very carefully grind away the discoloration. Or, since an India stone is supplied to remove burrs and sharp edges, it would be a simple matter for the craftsperson to stone the punch body to remove any remaining discoloration.

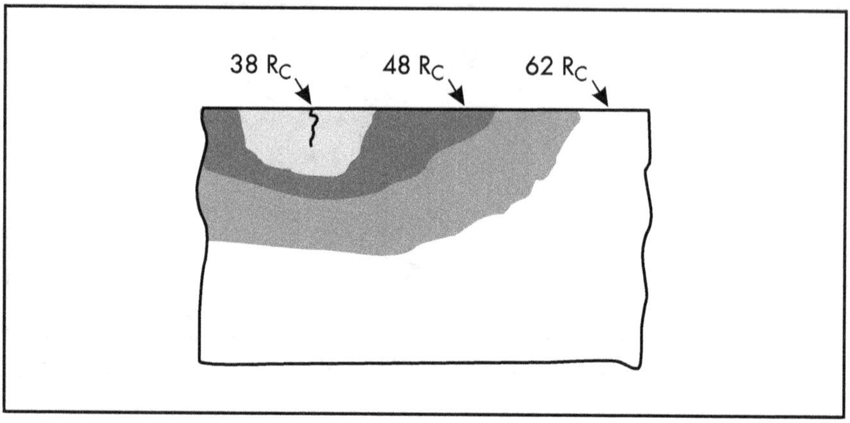

Figure 7-9. Close-up view of the area shown in Figure 7-8. A color change indicates how much overheating has occurred.

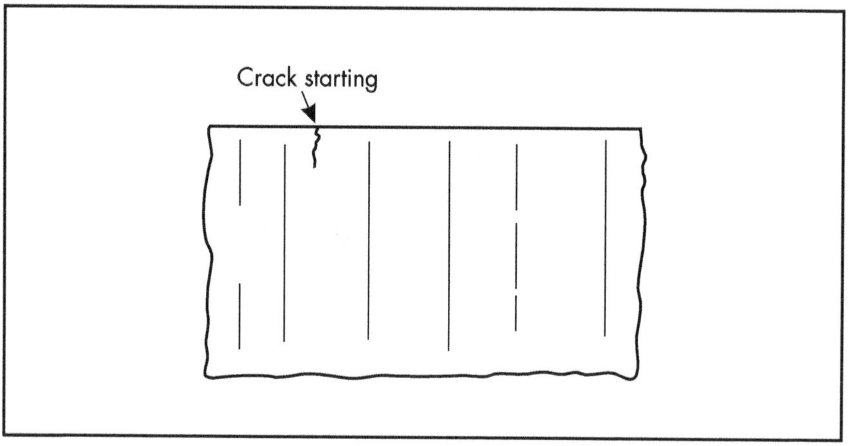

Figure 7-10. A small stress crack has formed because the punch has overheated while grinding.

However, as a result, the punch will have a soft spot that will dull quickly. The crack may grow and cause punch failure. Removing the discoloration will only hide the problem.

Cracks will grow. Once a crack in a highly stressed piece of steel such as a punch is formed, it tends to grow as shown in Figure 7-11. The rate of growth depends on how stressed the punch is when in use.

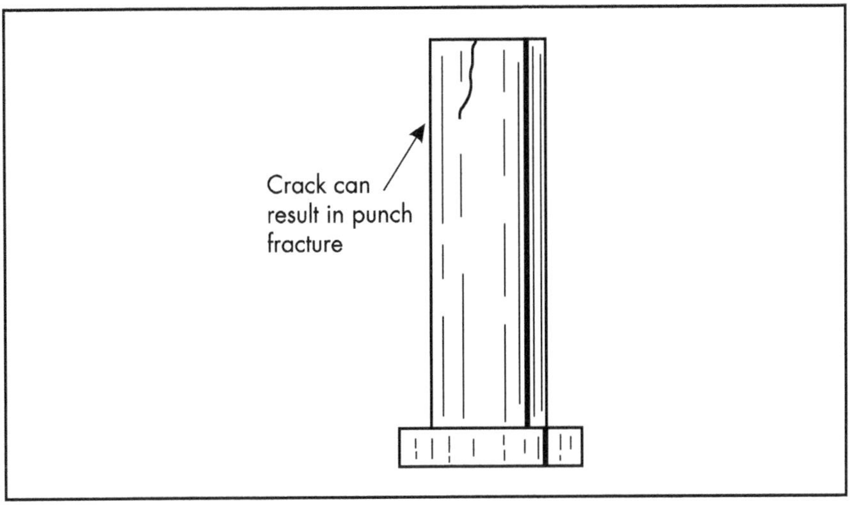

Figure 7-11. Crack propagation shown in punch body during use.

Figure 7-12 shows the final failure of the punch. A piece can break away. The result is unexpected downtime, scrap, and quite possibly, serious die damage.

STRESS-RELIEVING CUTTING TOOLS

Tool-steel sections that are subjected to frequent shock and/or high pressures will accumulate internal stresses. These stresses are much like the residual stresses that are relieved by the drawing process that follows quenching during heat-treatment. Hard-worked sections that tend to fail from fatigue fractures can benefit from redrawing at approximately 25° F (14° C) below the final-drawing temperature used in the heat-treatment process. Exceeding the drawing temperature will reduce the hardness of the section and may result in size change.

This stress-relieving procedure is a normal practice for severely worked, tool-steel details such as coining blocks. Experience obtained from coining high relief in 316-stainless-steel medallions using S7-tool-steel coining blocks is a good example. The coining pressures approach and may exceed 300 ksi (2,068 MPa) at the form block working faces. The typical failure mode is a fatigue

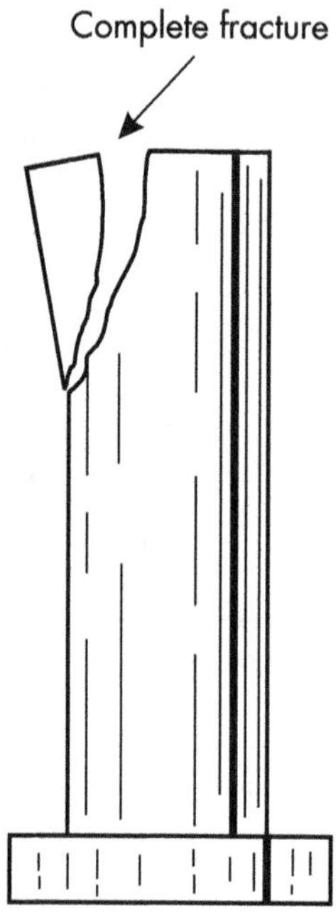

Figure 7-12. As the crack weakens the punch, failure will occur.

fracture of the raised-coining detail subjected to lateral metal flow. A typical amount of hits before failure is 2,000 pieces.

Stress relieving the form blocks after each 1,000 hits will extend the life to approximately 10,000 pieces. Placing the form blocks in a draw oven at approximately 25° F (14° C) below the final-drawing temperature does this. Shop practice typically is to leave the blocks in the oven overnight at the stress-relieving temperature.

Cutting sections are subjected to repeated pounding during the production run. This can also create excessive levels of residual stress, as well as work hardening. This will result in the Rockwell-C hardness increasing by a point or more. Gentle grinding of cutting sections will introduce or change the amount of stress in the die section. If chipping or cracking is a problem, the stress-relieving procedure carried out after grinding will probably increase tooling life.

SHIMMING DIE COMPONENTS

The use of steel shims to adjust the position of die components is a common toolroom activity. To compensate for sharpening, it is important to design dies that can be shimmed when necessary. Designing for ease of shimming is especially important in the case of progressive dies.

Shimming Progressive Dies

Progressive dies normally combine cutting and forming stations. The forming stations require little maintenance, except oil stoning and polishing to correct minor-metal pickup or scoring problems. However, the cutting stations require periodic sharpening to control burr height to within acceptable limits.

The normal practice is to shim the cutting details after sharpening. This will maintain proper cutting entry and the relationship between the forming- and cutting-die details. In the case of a forming detail that incorporates one or more details, the details should be separate pieces that are easily removed for sharpening. To compensate for the ground-off material, shims are then added.

Proper Shimming Materials

Clock-spring-tempered shim stock should be used for shims, except for use with low-grade tools for short-run work where working pressures are low. Spring-tempered stock has a hardness of approximately 45–48 R_C. Die-room supply companies stock this shim material in a variety of thicknesses. The material may be supplied in a handy dispenser box. If you make the mistake of opening the box incorrectly, the roll of shim stock may unwind violently, posing a cutting hazard. This is one reason the material is known as clock-spring shim stock. Soft-steel shims should not be used because they will extrude under the repeated-cyclical loading of many stamping operations.

Laying Out the Shim

New shim layout is usually accomplished by placing the die detail on the shim stock. The next step involves scribing the features, such as the outline and dowel, screw, and jackscrew holes.

If the material is purchased with a blue-oxide coating, it should be ready to use directly from the box. If the shim stock has no oxide coating, it should be painted with layout ink before scribing. It is important to punch holes for all items that will contact or go through the shim. For example, a spring-loaded pressure pin should not allow the spring to bear against the shim. Instead, it should make contact beneath the surface. This process will help maintain correct spring and pin travel.

Cutting Out the Shim

Spring-tempered stock can be cut with high-quality, sheet metal shears that have hardened-steel or carbide-jaw inlays. A conventional shim punch can be used to make the holes. Both tools require periodic resharpening when working with spring-tempered stock.

If a number of shims are to be produced, and machine time is available, a wire-cut EDM machine will produce accurate, burr-free shims if correct cutter-path information is available. This is especially true when thick shims are needed. The shim cutter-path information can be supplied with the die design. This information should figure in allowances for screw hole and other clearances.

Removing Burrs

If the shim is cut with shears and a shim punch, some edge burrs will remain. These burrs should be oil stoned off. The sharp edges and corners of wire-cut EDM shims also should be oil stoned. A good shim is flat, fits perfectly, and is safe to handle.

How Many Shims?

No more than three shims should be used under any one detail. It is an incorrect procedure to have a large stack of shims under a die detail. Figure 7-13 shows two thin shims and one thick shim in place on a progressive die. Any irregularities in the shims will be magnified if a large shim stack is used.

Fabrication of Solid Shims

As die details are repeatedly ground, more shim height is needed. The correct procedure here is to make a solid shim of tool steel. This is intended to replace the separate shims. Oil-hardening gage stock is a good material for solid shim construction.

To make solid shims quickly, heat-treating can be done in-house, provided a small heat-treat furnace is available. The soft shim may be wrapped in stainless-steel foil to avoid decarburization.

Figure 7-13. A stack of three shims is used to correct the height of a reground progressive-die cutting section.

The recommended hardness, depending on the amount of compressive force the shim must withstand, is 45–55 R_C.

It is advised to line ream or carefully drill and ream the dowel holes on location. Any movement in heat-treatment may be compensated for with careful hand grinding of the hardened-shim dowel holes. This should be done with a high-speed pneumatic grinder to provide a nice slip fit. The shim must be closely aligned, especially if slug drop through the holes occurs. Figure 7-14 shows how proper shims fit snugly on the dowels and have clearance provided around the screw holes. A sloppy shim can cause a die to become slugged.

Storage of Spare Shims

Extra shims can be identified with a scriber. They can then be stored on a pegboard provided for that purpose in the die room. Another method is to store them in a watertight, electrical-junction box attached to the die.

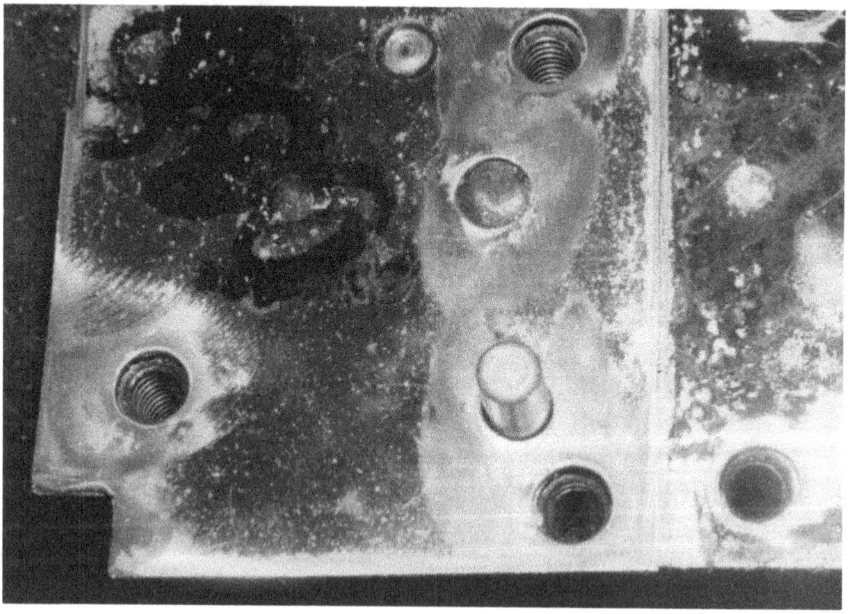

Figure 7-14. A properly made shim fits tightly around the dowel pins and has proper clearance for other holes.

BALL-LOCK PUNCHES AND DIE BUTTONS

Ball-lock punches and die buttons are popular for a number of reasons. These include:

- The punch and/or die button can be removed and replaced quickly to allow for rapid-style changes in the part.
- Should a misfeed cause a punch to break, it can be replaced quickly in the press.
- Most ball-lock punches are low-cost, perishable details and, in the smaller sizes, replacement cost is less than that required to grind a punch with a conventional head and shim punch retainer.
- In some cases, symmetrical punches that cut on one side only can be provided with dual teardrops 180° apart to permit rotation to expose the unused edge.

Figure 7-15 shows how punch and die buttons are retained by the ball-lock method. A ball-and-ramp locking principle is employed. As the punch or button is pulled, the ball locks more tightly on the tapered ramp. The ball has a spring to facilitate constant contact during normal operation, ensuring against vibration and shock.

A ball-lock punch shank has no head. Instead, a teardrop-shaped seat for the ball is ground into the punch shank near the middle of the retainer. The exact size and shape of this seat is very important for proper retention.

In the design shown in Figure 7-15, the in-line dowel and backing plug are used. This is one of several patented features of the product illustrated. Several manufacturers make ball-lock tooling in both light- and heavy duty designs. Although there are industry standards, a wise worker always double-checks the ball-seat geometry, because, in some cases, the ball may fail to retain the punch or button.

Teardrop Grinding

In some instances, the only punch available is an irregularly shaped punch that may be satisfactory, except for a teardrop in the wrong location. It is generally permissible here to jig grind or

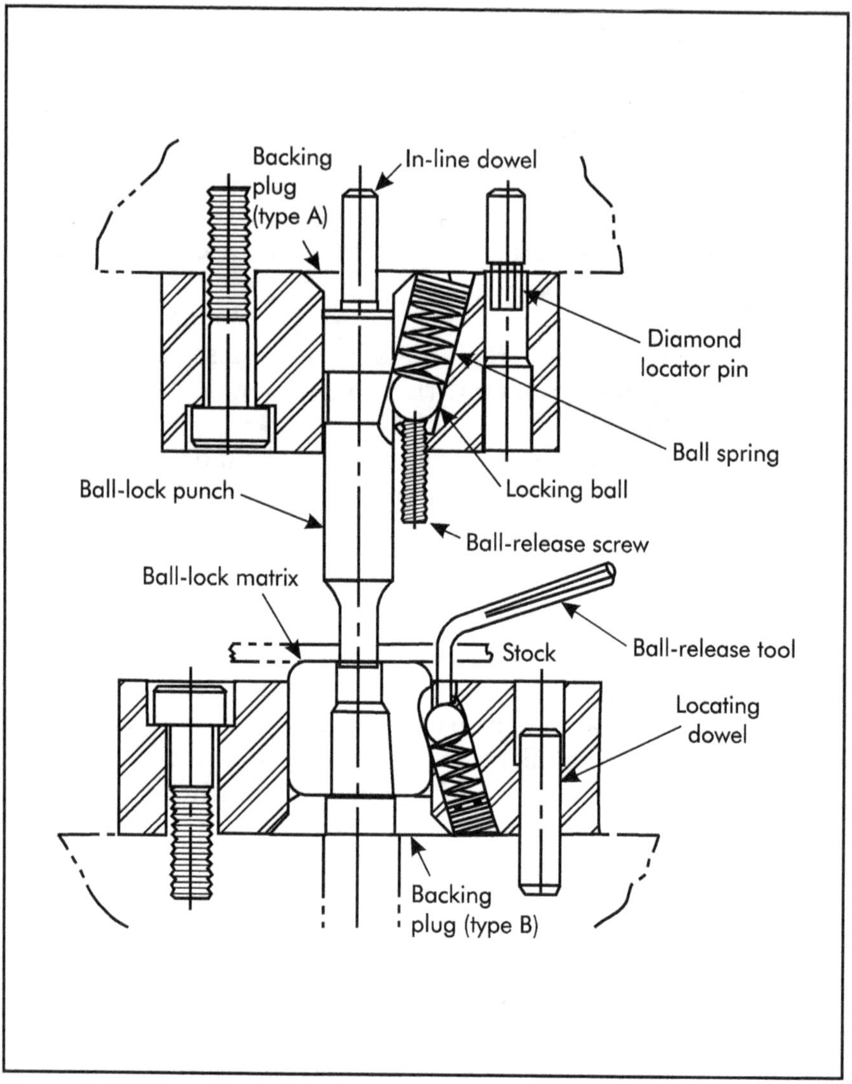

Figure 7-15. View of a ball-lock punch shows a ball-lock die button together with the retainers used. (Courtesy Dayton Progress Corp.)

use conventional EDM to make a new teardrop at the correct location. This is true only if the manufacturer's recommendations regarding the exact location and geometry can be duplicated. Incorrect teardrop location will result in unsafe conditions such as the ones shown in Figures 7-16 and 7-17.

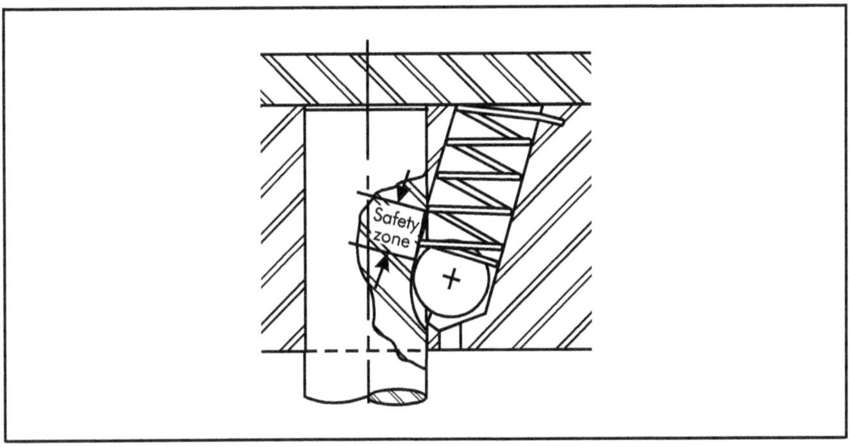

Figure 7-16. An unsafe lock condition. The ball is too low to take advantage of the wedge and ramp locking principle. (Courtesy Dayton Progress Corp.)

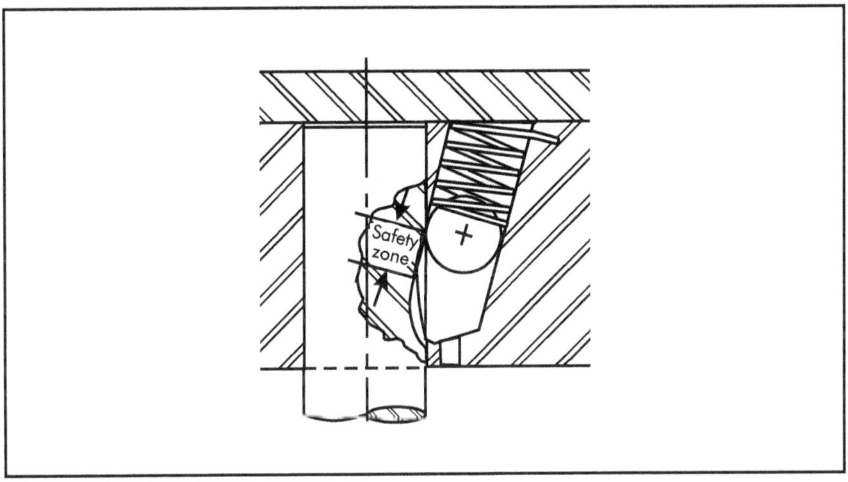

Figure 7-17. An unsafe lock condition—the ball is too high to securely lock. It is nearly at the top of the ramp. (Courtesy Dayton Progress Corp.)

It is essential for safety to plug the old teardrop securely with solder or a fast-setting resin compound. This procedure should eliminate the possibility of incorrectly installing the reworked punch in the retainer.

Make Sure the Ball Locks

When the punch shank is inserted, it is important to first depress the ball fully. The punch shank should be inserted at an angle to the correct position and be fully bottomed in the retainer. The ball depressor tool is then released and the punch is turned until the ball snaps into place. When the punch is in place, the punch depressor tool should again be inserted and checked for proper engagement in the teardrop as illustrated in Figure 7-18. When released, the punch should be firmly held in place with no discernable movement.

REFERENCE

ANSI B74.13-1990. "Markings for Identifying Grinding Wheels and Other Bonded Abrasives." New York: American National Standards Institute.

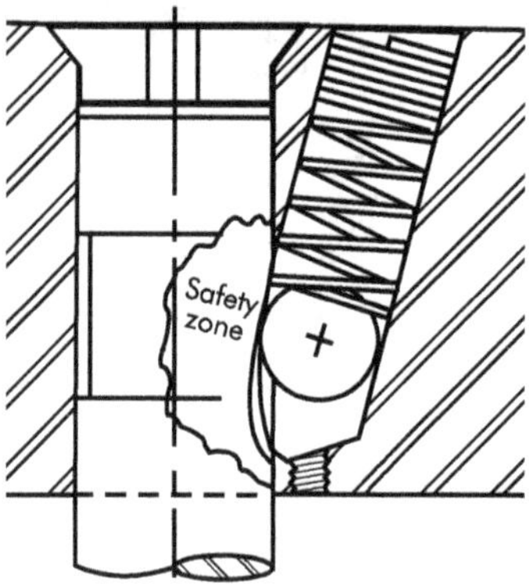

Figure 7-18. The ball is in the correct position on the ramp and securely locked. (Courtesy Dayton Progress Corp.)

8

Compound Dies

A *compound die* usually refers to a one-station die designed around a common vertical centerline where two or more operations are completed during a single press stroke. Usually, only cutting operations are done, such as combined blanking and piercing.

A common characteristic of compound-die design is the inverted construction, with the blanking die on the upper die shoe and the blanking punch on the lower die shoe. The pierced slugs fall out through the lower die shoe. The part or finished blank is retained in the female die, which is mounted on the upper shoe.

COMPOUND BLANK AND PIERCING DIES

Compound dies are widely used to produce pierced blanks to close dimensional and flatness tolerances. Generally, the sheet material is lifted off the blanking punch by a spring-actuated stripper, which may be provided with guides to feed the material. If hand-fed, a stop is provided to position the strip for the next stroke.

The blank normally remains in the upper die and is usually removed by a positive knockout at the top of the press stroke. Ejection of the blank from the die by spring-loaded or positive knockout occurs at the top of the stroke. Because of this feature, the die does not require angular die clearance. Not providing angular die clearance simplifies die construction and assures constant blank size throughout the life of the die.

A compound die for making a washer is shown in Figure 8-1. The center hole is cut and the outer diameter trimmed in a single-die station in one press stroke. The material is 0.015-in. (0.38-mm) cold-rolled steel strip. The piercing punch is attached to the upper die shoe. The blanking punch is attached to the lower die shoe. In this design, the piercing punch contacts the material slightly ahead of the blanking die. The part is stripped from both the blanking die and piercing punch by a positive knockout. The blanked strip is lifted off the blanking punch by a spring-loaded pressure pad.

Part Removal

A potential disadvantage of compound dies is that the part must be removed from the upper die at the top of each stroke. The part is usually knocked out at the top of the stroke by means of a press-actuated knockout bar.

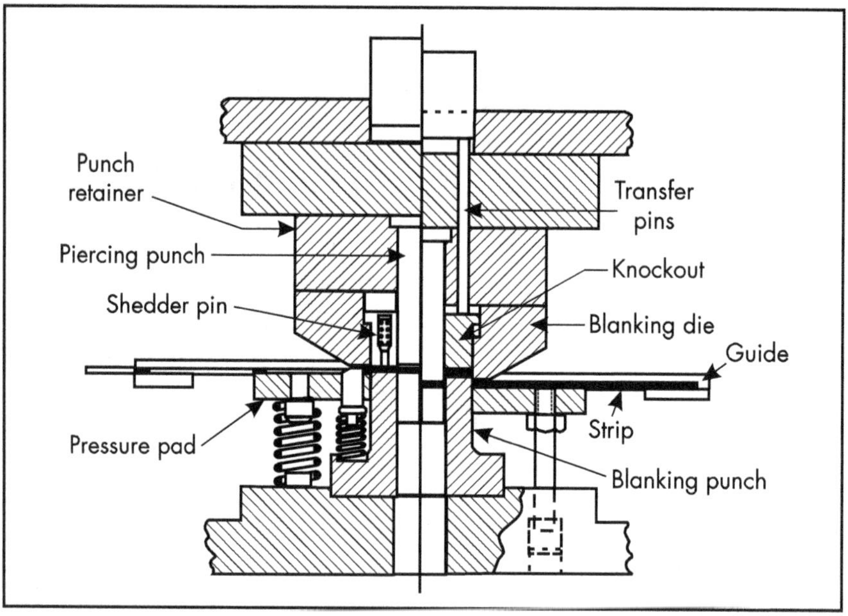

Figure 8-1. A compound blanking and piercing die used to produce a washer. Dies of this type are widely used to produce accurate flat blanks. (Courtesy Cousino Metal Products Company)

In the case of small parts, once knocked out of the upper die, they may be ejected by a timed blast of air. Larger parts can be removed by means of a shuttle unloader that enters the die opening as the ram ascends. The press ram normally drives the unloader, although air, hydraulic, or servomotor-driven units may be used. Accomplishing positive part removal during each press stroke may limit the speed of the operation. For low-production jobs, manual removal with appropriate safeguarding precautions may suffice.

Cut-and-carry or Pushback Operations

Cut-and-carry operations are very useful for producing parts with high flatness requirements. These operations are actually performed by an upside-down compound die.

The feature of pushing the part back into the carrier strip or scrap skeleton provides a positive means to get the part out of the die without an auxiliary unloading means as required with a conventional compound die. Cut-and-carry operations are useful as a station in progressive dies.

The cutting station makes use of a counter-force pad. Springs or nitrogen cylinders usually supply the pressure. Figure 8-2 illustrates a typical cut-and-carry operation. The part may progress through several other stations before being pushed out of the carrier strip in a knockout station. The knockout station must have approximately double the clearance of the cutting station to accommodate normal part expansion and pitch growth errors. Otherwise, the production of slivers could be a problem.

Cut-and-carry or pushback dies are a proven method to enhance part flatness while providing a convenient means to carry the part out of the die. Complete fracture must occur, and the punch and counter-force must be configured to prevent the part from leaving the carrier in the cut-and-carry station.

Success Factors

Successful cut-and-carry stations must completely fracture the part from the material and always contain the part within the carrier

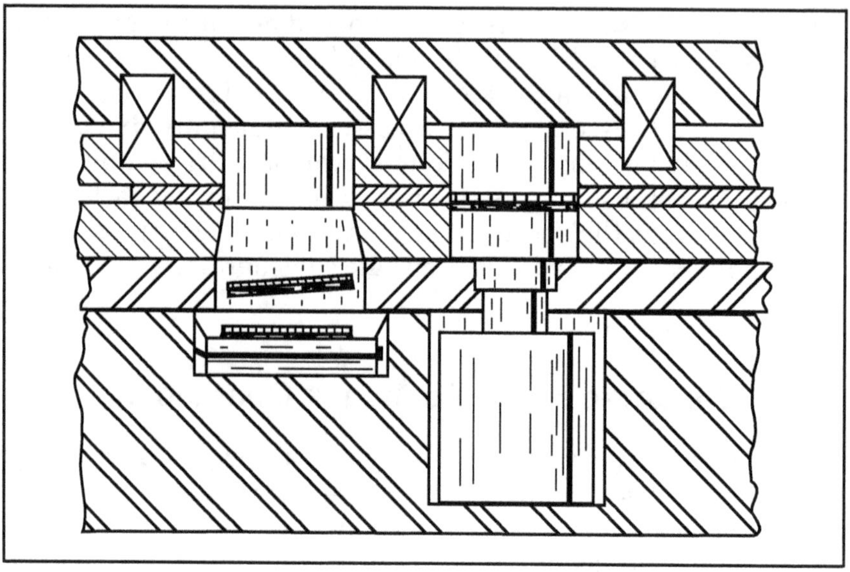

Figure 8-2. A section through a typical cut-and-carry die is shown.

opening. Figures 8-3 through 8-7 show close-up views of the fractured edge in the cut-and-carry station. Should the part not completely fracture, a double-breakage condition occurs in the push-out station.

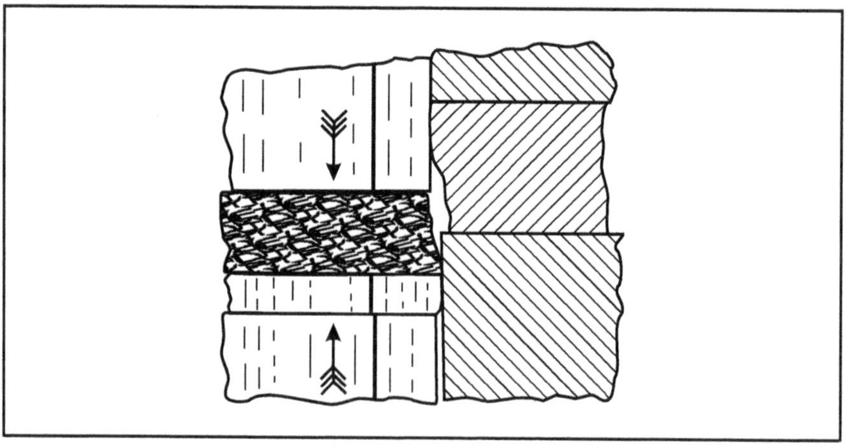

Figure 8-3. A close-up view of the fractured edge in the cut-and-carry station illustrates proper fracture.

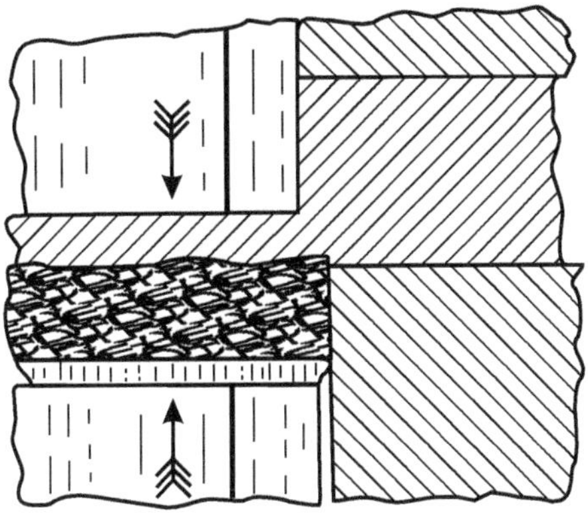

Figure 8-4. A close-up view of the fractured edge in the cut-and-carry station illustrates incomplete fracture. This will result in a secondary or double breakage in the knockout station.

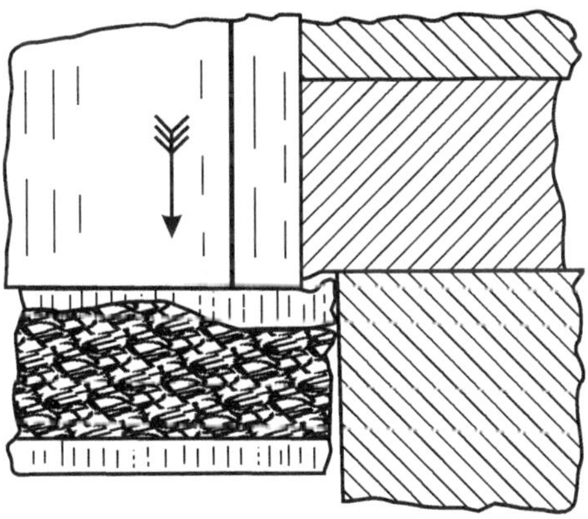

Figure 8-5. Close-up view of the fractured edge illustrates a secondary or double breakage in the knockout station.

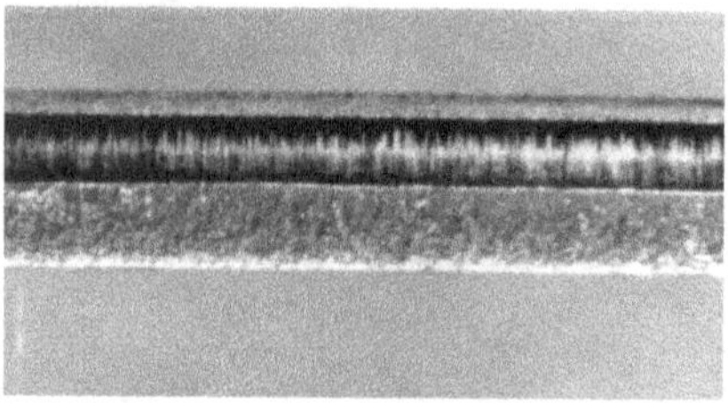

Figure 8-6. Shown is a complete clean fracture in a pushback station. The part is made of 0.108-in. (2.74-mm) hard HSLA steel.

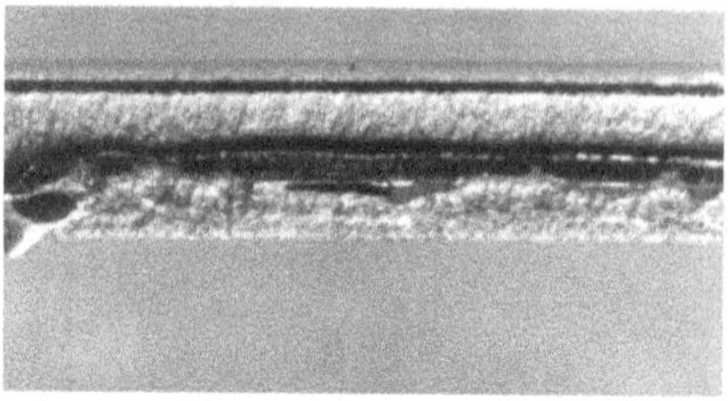

Figure 8-7. Complete double breakage is shown in the knockout station. The material is gummy 1020 high manganese steel.

If the part is shoved completely out of the carrier, it will expand and the carrier opening will shrink due to normal material elasticity. Forcing the part back into the opening normally results in distortion and possible cold-welding problems.

The main cut-and-carry success factors are:

- completely fracturing the part from the material; and
- always containing the part within the carrier opening.

Process Control Variables

Three main variables are available to control the process. The variables are, in order of effectiveness:

1. controlling the depth of punch entry;
2. amount of die clearance; and
3. control of material properties.

Variation of the spring or nitrogen pressure used to supply the counter-force is not good practice. If too low of a pressure is used, the part flatness will be affected. The depth of punch entry is a means to accommodate a range of material properties. Changing die clearances can only be accomplished by modifying or changing the die details. The correct clearance is often determined by experimentation in critical operations. Minimizing cutting forces with optimal clearances normally results in correct operation.

Material properties will have some effect on the ratio of shear to fracture. In general, harder materials will fracture more easily than softer ones.

9

Pressure Pad and Stripper Selection

In the case of cutting dies, the function of a stripper is to strip or remove material from the punches. A pressure pad is often used for this purpose. Pressure pads take many forms, depending upon the amount of stock control required.

For high-precision dies, especially those involving long slender punches, the stripper is guided with pins and bushings. The stripper can guide the punches with hardened bushings.

The simplest type of stripper is the tunnel or fixed stripper. Its two main advantages are economy and reduced force requirements. The force requirements must be added to the cutting-force requirements when springs or nitrogen cylinders are used.

POSITIVE OR TUNNEL STRIPPERS

Figure 1-1 illustrated a simple hole-punching die employing a positive or tunnel stripper. Close-up views are illustrated in Figures 9-1 and 9-2. Before stripping action can occur, the punch must lift the stock against the top of the stripper. The stock then falls to the die surface as the punch is withdrawn.

The tunnel or fixed stripper is popular for simple heavy punching operations as well as low-cost short-run dies. Fixed strippers are a good choice for long-run dies, provided the stripping action and stock feeding is satisfactory. One problem is that distorted

Figure 9-1. Close-up view of the tunnel stripper illustrated in Figure 1-1.

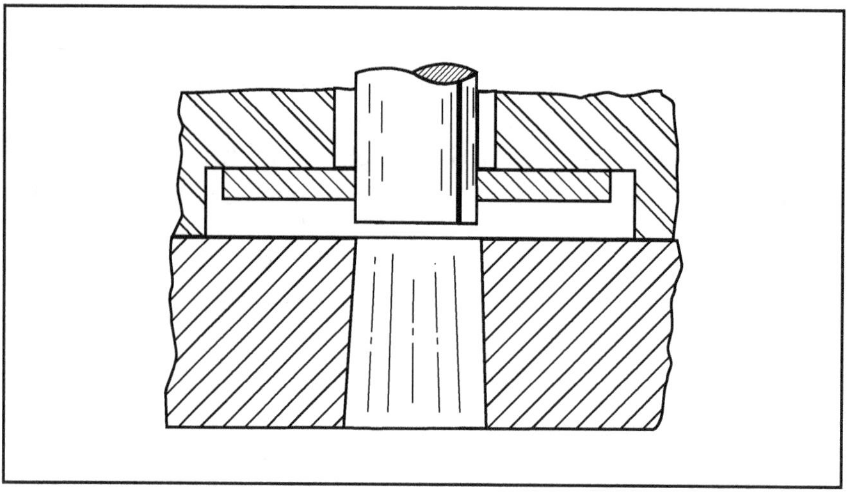

Figure 9-2. The stock is stripped upon withdrawal of the punch.

stock will tend to deflect the punch upon withdrawal, which may break or chip the punch face as shown in Figure 9-3 (Smith 1990, 1994). Figure 9-4 illustrates a close-up view. Figure 9-5 shows the effect of a "finger lock" on the punch. A combination of distorted stock, high stripping loads, and poor stock control results in a chipped punch face (see Figure 9-6).

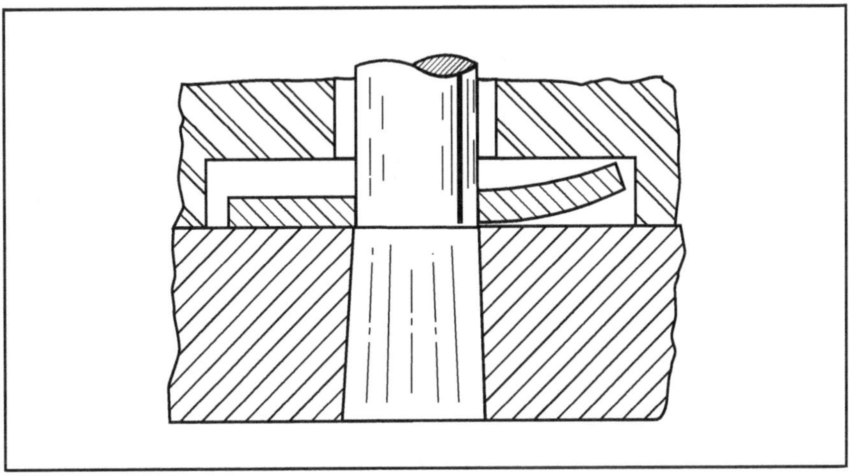

Figure 9-3. Distortion of the stock can break or chip the punch face.

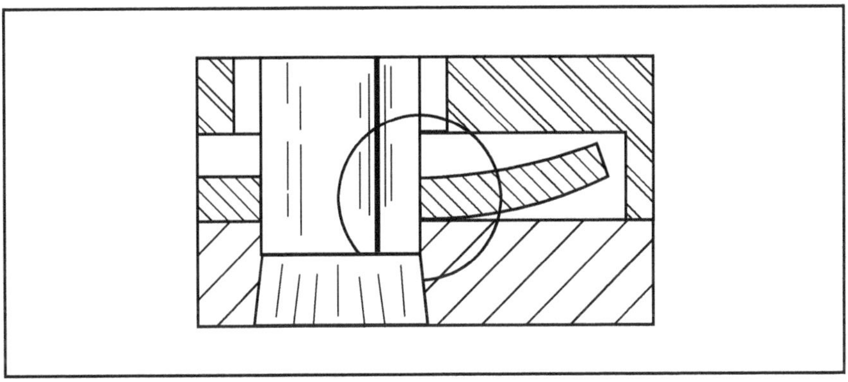

Figure 9-4. Close-up view shows distorted stock problem.

PRESSURE SYSTEMS FOR PADS AND STRIPPERS

Spring Selection

For low to moderate stripper or pad pressures and short travel distances, steel coil springs are normally used. Coil springs are economical and offer long service life when the manufacturer's recommendations for allowable compression or deflection are followed. When higher forces are needed, other pressure systems are used (Smith 1990).

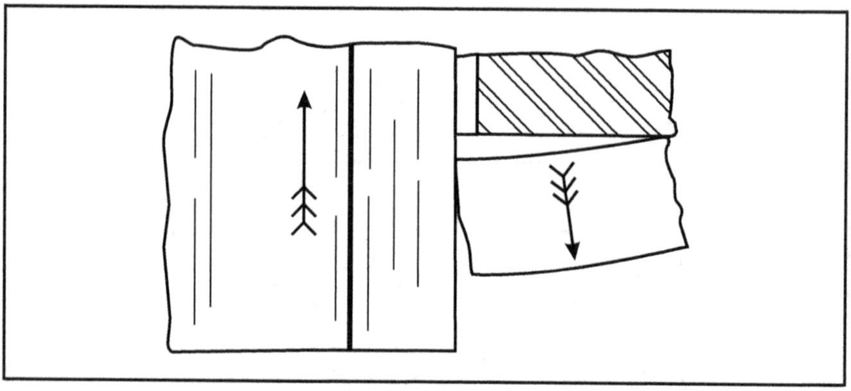

Figure 9-5. The distorted stock grips the punch like a "finger lock," which results in an extremely high stripping load.

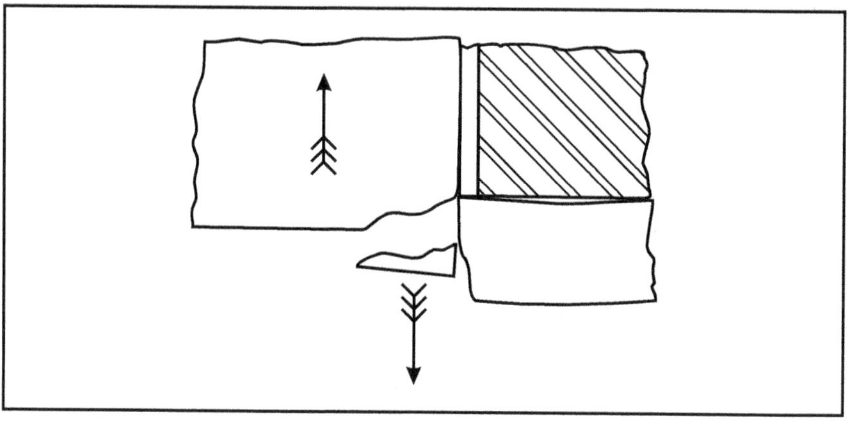

Figure 9-6. A combination of distorted stock, high stripping loads, and poor stock control results in a chipped punch face.

Mechanical Springs

Mechanical springs usually provide the lowest-cost stripper or pad-pressure source. The amount of maximum allowable compression is stated as a percentage of the undeflected spring length. A coil spring develops zero force unless deflected. Therefore, if pressure is needed upon initial pad or stripper contact, the spring must be preloaded by the pad- or stripper-retention system. The amount of preload must be subtracted from the allowable total spring deflection.

Mechanical springs are available in a variety of lengths and diameters. In addition, several duty classes are available from various die-spring suppliers. For example, a manufacturer of die springs typically provides light, medium, heavy-duty, and extra-heavy-duty springs.

Spring manufacturers normally color code the duty class of springs with paint. There is no uniform system of color-coding the various duty classes of springs by the manufacturers. Therefore, it is important to cross-reference color codes if springs from more than one manufacturer are used. Cross-reference information is normally available in the various manufacturers' catalogs. An alternative is to use springs that are color-coded according to duty class with the International Organization for Standardization (ISO) standard system, as outlined in Table 9-1. These standards, adopted by the North American Automotive Metric Standards Group, are applicable to springs made to metric standards.

Table 9-1. ISO standard spring color coding

Spring Load Rating	ISO Color Code
Light duty	Green
Medium duty	Blue
Heavy duty	Red
Extra heavy duty	Yellow

Manufacturers also rate the allowable deflection of the spring in terms of life expectancy. Springs can function without danger of breaking if the deflection is limited to that recommended for maximum life. Higher deflections will provide more force if spring failure and periodic replacement is acceptable. In no case should the manufacturer's maximum deflection rating be exceeded.

System Selection

Pad- and stripper-pressure systems are rated in several ways. It is important to determine several facts before choosing any stripper- or die-pad-pressure system. These include:

- the pressure required at the start of travel;
- the force needed during pad or stripper travel; and
- how much force is required at the maximum travel of the pad or stripper.

The spring-loaded strippers have several advantages. A sectional view of a die with a spring-loaded stripper is shown in Figure 9-7.

Nitrogen Cylinders

If more pressure is needed than can be supplied by mechanical springs, nitrogen cylinders are a good choice. Nitrogen cylinders have an additional advantage in that the full-rated capacity is available without the need to preload the pad or stripper.

Nitrogen cylinders are available in self-contained units that are size-for-size compatible with some die springs. Self-contained cylinders typically provide about 40% more pressure at maximum travel than the rated force at initial contact.

There is no industry standard to rate the force capacity of self-contained nitrogen cylinders at initial contact, or for pressure buildup throughout the stroke. It is advisable to have the manufacturer or supplier provide a table listing force V travel for their cylinders to have accurate data when designing or modifying dies.

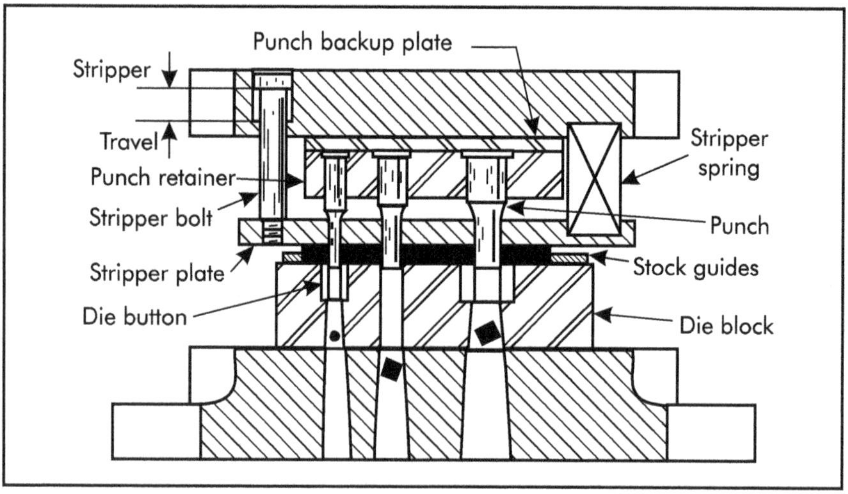

Figure 9-7. Sectional view of a die features a spring-loaded stripper.

Nitrogen Manifold Systems

Nitrogen manifold systems are often built into dies. The manifold is made of steel plate. It is advisable to test the plate ultrasonically to inspect for any slag inclusions that cause leaks.

The manifold can serve as a structural part of the die. For example, it can serve to stiffen the die shoe, permitting the use of a thinner die shoe than would otherwise be necessary.

The cylinders screw into most modern manifold systems. A carefully machined O-ring and O-ring groove accomplish sealing. Manufacturer's instructions for torque values must be followed carefully.

The manifold is cross-drilled to provide surge capacity. For draw-ring applications to avoid fractures, it is undesirable to have the pressure increase more than 10% during the stroke. This requirement may make one or more surge tanks necessary.

The operating pressure of most self-contained cylinders is 2,000 psi (13,789 kPa), although it is permissible to operate them at lower pressures to obtain the correct force for the job. Modern nitrogen manifold systems operate at a maximum pressure of either 1,500 psi (10,342 kPa) or 2,000 psi (13,789 kPa).

Air Cylinders

The use of air cylinders in dies is currently limited to automation applications such as part lifters and kickers. Years ago, air cylinders were extensively used in die cam return, stripper, and pressure pad applications. Nitrogen cylinders have almost entirely replaced air cylinders in such applications.

Hydraulic Die Pressure Systems

When extremely high forces are required, a hydraulic pressure system is capable of providing more pressure per unit area than any other system.

There are several configurations of hydraulic die pressure systems. A popular system uses an external oil tank under regulated air pressure. The air pressure fills the cylinders when the die is in the open position. The amount of pressure developed on die closure is determined by the setting of a pressure-relief valve through which oil is returned to the external air-over-oil tank.

Other systems make use of air-over-oil intensifiers. An important feature of such systems is the ability to delay the return of a die pressure pad. The feature permits the forming of flanges and other stamped-part features that would be distorted during the die opening if the pad return were not delayed. Some models permit pad-return initiation by an electrical signal to a solenoid-actuated valve.

ELASTOMER STRIPPERS

Special die rubber or polyurethane strippers are frequently used on low-cost tooling. For simple punch shapes having round shanks, a variety of styles are commercially available. Figure 9-8 shows one popular design installed on a round punch.

Shapes of elastomer strippers and pressure pads can be cast from two-part elastomer compounds. Often a prototype part can serve as a part of the mold. These materials are also useful for fabricating non-marking automation grippers and other parts-transfer components.

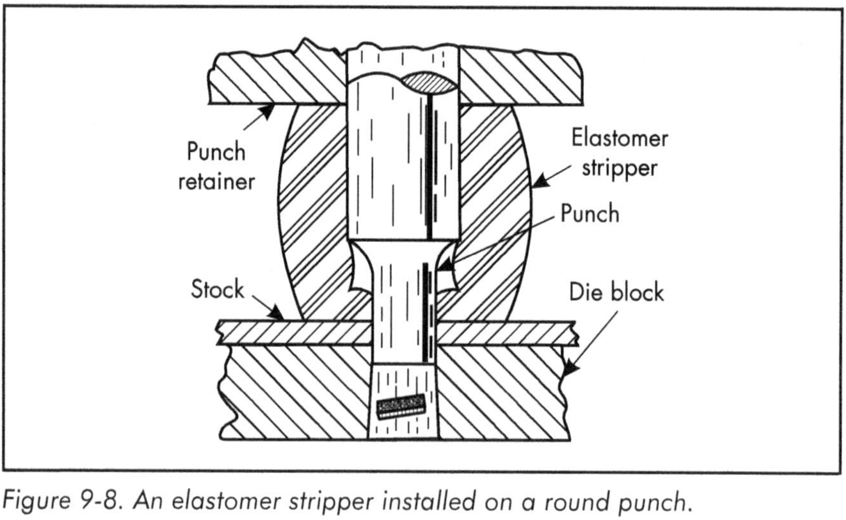

Figure 9-8. An elastomer stripper installed on a round punch.

STRIPPING FORCES

The amount of force required to strip the stock from punches varies greatly. The stripping requirements can vary from 1–2% percent of the cutting force to 20% or more. Stripping loads tend to increase with tighter cutting clearances. This is because there is a higher ratio of burnish to fracture and, therefore, more area gripping the punches.

Tight cutting clearances result in more residual stress in material around the punches than when greater clearance is used. One result is that the stock will grip the punch tightly. More hole-size closure upon punch withdrawal may also occur.

Proper die clearances serve to reduce cutting and stripping forces. Sufficient lubrication is also helpful. If cold welding and galling occurs on the punch surface, stripping forces will be greatly increased, and die resharpening will be required much sooner. In some cases, punch coatings will prevent galling and extend the resharpening intervals.

REFERENCES

Smith, D. 1990. *Die Design Handbook*. Section 5, Cutting Dies. Section 22, Die Sets and Components. Dearborn, MI: Society of Manufacturing Engineers.

——. 1994. *Fundamentals of Pressworking*. Chapter 6, Basic Die Operations. Dearborn, MI: Society of Manufacturing Engineers.

10

Anti-slug Pulling Methods

RESOLVING SLUG-PULLING PROBLEMS

When the punch is withdrawn after cutting a slug, a vacuum is created on the upstroke that will tend to pull the slug out of the hole. If heavy lubricants are present, this tendency is increased.

One solution is to build the die with tight punch-to-die clearances. This will stop most slug-pulling problems at the expense of accelerated wear, and shorten punch and die life. Other solutions also work successfully.

Spring-loaded Pusher Pin

Figure 10-1 is an example of a spring-loaded pusher pin to force the slug to remain in the die opening. It permits the punch to be sharpened without removing the pusher pin as shown in Figure 10-2.

Blow-off Device with Hole Through the Punch

Figure 10-3 shows how compressed air may be supplied through a drilled passage in a punch to blow the slug out of the die opening. A timed burst of air is advised, rather than a continuous flow.

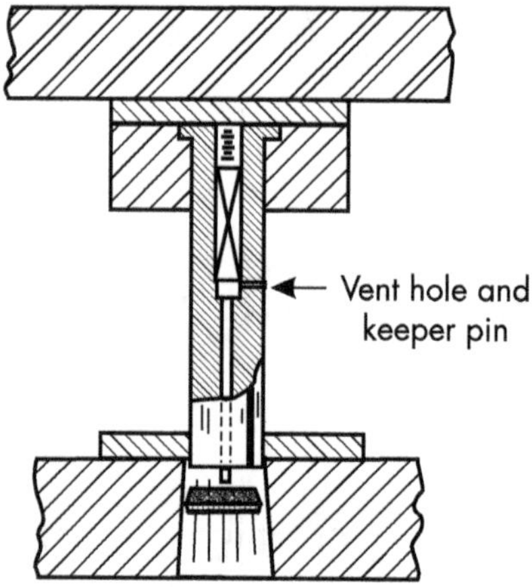

Figure 10-1. An example of a spring-loaded pusher pin to force the slug to remain in the die opening. (Courtesy Dayton Progress Corp.)

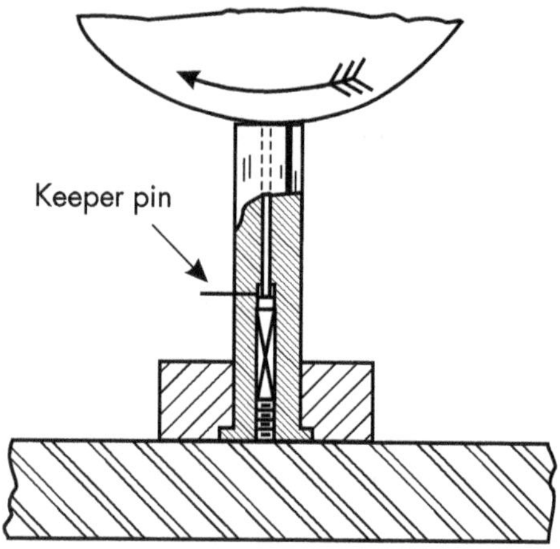

Figure 10-2. Inserting a keeper pin into the vent hole in the side of the punch will hold the pusher pin at the correct retraction amount to permit sharpening the punch, and shorten the pusher pin the correct amount at the same setup. (Courtesy Dayton Progress Corp.)

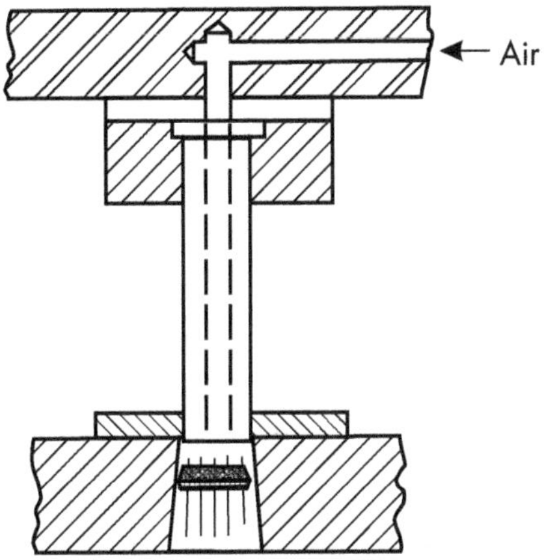

Figure 10-3. Example of compressed air supplied through a drilled passage in the punch to blow the slug into the die opening.

This method is seldom used, but is excellent for thin materials and larger slugs—it provides a large effective pressurized area.

A continuous flow, as shown in Figure 10-4, should be avoided because it will:

- waste costly compressed air;
- create excessive noise, which can damage human hearing;

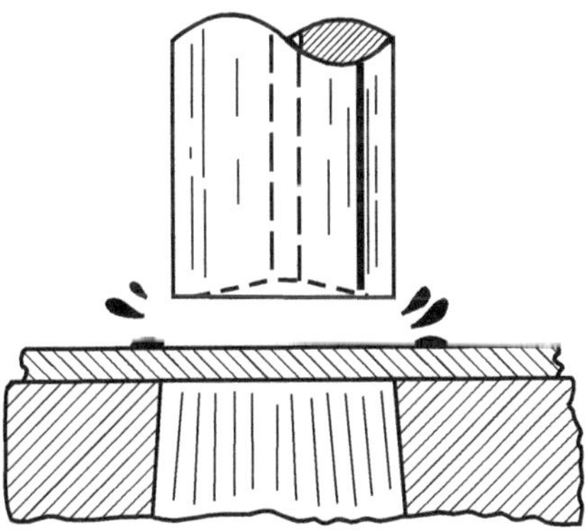

Figure 10-4. An example of how a continuous blast of air through the punch blows the lubricant from the stock surface.

- blow the lubricant off the surface of the stock before contact occurs;
- atomize lubricant, creating health concerns and housekeeping problems.

The use of a timed burst of air is advised for several reasons. These include:

- reduction of air consumption;
- the lubricant is not atomized, resulting in more healthful conditions in the pressroom and less lubricant waste;
- the timed air burst has maximum effectiveness.

Figure 10-5 illustrates a ground conical depression in the face of the punch. This depression permits the compressed air to act against most of the slug surface since it is fractured from the part. This effectively increases the available piston area and serves to improve slug blow-off action.

Reverse Taper in the Die Button

In most cases, slug pulling can be stopped by grinding the die for approximately one-half to one-and-one-half the metal thickness with a reverse taper. This will cause the slug to jam into the

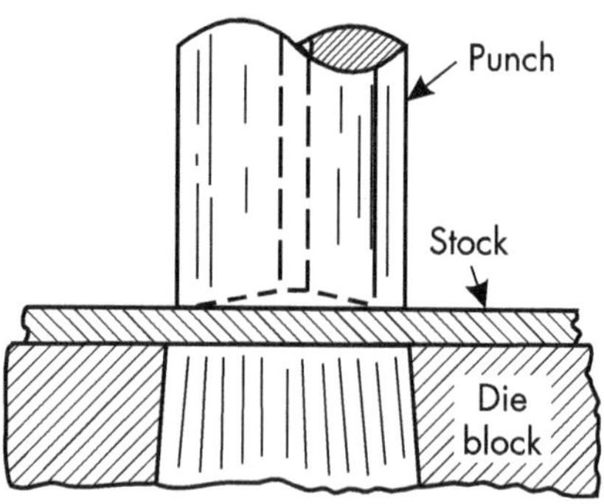

Figure 10-5. A ground conical depression in the punch face effectively increases the available piston area and serves to improve slug blow-off action.

die opening tightly so it cannot pull back. As shown in Figure 10-6, the punch entry is set to jam the slug just cut into the taper. This will dislodge the previously cut slug, which is shown being discharged.

There are some disadvantages to using reverse taper in the die button. Punch, and especially die wear, are increased. In addition, the cutting pressure is increased, tending to cold weld the slugs together in a string or solid stick-like continuous mass, which can cause severe slug jamming.

When reverse taper is used, (Figure 10-6), it is helpful to keep the tapered portion of the die opening short and provide a stepped or normal tapered relief beneath the reverse-tapered portion. While this will aid slug shedding and cold-welding problems, it tends to weaken the die section. In addition, die life is reduced. This method is not considered good die-making practice. Rather,

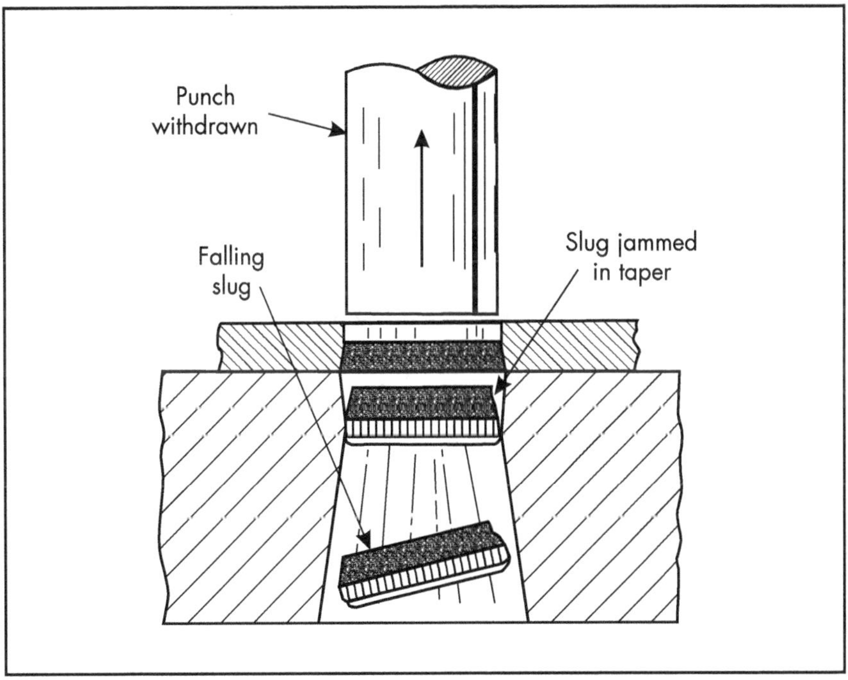

Figure 10-6. The use of reverse taper in the die opening will cause the slug to jam into the die opening tightly so it cannot pull back.

it is a method of dealing with slug pulling at the expense of die life and maintenance costs.

Emergency Anti-slug Pulling Measures

In an emergency, many anti-slug pullback methods have been tried and found helpful. These include everything from deliberately nicking the punch and/or button to set up a small burr, to applying electrical tape or Velcro™ to the punch face.

Electrical Etching and Metal Deposition in the Die Opening

One method that works in an emergency is to etch the inside of the die button with an electrical etching marker. If etching equipment is not available, arc-welding equipment set for low amperage can prove effective for larger die openings in an emergency.

A similar technique is to deposit hard wear-resistant material with an electrical deposition tool within the die opening. This type of equipment and deposition technique may give better results than etching equipment intended for tool marking or low-amperage, electrical-weld deposition. Several companies provide equipment for depositing wear-resistant materials on tool steel.

Punch Shear Angles

Grinding a shear angle on the punch can stop slug pulling by curling the slug into the die opening as the cutting action occurs. One example of a punch shear angle is shown in Figure 10-7. A disadvantage of this method is a tendency for the punch to deflect to one side.

Unbalanced shear as shown in Figure 10-7 is useful when the die opening is on an angle. This condition is frequently a problem with dies for trimming and piercing holes in large irregularly shaped parts. Here, the point should be ground to a slightly greater angle than that of the die opening. The sharp edge should begin the cut. In this way, the punch portion that heels up the punch

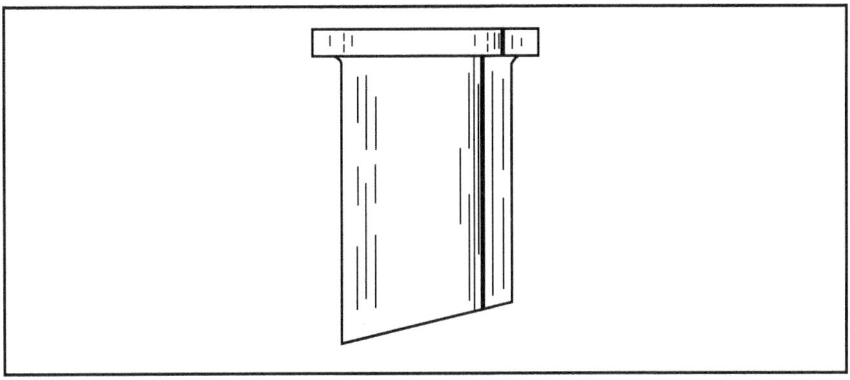

Figure 10-7. Grinding a shear angle on the punch curls the slug into the die opening to stop slug pulling. A disadvantage is a tendency for the punch to deflect to one side.

limits any tendency for the punch to deflect. Developing proper punch-point geometry may require some trial and error. However, once the best point form is found, it can be reproduced as needed from die-maintenance data.

Most hole-cutting operations do not involve angular punch entry. To curl the slug into the die opening with straight entry, a punch geometry of the type illustrated in Figure 10-8 is useful to prevent slug pullback. Angular shear is good to curl the slug to avoid slug pullback. An added advantage is reduction of peak cutting force and snap-through energy release.

Vacuum Slug-ejector Systems

A very useful slug- and part-ejection system makes use of a commercially available air-powered-vacuum ejector device known as a *bazooka*™. The bazooka is easily fitted under die openings. The blast of air may be timed to avoid waste. Like all vacuum systems, it is best suited for thin, light materials (see Figure 10-9).

Figure 10-9 illustrates one of several designs of compressed-air-powered vacuum slug-ejector devices. This two-piece design fits into precision-machined openings in die shoes and subplates. The O-rings seal the bored openings. Compressed air is supplied by drilled holes in the shoe or die subplate.

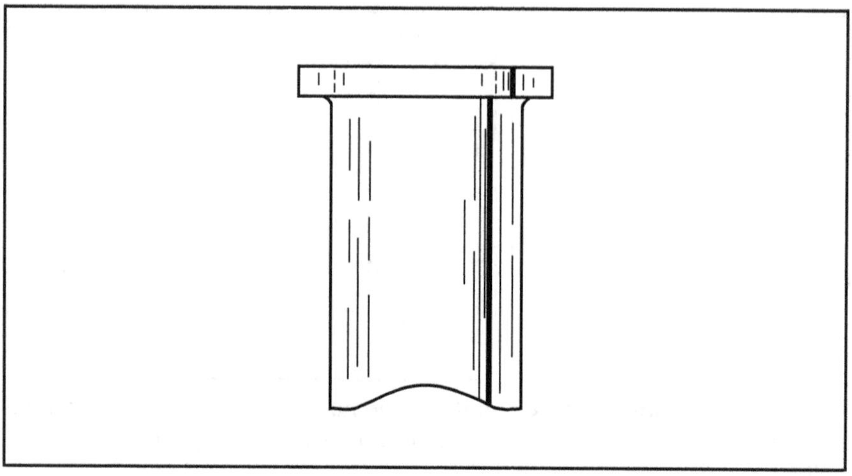

Figure 10-8. Balanced V-shear is used to curl the slug into the die opening. The flats at the edges of the punch together with the radius in the center increase punch's wear life and lessen the tendency for the punch to split in the middle.

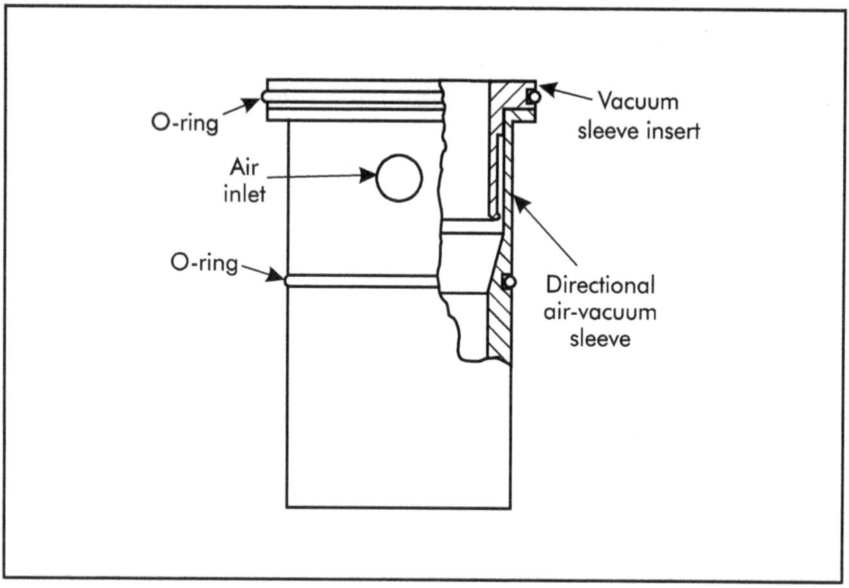

Figure 10-9. An air-powered-vacuum slug-ejector design. (Courtesy Koppy Corporation)

External Vacuum Systems

External vacuum systems are suited for removing large, thin slugs in high-speed work, such as semiconductor lead-frame production. If scrap is discharged through the bed, the entire chamber can be sealed so that an external vacuum source can be connected.

The use of a compressed-air-powered vacuum-ejector device is not recommended due to the high air consumption needed to achieve satisfactory results. A high-volume, dry-vane or roots-blower pump is advised for such applications. Of course, a filter must be used on the pump inlet to prevent the entry of slugs and debris.

One consideration of a vacuum system is that the vacuum force acting on a slug in a die opening cannot exert more than 14.5 psi (750 mm Hg), which is normal atmospheric pressure at sea level. In addition, some airflow is necessary since a vacuum, by definition, contains no air with which to convey slugs.

Slug-retention System

A unique patented die-opening geometry has shallow helical grooves cut into the die opening. A slug is retained in the die opening as the cutting action occurs. This principle is illustrated in Figure 10-10.

In a round die opening, the grooves are helical or straight. If helical, then the grooves have opposite helixes. If straight, they are terminated at the depth of punch penetration. Non-round shapes have angular grooves or straight grooves similar to the round shapes. The inability of the small discontinuity on the slug to follow the groove locks the slug into the die section. This method may leave a slight burr on both the slug and opening that may be objectionable in some cases.

OVERVIEW OF METHODS

Slug damage to the die and parts are a result of failing to retain or properly eject slugs each time. There are three basic ways to retain or properly eject slugs from die openings. These methods are:

Figure 10-10. A small carbide die button is shown with grooves in the sectioned button halves. (Courtesy Oberg Industries)

1. pushing the slug into the die opening with either a spring-loaded pin within the punch or a blast of compressed air through the punch;
2. sucking the slug into a cavity beneath the die opening with some form of vacuum system—this method works best with thin materials; and
3. modifying the die-opening clearance and geometry to retain the slug within the die opening—there are many methods, including a recently patented system that is unique, compared to any past shop practices.

The most common die-opening deviation to control slug pulling is to reduce the amount of die clearance from values known to produce long die life. While this often does stop the slug-pulling problem, cutting pressures and tooling wear are increased. In extreme cases, a reverse die taper is used. This also greatly increases cutting pressure and die wear.

Some emergency measures, such as nicking the die cutting edge with a carbide scriber, etching the die opening with an electric marker, or depositing globules of material with an electric welder, are effective. However, these do not reflect good die-making practices. They are emergency measures that are wasteful of time and tooling life.

11

Cutting Force and Snap-through Reduction

Improvement of die and press operations frequently requires a skilled die-maintenance technician to optimize timing for smooth operation with minimal tonnage requirements. The peak pressure required to cut through material can be reduced by grinding one or more shear angles on the punch or die.

TYPES OF PUNCH SHEAR

Figure 10-6 illustrated a method of reducing cutting forces by grinding angular shear on the punch point. A shear angle will reduce the peak cutting force. If a cutout is discarded, such as a slug, the shear is placed on the punch. The slug or cut-out portion will be curved or bent.

Simple Angular Shear

Grinding a shear angle on a punch point will reduce peak punching forces. However, the shear is unbalanced and will result in a side or lateral forces. For thin stock, this method may be satisfactory. The punch and die guiding system must limit the side or lateral movement of the punch relative to the die. In no case should so much deflection occur that the punch and die come into direct contact.

If the punched-out portion is a part that must be flat, the shear must be ground on the die, rather than the punch, to limit part distortion. Shear on the punch will curl or bend the cut-out part.

Balanced-vee Shear

Balanced-vee shear is an improvement over simple angular shear. This method has two advantages. By balancing the load with symmetrical shear, lateral force is avoided. The hollow-vee designs start cutting at the edges of the slug and finally complete the fracture at the center. This may help reduce the spreading force on the die.

A disadvantage of a balanced-vee design is that pressure is concentrated in the center of the punch. Cutting thick, high-strength materials can develop sufficient tensile stress to cause this punch design to split at the center of the vee. See Figure 10-8 for an example of a balanced-vee shear.

Balanced Pointed Shear

For very heavy punching operations, a pointed punch, shown in Figure 11-1, can be used to reduce peak punching loads. If this design is used, the die block may need to be larger to withstand any increase in lateral or side loads caused by the slug's bending action.

In both punch styles shown in Figures 10-8 and 11-1, the total energy required is greater due to the addition of bending forces. However, the peak force is less, which is the goal of both force and snap-through energy reduction.

PUNCHING THICK CARBON STEEL

Figure 11-2 is a photograph of a large chain side bar made of 0.625-in. (15.9-mm) thick AISI-SAE 1039 fine-grained carbon steel. A 6-in. (152.4-mm) scale is shown in the photograph for size comparison.

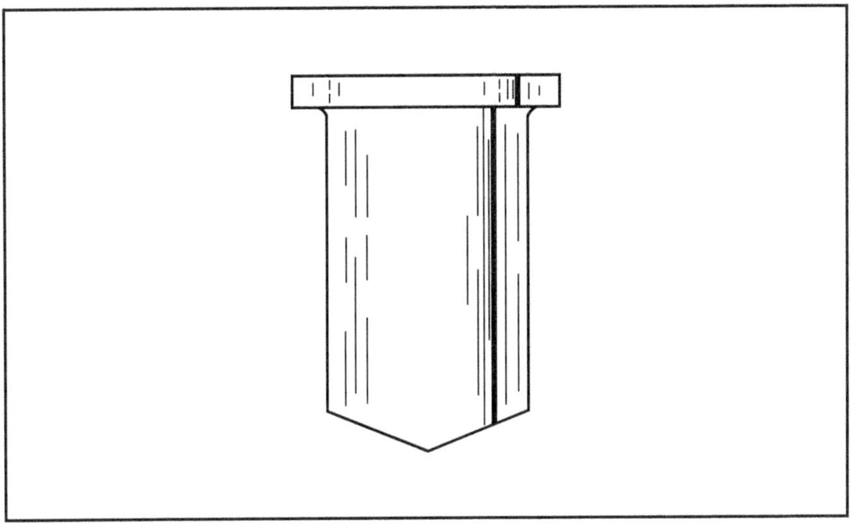

Figure 11-1. A balanced pointed shear is useful for heavy punching operations, because it helps reduce peak punching loads.

Figure 11-2. Shown is a chain side bar made of 0.625-in. (15.9-mm) thick SAE-ASTM 1039 fine-grained carbon steel. A 6 in. (152.4 mm) scale shown in the photograph gives size comparison. (Courtesy Webster Industries)

Figure 11-3 is a close-up view of the chain side bar shown in Figure 11-2. Note the curvature in the slug. The punch has a pointed angular shear optimized for the material (Smith 1990).

Figure 11-4 illustrates the waveform signature of the stress-strain relationship when cutting off and piercing two holes in a chain side bar. The part is slightly smaller than the side bar shown in Figures 11-2 and 11-3. The material is AISI-SAE 1039 fine-grained steel 0.500-in. (12.7-mm) thick by 3.0-in. (76.2-mm) wide.

The data was taken with a chart recorder speed of 7.874 in. (200 mm) per second.

IMPORTANCE OF TIMING

In timing punch entry or die shear, care must be taken to provide for a gradual release of the force developed. With the exception of high-speed applications, a shock load is not normally generated by the impact of the punch on the stock. In fact, when the punch first contacts the stock, the initial work may be done by the kinetic energy of the slide. To complete the work, energy must

Figure 11-3. Shown is a close-up view of the chain side bar shown in Figure 11-2. Note the curvature in the slug. (Courtesy Webster Industries)

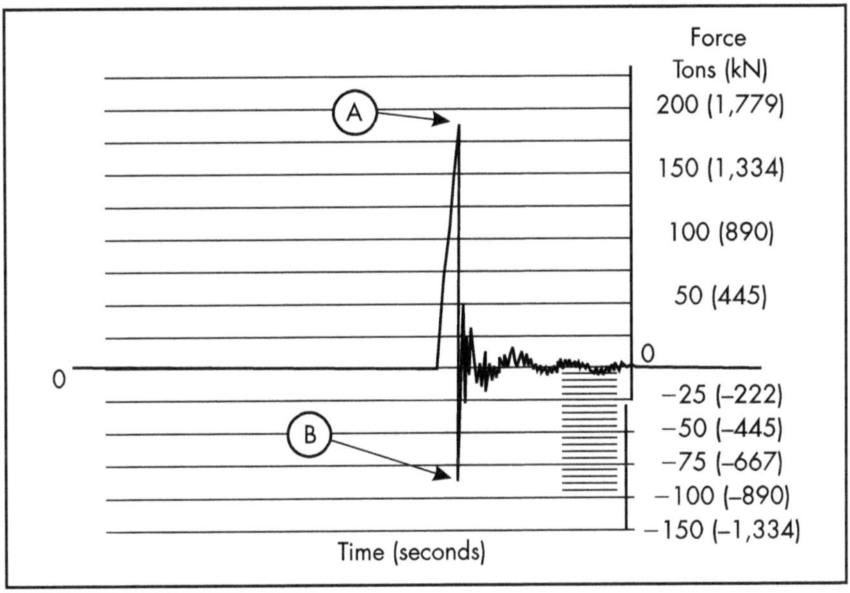

Figure 11-4. Actual waveform signature analysis of a combined piercing and cut-off operation with excessive snap-through or reverse load is illustrated. (Courtesy Webster Industries)

be supplied by the press flywheel or hydraulic pump. As this occurs, the press members deflect. An analysis of the quantity of energy involved will show why a gradual reduction in cutting pressure before snap-through is very important.

PRESS REVERSE-LOAD CAPACITY

The reverse or stripping load that most presses can withstand without sustaining damage is 10% of the rated press force or tonnage. Some presses designed for difficult cutting work have higher reverse-load capacities. Exceeding the reverse or snap-through rating of any machine can damage components. An especially critical part is the slide connection. The connection is the attachment of the pitman to the slide. Should the connection fail, the slide may fall unexpectedly. Some presses are designed to withstand higher reverse loads. For example, some presses can be designed to withstand repeated 50%-reverse loads.

Square-law Relationship

The press speed of the waveform signature in Figure 11-4 is 60 strokes per minute (SPM). Even at a chart speed of 7.874 in. (200 mm) per second, the waveform trace distance from initial contact of the punch on the work until it breaks through is short. The portion of the waveform from initial punch contact to break-through occurs in 0.20 in. (5 mm) or 0.025 seconds.

The punching waveform exhibits a sharp negative spike below the zero trace at break-through. This is due to a sudden release of the energy stored in the press and die in the form of strains or deflection. A simplified mathematical analysis is possible, based on Equation 11-1 (Smith 1990). The magnitude of the actual energy released increases as the square of the actual tonnage develops at the moment of final break-through. This is known as the *square-law relationship*. The actual energy is determined by:

$$E = \frac{F \times D}{2} \qquad (11\text{-}1)$$

where:

$E \times 166.7$ = energy (ft-lb)
$E \times 9.807$= energy (J)

F = force at moment of break-through (lbf [kN])
D = total deflection (in. [mm])

For example, if 400 tons resulted in 0.080-in. total deflection to cut through a thick steel blank, the energy released at snap-through, from the formula, is 2,667 ft-lbf. Careful timing of the cutting sequence results in a reduction of the tonnage as snap-through occurs to 200 tons. The reduction in shock and noise is dramatic. This is because half the tonnage produces only half as much deflection or 0.040-in. The resultant snap-through energy is only 667 ft-lbf or one-fourth the former value.

DIE TIMING IMPROVEMENT

Timing shear and punch-entry sequences to release force, before snap-through, reduces shock and noise. The simplified analysis of the square-law relationship can be applied to the following case study.

A 300-ton (2,669-kN) straight-side press is used for this operation. The allowable reverse load is 30 tons (267 kN). Point (A) of Figure 11-4 illustrates a peak load of 191 tons (1,699 kN), which is well within press capacity. The reverse load (B) is 87 tons (774 kN), which is nearly three times the allowable amount. The die is immediately taken to the repair bench and one punch is shortened by 0.312 in. (7.92 mm). Balanced angular shear is ground on the punches. Balanced shear is also ground on the parting punch.

Figure 11- 5 illustrates the improvement achieved by modifying the tool. The peak tonnage is reduced to 82.8 tons (737 kN), less than half the initial value. The reverse load is reduced to 22 tons (196 kN), or about one fourth of the former value.

The previous example and the documented results of many other tests show snap-through reductions that conform closely to the square-law formula. Simply stated, if the amount of force or tonnage released as punch break-through occurs can be reduced by one-half, the amount of stored energy that causes snap-through problems will be reduced to one-fourth the former amount.

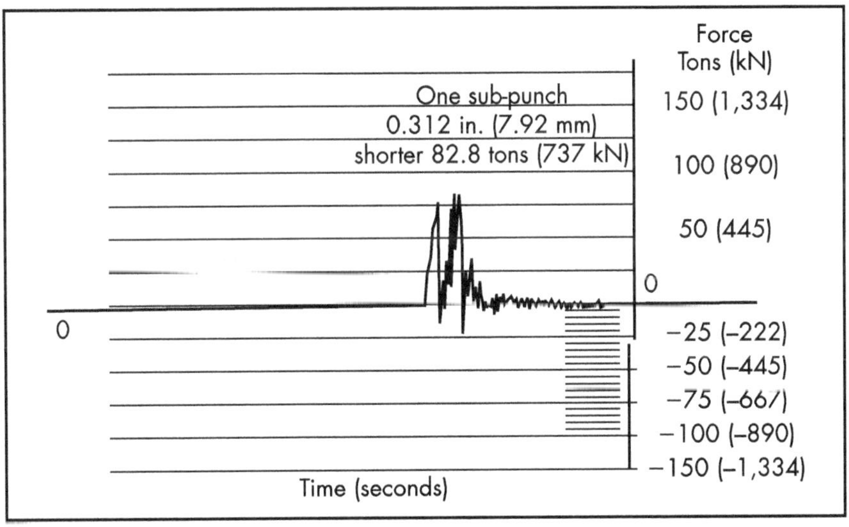

Figure 11-5. Waveform signature of the same operation illustrated in Figure 11-4 after modifying the die by adding timing and balanced shear. (Courtesy Webster Industries)

Combined Cutting and Bending

Figure 11-6 illustrates the waveform signature of a combined punching, cut-off, and joggle-bending operation. Here, AISI-SAE 1039 steel 0.500-in. (12.7-mm) thick by 2.000-in. (50.8-mm) wide has two holes punched, and a 0.562-in. (14.27-mm) joggle formed. The part, an engineering-class chain side bar, is also cut off in this combined operation.

The die is correctly timed. The snap-through energy release is well under 10% of the capacity of the 300-ton (2,669-kN) straight-side press used for this operation.

PROVIDING ANGULAR SHEAR ON THE DIE

All types of shear ground on the punches will result in curled or bent slugs. If the slug is the part, this may be unacceptable. In such cases, angular shear may be provided on the die. Distortion

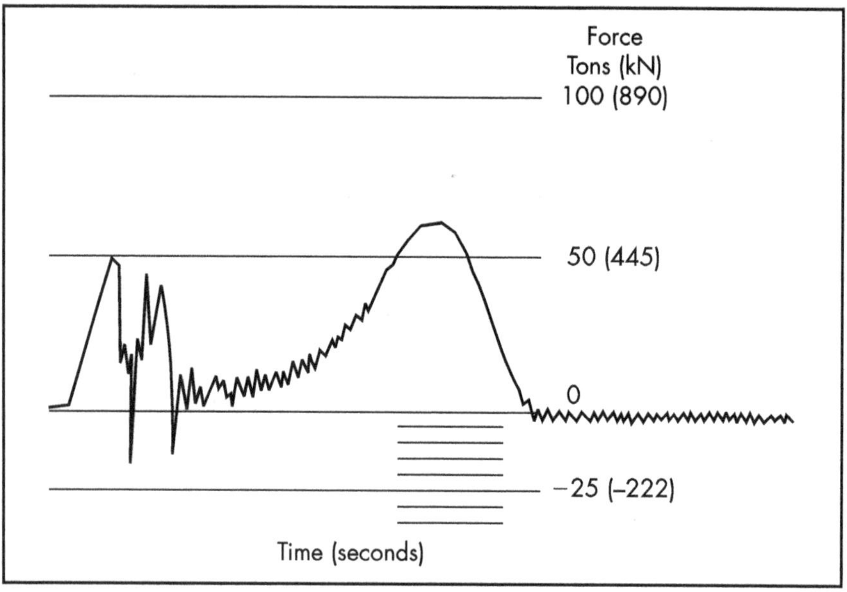

Figure 11-6. The waveform signature of a combined punching, cut-off, and joggle-bending operation is shown. The part, an engineering-class chain side bar, is also cut off in this combined operation. (Courtesy Webster Industries)

of the scrap skeleton is usually not a concern. The goal is to produce flat parts.

In the die shown in Figure 11-7, flat areas around the corners at *D* provide a level support for the stock and prevent slippage at the start of punch entry. With the amount of shear *B* and angle of shear *A*, cutting progresses from the outside corners to the center, producing a flat blank. Typically, the amount of shear *B* is approximately one to three times stock thickness.

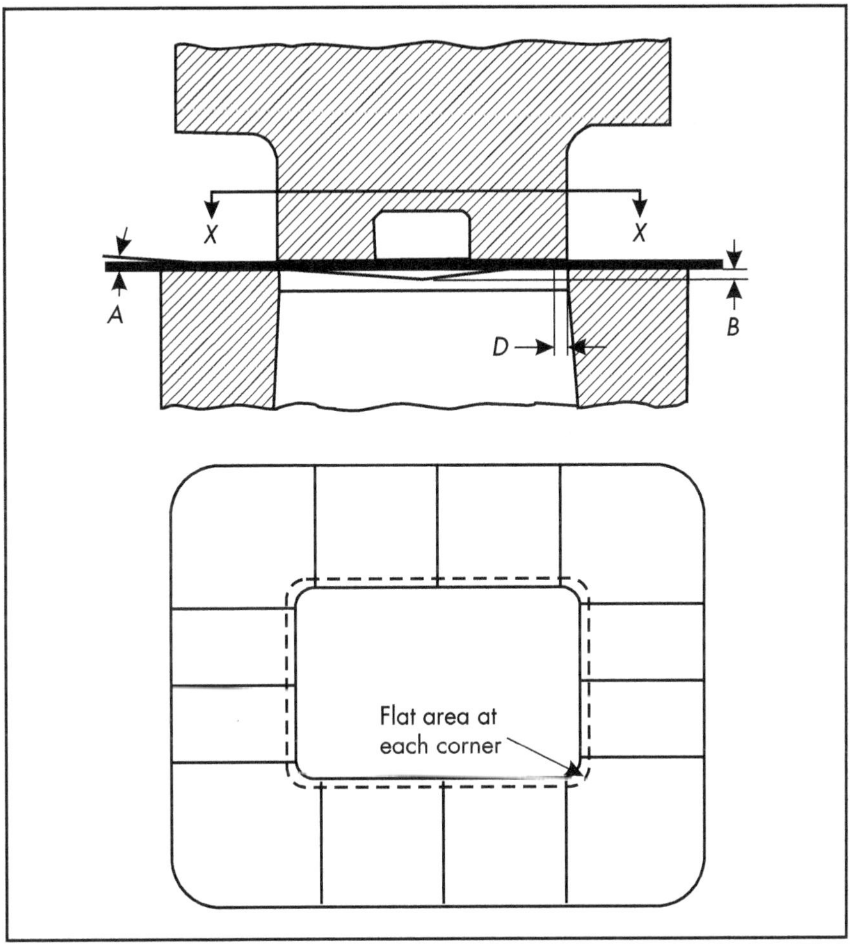

Figure 11-7. Concave shear applied to the face of the die: the cutting progresses from the outside to the center, leaving a flat blank (Smith 1990).

SUMMARY

In addition to providing angular shear on the punch and/or die, entry of individual punches may be timed to reduce cutting forces. When using typical cutting clearances, the punches penetrate the stock one-third of stock thickness when fracture occurs. Therefore, the entry of the punches is usually stepped in increments of approximately one-third stock thickness.

In many cases, optimizing punch and die shear, together with stepping-punch entry, reduces peak-cutting forces by one-half to two-thirds of the untimed value. The total press energy required per stroke is not reduced. The process of optimizing cutting forces is aided by force monitoring and waveform-signature analysis. These methods are valuable process-control tools.

It is very important to maintain records of the optimum die timing for each job. This information will save trial-and-error work when the die is sharpened. Proven timing data is also invaluable in avoiding trial-and-error adjustment of new dies.

REFERENCES

Smith, D. 1990. *Die Design Handbook*. Section 4, Shear Action in Metal Cutting. Dearborn, MI: Society of Manufacturing Engineers.

——. 1990. "Using Waveform Signature Analysis to Reduce Snap-through Energy." Technical Paper MF90-11. Dearborn, MI: Society of Manufacturing Engineers.

12

Bending of Metals

Bending is frequently used to increase the rigidity of shaped parts in pressworking operations. The simplest bending operation is air bending, so called because the die does not touch the outside of the bend radius. The part to be bent is supported on each side of the bend and force is applied to the forming punch in the center.

Figure 12-1 illustrates a metal beam supported at two points, with a load applied at the midpoint. The load produces compressive stresses in the material on the inside of the bend as it is forced into compression. Tensile stress or stretching occurs on the outside of the bend.

To produce a bend in a finished part, the yield point of the material must be exceeded. If the bending force applied does not exceed the yield strength of the material, the beam will return to its

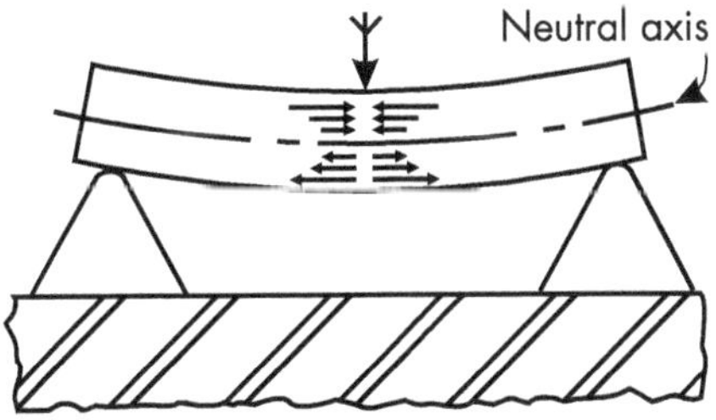

Figure 12-1. A metal beam supported at two points, with a load applied at the midpoint, resulting in bending or deflection.

original shape upon removal of the load as shown in Figure 12-2. However, if the stress exceeds the material-yield strength, the beam will retain a permanent set or bend when the load is removed, as illustrated in Figure 12-3.

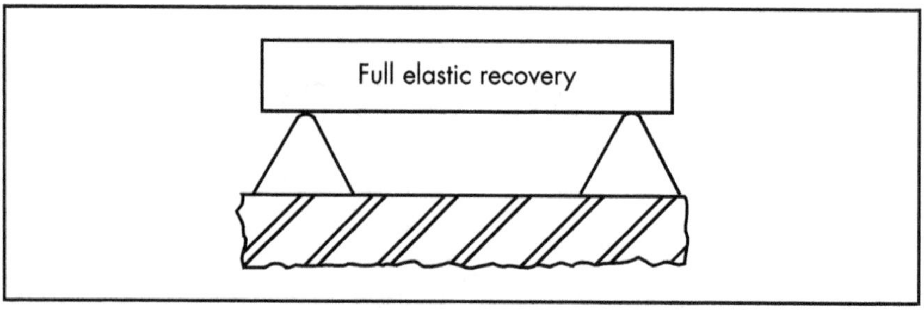

Figure 12-2. If the applied force does not exceed the material-yield strength, the beam returns to its undeflected shape.

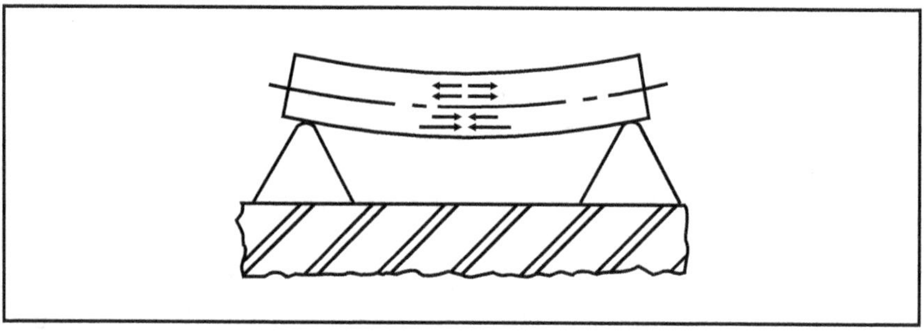

Figure 12-3. Simple beam deflection occurs in air bending. If the applied force exceeds the material's yield strength, the beam retains a permanent set or bend when the load is removed.

The goal of the process is to bend the material the correct amount. Springback or elastic recovery will then occur until residual stresses in the bend are equal to the stiffness of the material. This concept is illustrated in Figure 12-4.

Not all of the material in the bend zone is stressed equally. The material in the inner and outer surfaces is stressed the most, and the stress gradually diminishes toward a neutral axis between the two surfaces. At that point, the stress is zero, and there is no length change.

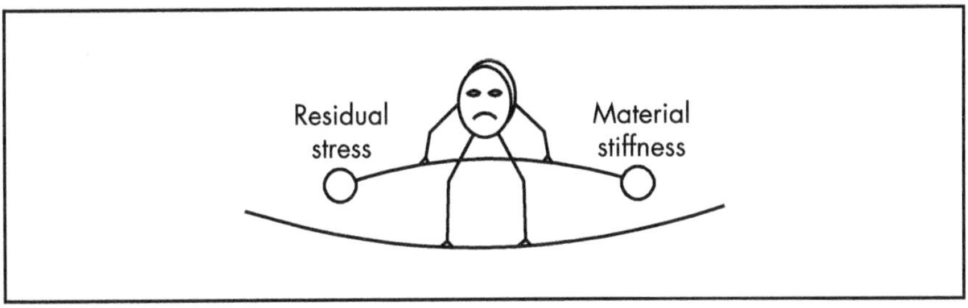

Figure 12-4. Springback occurs until the residual stress forces are balanced by the stiffness of the material.

BEND ALLOWANCES

The exact length of a bend is determined by trial and error. The assumed neutral axis varies, depending upon the bending method used, the location in the bend, and the type of stock to be bent. The direction of grain in a steel strip relative to the bend has a slight effect on the length of metal required to make a bend. Bending with the grain allows the metal to stretch more easily than bending against the grain. However, this results in a weaker stamping.

Bend allowances depend upon the physical properties of the metal, such as its tensile and yield strength and ductility.

Empirical Rules

The exact bend allowance is the arc length of the true neutral axis of the bend (metal is stretched above the neutral axis; below it, metal is compressed). The neutral axis only can be approximated. Many manufacturers assume the neutral axis is 1/3-stock thickness from the inside radius of the bend for inside radii of less than twice the stock thickness. For an inside radii of two times the stock thickness or greater, the neutral axis is assumed to be 1/2-stock thickness from the inside radii. One reason for the requirement of relatively less metal to make a tight bend is that the sharp radius tends to be drawn or stretched slightly.

Many experts believe that the location of the true neutral axis from the inside radius varies from 0.2–0.5 times the stock thickness. An important factor that determines the neutral axis is how the bend is accomplished. Less metal is required for a bend made by a tightly wiped flange than for an air bend on a press brake. Wiping the flange tends to stretch the metal.

Formulas and Data

Tabular data is available to determine the inside of the bend radii, from 0.015–1.250 in. (0.38–31.75 mm), including metric equivalents for various materials and types of bends. The formula used to develop these tables is based on extensive experimental data. Such tables were very useful before the availability of the electronic calculator. For 90° bends, the coefficient of 1.57 (the number of radians in 90°) determines the assumed amount of metal necessary to make the bend, which is multiplied by the assumed neutral axis. For bends that are not exactly 90°, it is necessary to multiply the number of degrees of bend times 0.0175 (the number of radians in 1°), and to substitute the result for the coefficient 1.57.

Springback or Elastic Recovery

Whenever forming metals, some springback occurs. The cause of springback is the residual stress that is a result of cold-working metals. For example, in a simple bend, residual compressive stress remains on the inside of the bend, while residual tensile stress is present on the outside radius of the bend. When bending pressure is released, metal springs back until residual stress forces are balanced by the material's stiffness, which resists further strain. The most common method of correcting for springback is to overbend material to obtain the desired shape after forming.

Stiffness is a function of the material's modulus of elasticity. This explains why materials, such as mild steel with a high modulus of elasticity (as compared to tensile strength), spring back less than materials with a lower modulus. However, some materials do have comparable tensile strength, such as hard aluminum alloys.

Springback Compensation

The most common method used to compensate for springback is to bend material a sufficient amount, beyond the desired angle, and allow it to spring back to the desired angle after elastic recovery occurs. This method of springback compensation is termed *overbending*. Because of the uncertainty of the exact location of the neutral axis, it is best to use trial-and-error methods when developing close-tolerance stampings.

Many complex factors determine the amount of springback that will occur in a given operation. Because the exact amount of springback is difficult to predict, data for a specific material and forming method is often developed under actual production conditions to aid process control and future product development. If the die designer and builder fail to include correct springback compensation in the die, correction will need to be done by the repair facility of the press shop that uses the tool.

Factors Affecting Springback

Some factors that increase springback are:

- higher-material strength,
- thinner material,
- lower Young's modulus,
- larger die radius,
- greater wipe-steel clearance,
- less irregularity in part outline,
- and flatter part-surface contour.

If a flanged part is irregular compared to either the outline or surface contour, the springback will be slight. The springback for large-wipe-steel clearances can be several degrees or more.

PRESS-BRAKE OPERATIONS

Press-brake tooling for air bending (see Figure 12-5) is quite simple. Air bending is one of the most common press-brake operations.

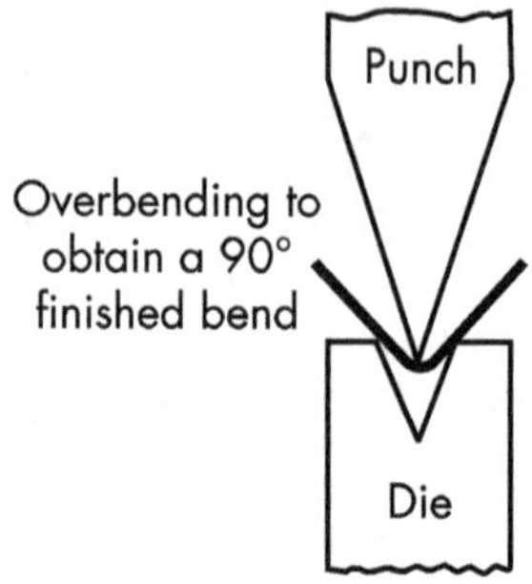

Figure 12-5. Simple tooling of the type used to air-bend sheet metal parts in press brakes. The upper die is lowered and a hit is made until the desired bend angle is obtained.

This method of bending requires minimum tonnage for the work performed. Exact repeatability of ram travel is required to maintain close repeatability of the bend angle. The amount of over-travel is determined experimentally to compensate for springback.

Causes of Bend-angle Variation

There are several causes of bend-angle variation when bending materials in pressworking operations. These include:

- changes in the stock's yield strength;
- variation in stock thickness;
- machine variations due to temperate changes; and
- machine deflection, especially in long press-brake bending operations.

Compensation for any change of the conditions that affect the bend angle may require adjustment of the ram travel. With press-brake bed deflection, shimming also may be required. In addition, shims may be required to correct for additional machine deflection. Some press-brake designs have automatic-deflection compensating devices such as hydraulic cylinders built into the bed. If high force is required in press-brake bending because of an increase in stock thickness or hardness, a simple ram adjustment may not be enough to correct the problem.

Coining the Bend to Control Springback

Coining has the advantage of producing sharp, accurate bends with less sensitivity to material conditions than air bending. The disadvantages of coining compared to air bending are high-force requirements and accelerated die wear.

Figure 12-6 illustrates a press-brake die designed to coin the bend for a precise angle. This coining action eliminates the root causes of springback, including the tensile and compressive residual stresses on opposite sides of the bend. Coining action is accomplished with pressure that is sufficient to subject the metal to the yield point in the bend area.

The tonnage required for coining might be five to ten times that required for simple air bending. Higher forces increase machine deflection. Air-bending jobs that produce acceptable bend angles throughout the entire length of the bend may need shimming if coining is required. The amount of machine deflection increases approximately in proportion to the developed tonnage.

WIPE-BENDING DIE OPERATIONS

It is often not feasible to use V-bending tooling for bending. V bending is popular for press-brake work. The tooling is simple and a variety of work can be accomplished with it. Usually, only a single bend occurs per stroke. Accurate work requires that each

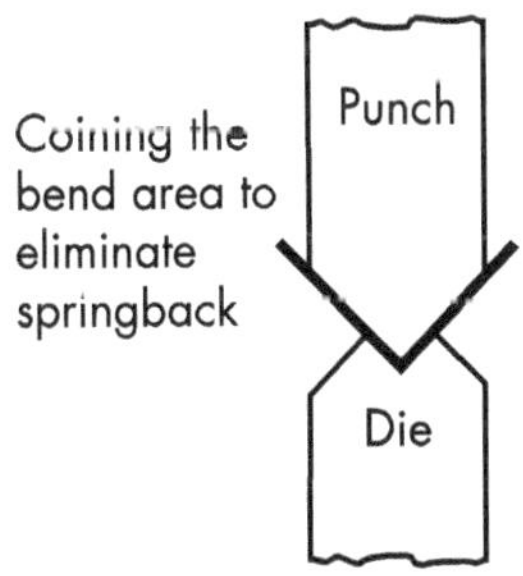

Figure 12-6. Coining the bend requires high tonnage to obtain a sharp, accurate bend.

previous step be accurate. Skilled and experienced operators are required. A limiting factor is the cost of press-brake work, because of low throughput and the high skill required to produce accurate work.

Wipe Flanging and Springback Control

Figure 12-7 illustrates a sectional view through a wipe-flanging die. In this design, the flange steel attached to the upper die wipes the metal around the lower die. A popular method for controlling springback is to coin the top of the bend with the flange steel. A disadvantage is the limited springback compensation that can result in a distorted bend-angle condition. A close-up view of this is shown in Figure 12-8. The top thickness of the bend can be squeezed beyond the material yield point by careful adjustment of the die's shut height. Only the top portion of the bend is coined.

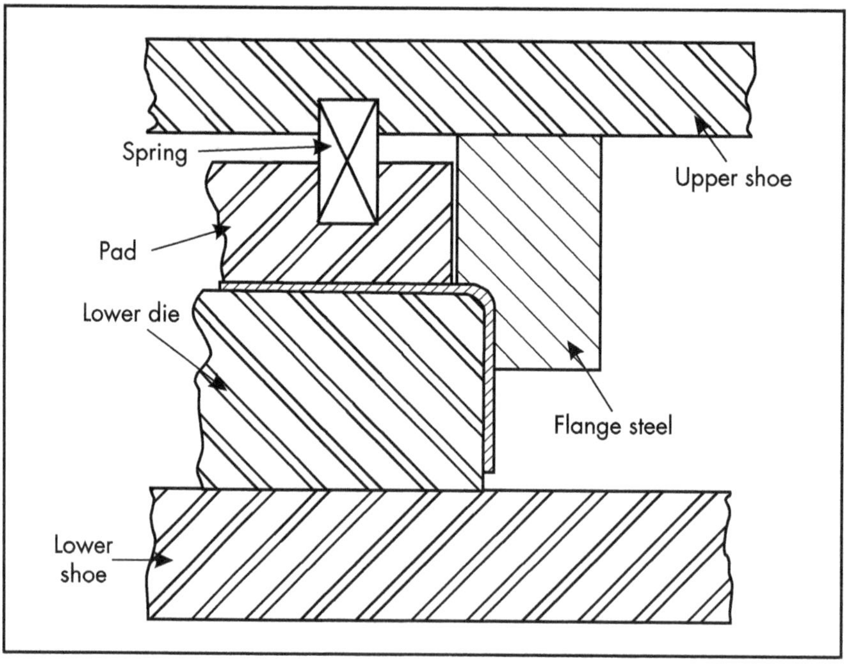

Figure 12-7. Sectional view is shown of a wipe-flanging die.

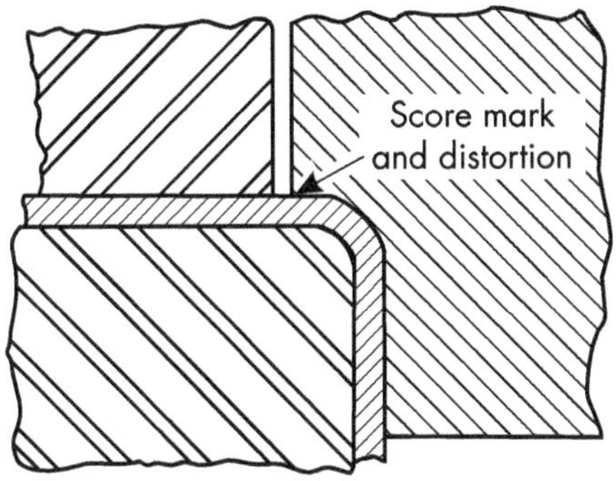

Figure 12-8. Close-up view is shown of the point of flange steel contact on the bend radius in a wipe-flanging die.

This can result in a score mark that might weaken the stamping and extrusion of the metal being coined.

If excessive coining pressures are applied, the metal at the top of the bend will extrude and result in a weak and distorted bend condition. An improved flanging method relieves the radius in the flange steel so it does not contact the top of the bend radius. One way to do this is to relieve the flange steel at an angle that is approximately 20° tangent to the radius. Another method is to machine the flange steel to a radius that is larger than the outside of the bend. The flange steel is positioned so the tightest point is 45–60° beyond the top of the radius. The side of the form steel is relieved 5° or more to permit the material to be overbent (see Figure 12-9). This method is more effective than coining the top of the bend. In addition, the improved bending process is not as sensitive to variation, due to press adjustments and material conditions.

ROTARY-ACTION DIE-BENDING OPERATIONS

A patented rotary-action bender known as the Ready™ bender combines the low-tonnage requirements of air bending with the accuracy and multiple-bend capabilities of wipe-flange tooling. Figure 12-10 illustrates a Ready bender making initial contact with the stock. As the upper die travels down, the stock is clamped

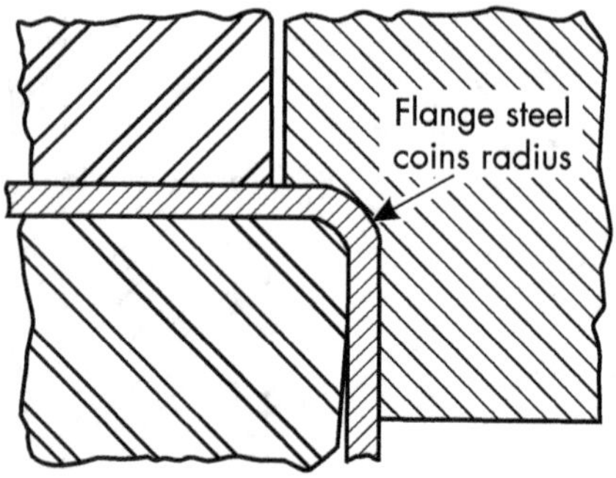

Figure 12-9. The side of the radius is coined and a relief angle is provided in the lower-die steel in this improved springback control method.

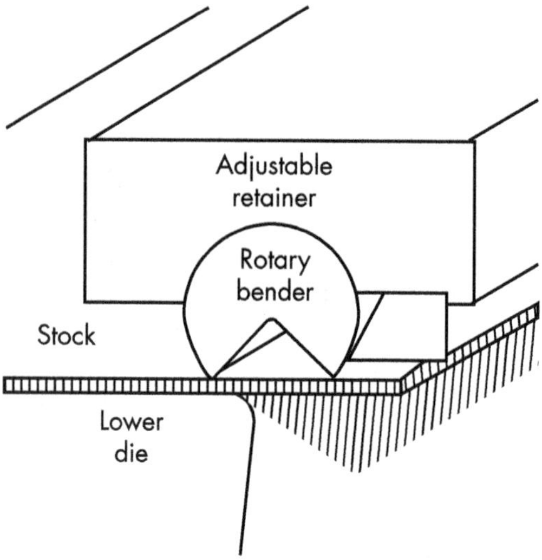

Figure 12-10. A Ready™ bender makes initial contact with the stock. As the die closes, the bender clamps and bends the stock. (Courtesy Ready Tools, Inc.)

and bent by the rotating bender. As the die closes, the rotary-bending action progresses (see Figure 12-11). An optional relief angle in the lower die permits the rotary member to overbend the stock at the bottom of the stroke to provide springback control, as shown in Figure 12-12.

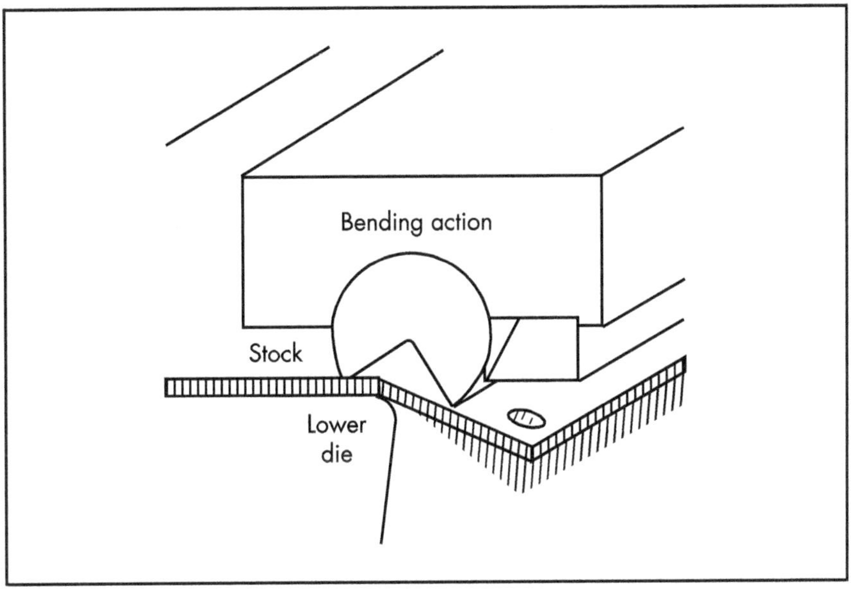

Figure 12-11. A Ready™ bender bends the stock through rotary action of the circular member. (Courtesy Ready Tools, Inc.)

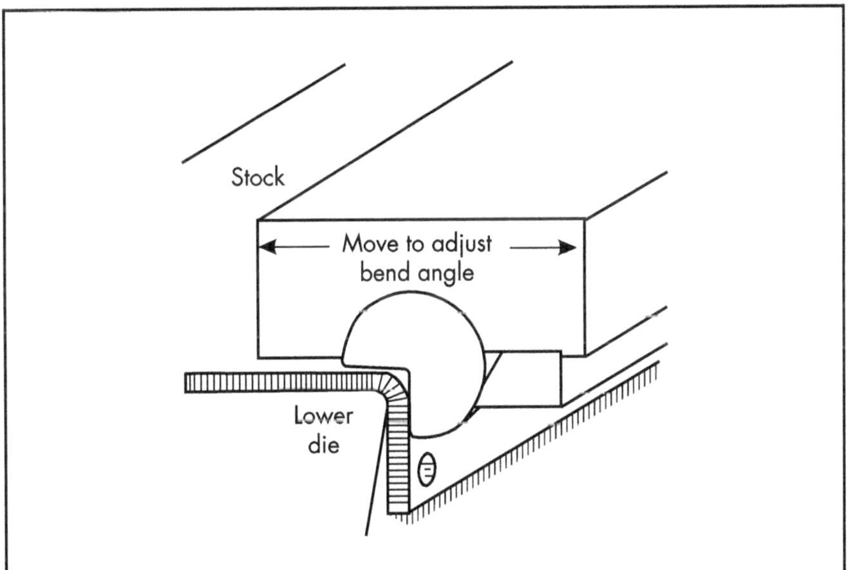

Figure 12-12. A Ready™ bender overbends the stock at the bottom of the stroke to compensate for springback. (Courtesy Ready Tools, Inc.)

Rotary-action benders can bend angles up to 120°. The rotary member is usually made of tool steel that is heat treated for long wear. The bending pressure is typically 50–80% less than that required for conventional wipe bending. The lower pressure permits many types of prepainted materials to be fabricated without damaging the finish. Rotary benders also can be constructed of nonmetallic materials, such as hard thermoplastics, for work with prefinished materials.

Fine Adjustment of Bend Angle

The bend-angle adjustment of conventional wipe-flange tooling is usually made by adjusting the flange steel up or down with shimming. In the case of rotary-action benders, the bend angle is adjusted by moving the assembly containing the bender in the horizontal plane, relative to the lower-die member or anvil. Attempts to obtain a tighter bend by excessive lowering of the press shut height can result in tooling damage.

CONTROL OF BEND ANGLE BY ADJUSTING PAD PRESSURE

One of the most frequent stamping variables affecting quality is that of incorrect bend angles. Press-brake tooling for air bending can easily compensate for bend-angle variation by making a shut-height adjustment. Many wipe-flange dies allow coining of the bend radius slightly to control springback. Often pressroom technicians try adjusting shut height immediately to attempt to correct a part's dimensional problems. Shut-height changes can affect other part features that are sensitive to shut height, such as embossments and identification stamps. Any shut-height adjustment should be thought out carefully before a change is made to avoid undesired results.

The wipe-bending sequence shown in Figures 12-13 through 12-16 shows how the bend angles of a wipe-flanging operation can be controlled by adjusting pad pressure. Technicians in a pressroom where this procedure is used have found that low pad pressures result in substantial overbend in a 90° wipe-flanging operation. The

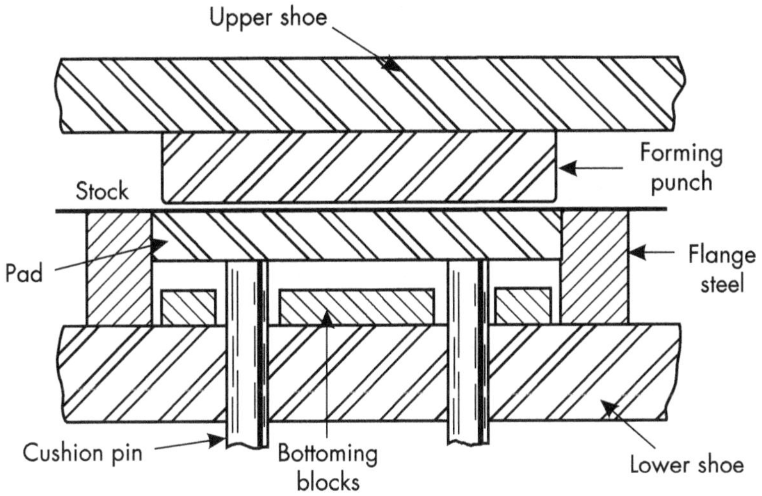

Figure 12-13. A wipe-flanging die closes with a blank to be formed in place. Note the die-cushion pressure pins and pad-bottoming blocks.

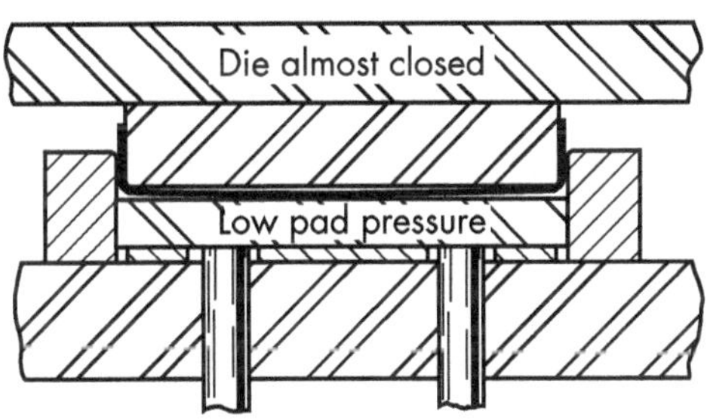

Figure 12-14. A wipe-flanging die has low cushion pressures nearing the bottom of stroke. Note that the blank has a crown or bow.

job operates in a 60-ton (534-kN) open-backed inclinable (OBI) press. Why this overbending condition occurs is unclear and requires further study and documentation.

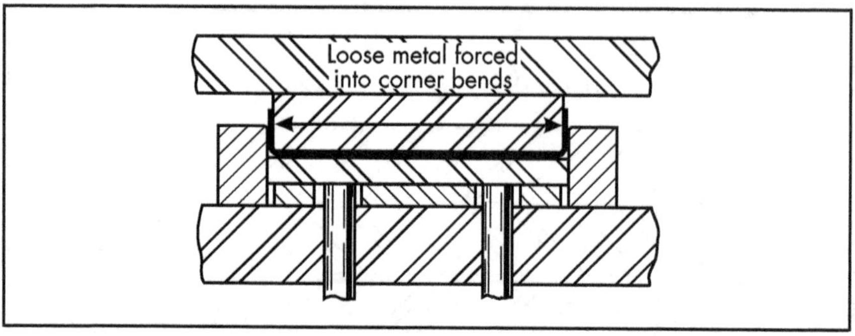

Figure 12-15. A wipe-flanging die shown at the bottom of the stroke. As the pad forcefully contacts the bottoming blocks, the loose metal is forced into the bend radii, resulting in a controllable overbending condition.

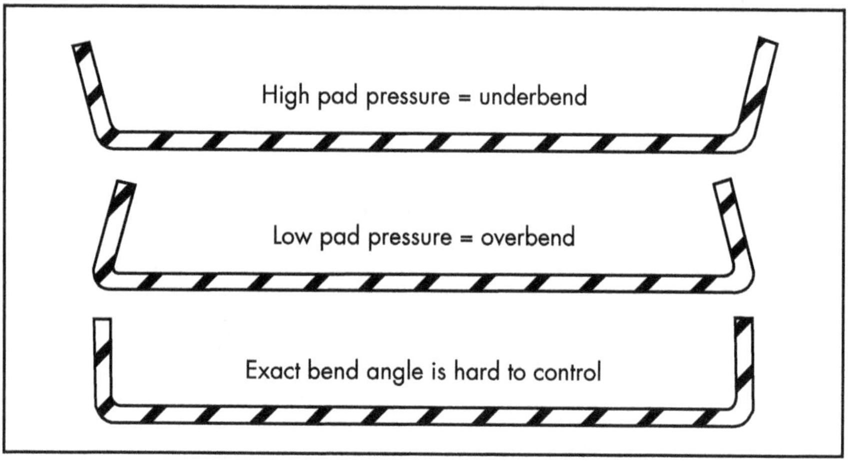

Figure 12-16. Examples of how bend angles can vary with pad-pressure settings.

Pad Pressures

Normal pad pressures typically are in the range of 5–20% of the forming force. However, in the following case study, pad pressures this high result in varying bend angles that are less than the specified 90°. In addition, coining or restriking the bend to control springback is not possible in the wipe-flanging detail design. However, carefully controlling pad pressure at low values has been found a workable method to obtain 90° bends on this type of job. In fact, several degrees of overbend can be obtained if desired.

The Effect of Low Pad Pressure on Overbending

In this case study, the pad-pressure adjustment is critical. While conventional overbending, bend coining, or Ready bender rotary-bending methods may be preferred, this particular method may be useful under some circumstances.

The operation involves wipe bending a 0.050-in. (1.27-mm), 6 × 9 in. (152.4 × 228.6 mm) rectangular steel blank on two sides. The cold-rolled steel material was checked and showed a minimum yield strength ranging from 45–50 ksi (310–345 kPa). Pressroom troubleshooters expect normal variations in material thickness and hardness to cause product variation. The finished part is a symmetrical U-shaped stamping with two straight flanges of equal length. A sectional view of the die with the blank in place is shown in Figure 12-13.

The die cushion that supplied pad pressure with cushion pins operated in the 5–7 psi (35–48 kPa) range. The available cushion force at 80 psi (552 kPa) was 5 tons (45 kN). This provided a pad force of only 625 lb (2.8 kN) over a 6-in. (152.4-mm) square pad. This pressure was quite low for accurate stock control. However, the symmetry of the part, with good press and die alignment, avoided part slippage and product variation.

Figure 12-14 illustrates the flange die near the bottom of the stroke. The excess metal in the crown or bow is deformed, but the yield point of the material is not exceeded. As the die bottoms out on the pad-bottoming blocks, the excess metal is forced into the bend radii in the corner of each flange. Having excess metal in the crown or bowed blank as the die closes is a result of using low pad pressure. This loose or excess metal is also the key to achieving controlled overbending.

Using Pad Pressure to Control Overbend (Smith 1990)

As pad pressure is increased, the amount of bowed or loose metal is reduced between the forming punch and pressure pad. This factor decreases the amount of overbend because there is less metal to force into the bend radius as the pad bottoms out. If pad pressure is high enough to hold the blank firmly against the forming punch throughout the press stroke, the part flanges will be formed at less than a 90° angle. The action of the metal being forced into the bend radii at the bottom of a stroke is shown in Figure 12-15. The product variation as a function of slight pressure variations is shown in Figure 12-16.

Practical Applications and Process Limitations

Using pad pressure to provide springback control is not recommended for the following forming application types and conditions:

- die cushions operated at very low pressures are apt to provide non-uniform forces due to temperature changes and a stick-slip, packing-adhesion phenomenon known as *stiction*;
- slight changes in stock thickness or temper will result in flange-angle variation;
- especially likely in the case of non-symmetrical parts, low pad pressures are apt to result in blank slippage and product variation.

Wipe-flange details that provide for bend-radii coining and the use of a Ready bender are effective and dependable methods of controlling the bend-angle variations caused by springback. This method can be useful under some difficult flange-bending conditions when a strong and dimensionally accurate part is required. One example of a difficult-to-bend stamping is a commercial truck or trailer spring-hanger bracket. Other parts include tilting-chair hardware and a variety of appliance stampings.

The main disadvantage of using low pad pressure to control overbend is the care required to deal with process variables. It is not considered the best method for springback control for the flanging operation in the previous case study. The factors that permit part geometry to be held within specifications include:

- The part is symmetrical, which balances the lateral forces resulting from the wipe-flanging operation.
- The flat blank is nested in accurately doweled gages that help prevent lateral shifting until the bend is correctly started as the forming punch contacts the work.
- The operator frequently checks the parts to assure conformity to dimensional tolerances. This includes catching the

parts in a small plastic bin as they are produced for inspection before putting them in the shipping container.

Lateral shifting of the blank is the greatest concern when using less pad pressure than that recommended by good design practice. Low pad forces that permit bowing of the stock, which is subsequently forced into the bend radius, can have practical applications for heavy parts that are difficult to flange. Recommended stock locating and control measures include:

- The blank should be nested in accurate robust gages to prevent lateral shifting until the bends are correctly started as the forming punch contacts the work.
- Two or more locating pins projecting from the forming punch should be used to pick up mating holes accurately located in the blank.
- The locating pinholes can be considered "nonfunctional" tooling holes if they do not affect the function of the finished part. The use of robust pins in the punch will prevent lateral shifting if the side forces are not great enough to distort the locating pin holes in the blank.

REFERENCE

Smith, D. 1990. *Die Design Handbook*. Section 6, Shear Action in Metal Cutting. Dearborn, MI: Society of Manufacturing Engineers.

13

Forming Flanges and Hems

Flanges on metal parts fall into three basic types. The simplest is the *straight flange*, which is a straight bend. Concave flanges are termed *stretch flanges* because the metal formed into the flange stretches. Convex flanges are called *shrink flanges* because the metal is compressed or shrunk. There are also combinations of these types of flanges that occur in a single operation.

FLANGING OPERATIONS

There are factors that place limits on the flanging process, including:

- the severity of deformation accomplished before fracture occurs;
- the amount of wrinkling or puckering permissible on a compression flange; and
- press energy or force available for flanging large areas with conventional wipe flanges high in the press stroke.

FLANGE TYPES

Figure 13-1 illustrates a number of different flanges. The straight flange is the most common type. The problems associated with straight flanges are springback and scoring.

Close tolerance bends should not be specified unless necessary to enhance the appearance or function of the part. A variation in

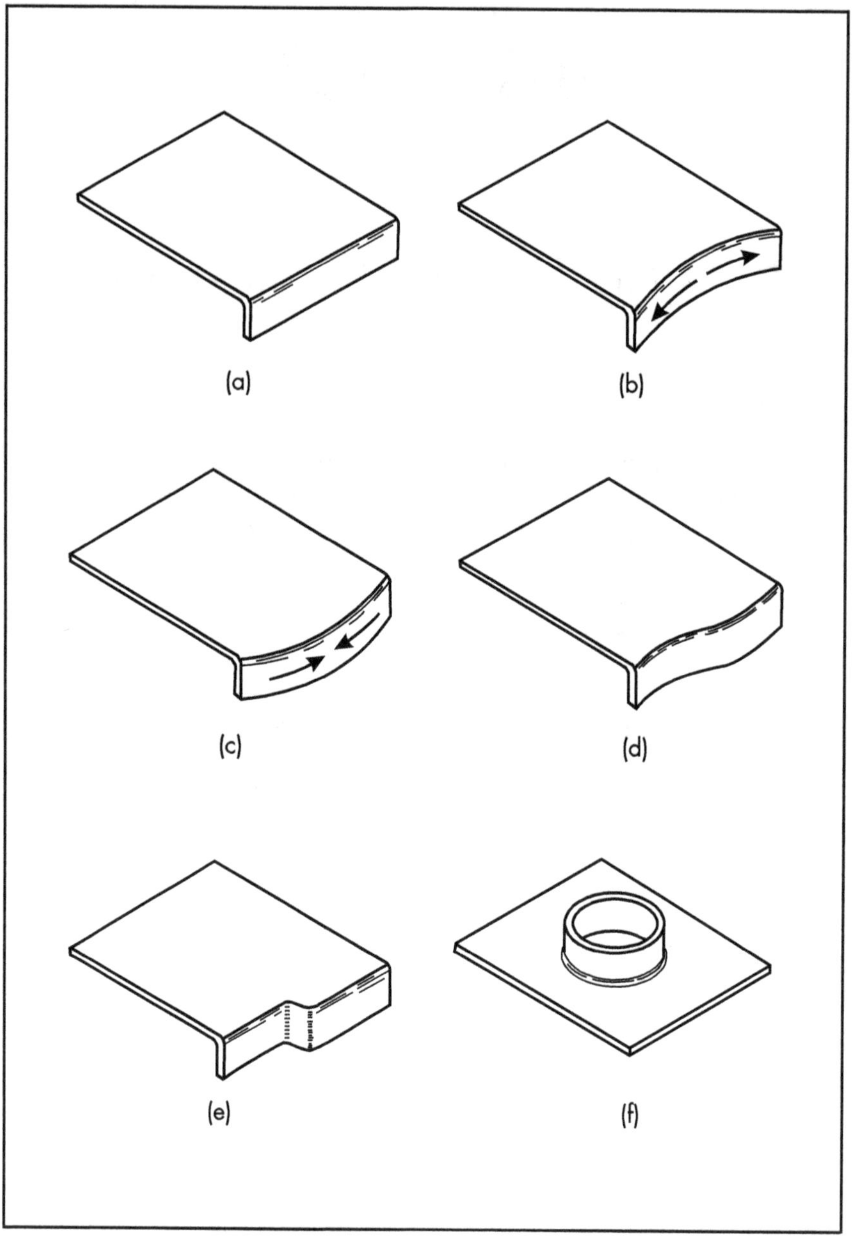

Figure 13-1. Examples of types of flanges: (a) straight bend or flange; (b) stretch flange; (c) in shrink flanges, the metal is compressed; (d) a reverse flange is a combination of a stretch and shrink flange; (e) a joggled flange; (f) a flanged hole is a type of stretch flange.

elastic recovery or springback is a problem. This results from variations in material properties.

Forming stretch flanges involves stretching the metal during the bending operation. The greatest amount of stretch occurs at the edge of the flange. The stretch amount is essentially zero at the bend radius.

Edge splitting can be a problem when stretch flanging (see Figure 13-2). The likelihood of splitting depends on the material properties and the edge condition resulting from shearing or trimming. Tensile stress can be reduced by shorter flange lengths or by providing notches or scallops. Notches will reduce the flange strength.

The metal in a shrink flange compresses and causes it to shorten in length. The amount of shrinkage is greatest at the edge of the flange and diminishes to zero at the bend radius.

Irregular or curved flanges tend to have less springback problems than straight flanges. A reverse flange is a combination of a stretch and shrink flange. Another combination flange is the joggled flange. While the majority of the joggled flange is straight, the corners are stretch and shrink flanges, respectively.

A flanged hole is a type of stretch flange. Some flanged hole applications are locating bosses, holes for tapped threads, openings for heat transfer tubes, and non-chafing passages for wires. Flanging an opening in a stamping can greatly increase the rigidity of the part.

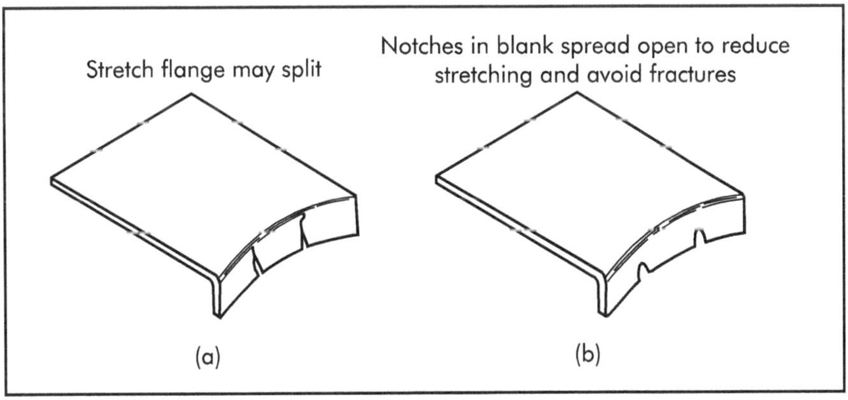

Figure 13-2. (a) Edge splitting can be a problem when stretch flanging. (b) Tensile stress can be reduced by providing notches or scallops.

MINIMUM BEND RADII

The minimum bend radii vary depending on the type of metal. Most annealed metals can be bent to a radius equal to the thickness, although some softer metals can be bent to an inside radius one-half metal thickness. Short bend-radius lengths reduce the minimum bend radius. This is not a practical consideration if the minimum bend-radius length is eight or more times the metal thickness.

The rolling direction in sheet or strip metal limits the minimum bend radius. An angle of 90° between the bend axis and the direction of rolling allows most metals to bend to the smallest possible radii.

Bend allowances are discussed in Chapter 12.

Strain

A simple general equation expresses the strain at the edge of the stretch flange, where most failures begin:

$$e_x = \frac{R_2}{R_1} - 1 \tag{13-1}$$

where:

e_x = strain at flange edge
R_1 = flange-edge radius before forming (in.[mm])
R_2 = flange-edge radius after forming (in.[mm])

Circle grid analysis (CGA) is an excellent method for determining the actual amount of strain at a flanged edge. Expanding a drilled, deburred hole with a lubricated conical punch to determine the forming limit is one type of comparative test performed on materials to be flanged.

Edge conditions such as burrs and rough fractured edges reduce stretch-flange formability. Such edge conditions result in excessive cold working of the metal.

STOCK THICKNESS VARIATION

Producing close-tolerance stampings requires stock thickness variations to be held to a minimum. Figure 13-3 illustrates sections through a typical bending or forming die. If the stock thickness is correct, the part will be properly formed as shown in Figure 13-4.

If the stock is too thin, angles will be under-bent as illustrated in Figures 13-5 and 13-6. Severe die damage can result from attempting to form stock that is too thick for the die clearances. Large lateral forces are developed that can greatly exceed the applied press tonnage. The force is multiplied by wedge-like action as shown in Figure 13-7. This type of damage often occurs due to operator inattention or an automation malfunction resulting in two blanks being formed in the die at one hit, as shown in Figure 13-8.

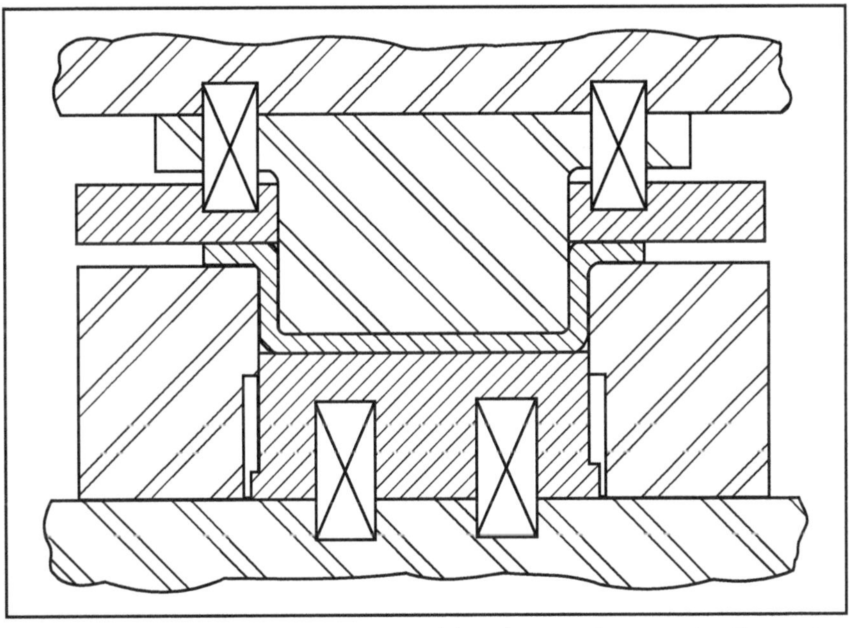

Figure 13-3. Section through a typical bending or forming die illustrating forming of the correct stock thickness.

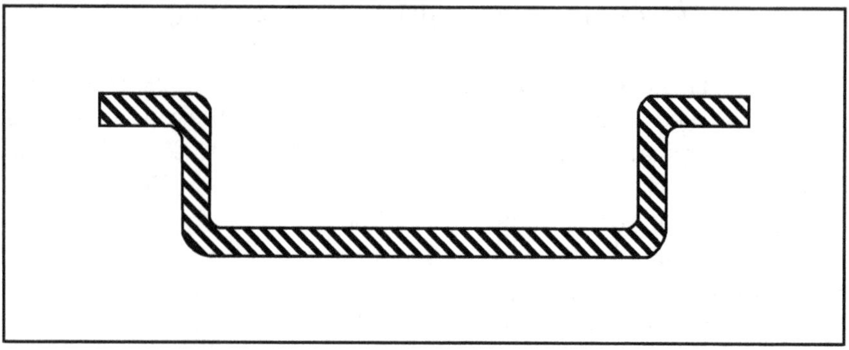

Figure 13-4. A correctly formed part produced in a die. Die clearances, die geometry, and stock specifications must be correct to produce properly formed stampings.

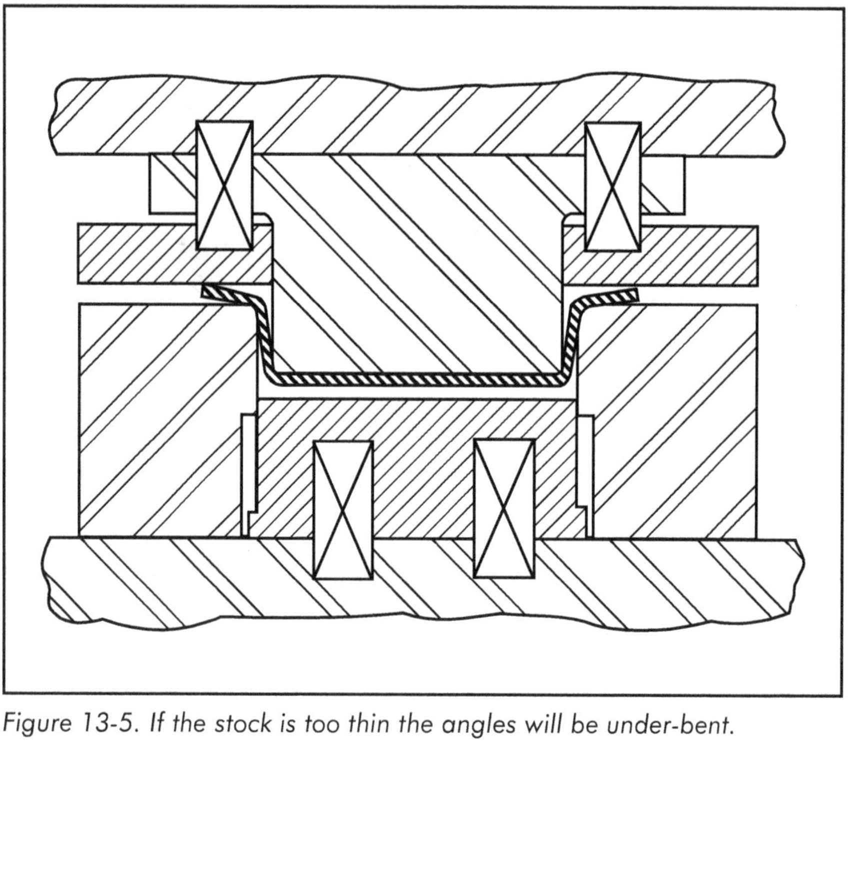

Figure 13-5. If the stock is too thin the angles will be under-bent.

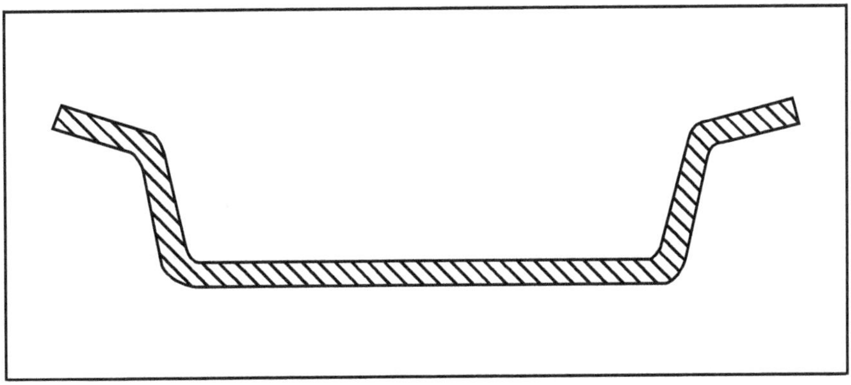

Figure 13-6. An incorrectly formed part produced from stock that is too thin results in the bend or form angles being less than required.

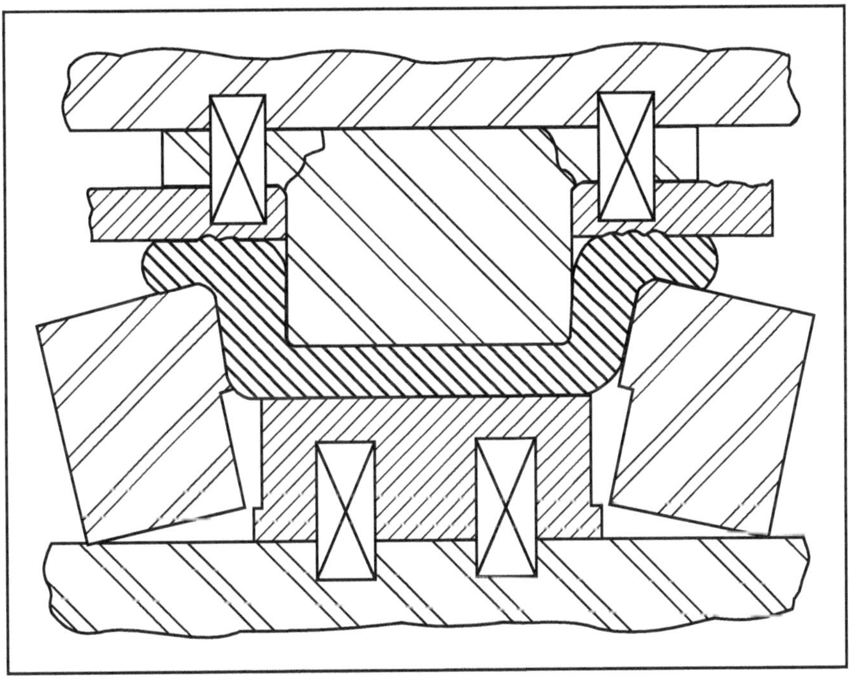

Figure 13-7. Section through a bending or forming die shows that if the stock is too thick, malformed parts and severe damage can occur.

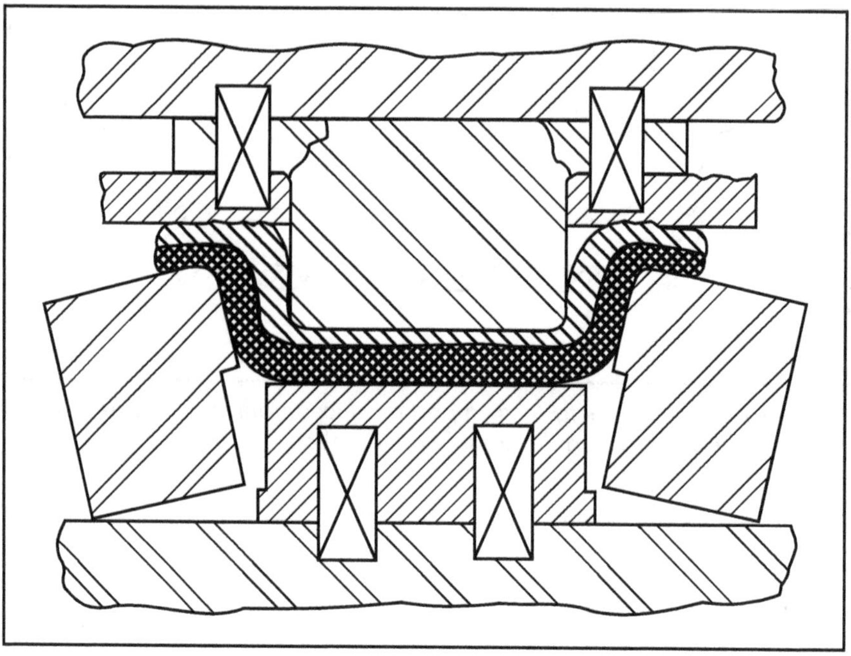

Figure 13-8. Section through a bending or forming die shows that severe damage can occur if more than one thickness of stock is in the die.

HEMMING OPERATIONS

Hems are primarily used to provide a smooth rounded edge and for the attachment of one sheet metal part to another. They are a very effective way to eliminate a dangerous sheared edge. Hems find extensive use in automobiles to join closure panels.

A sharply bent flattened hem requires materials with high ductility. Materials that do not have the ductility required to form a flattened hem may require a teardrop or rounded edge hem. The same minimum bend radii considerations based on material formability apply to hems used for attachment to other sheet metal parts (see Figure 13-9).

Tooling

Flanges and hems made in one operation require complex tooling. However, to simplify tooling, hemming a part that has been

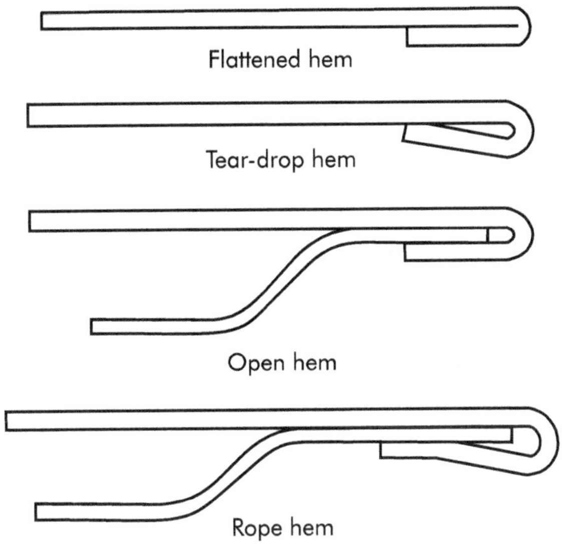

Figure 13-9. Four different types of hems are shown; the bottom two show hem types used to assemble stampings.

flanged in a previous operation may reduce overall cost. Figure 13-9 illustrates four different types of hems (Smith 1990).

Using Hems for Assembly of Stampings

The sequence of a typical hemming stamping assembly operation after bending a 90° flange in the outer panel is to place the inner panel in the flanged outer panel and bend the outer panel an additional 45°. The partially assembled panels are next transferred to the hem die, where the final assembly takes place by flattening the pre-bent hem.

PRESSURES

Typical hemming pressures, including seaming pressures, are generally seven times the forming pressure required for 90° bends, and may be as high as a ratio of 40:1. Variables are: stock thickness, tensile strength, size of area to be flattened or hemmed, and tightness of the hem.

The amount of pressure required depends upon the thickness of the stock, the length of the bend, the width of the die, whether a lubricant is used, and the amount of wiping, ironing, or coining present.

The use of Equation 13-2 for deriving bending pressures is valid for V-shaped dies only. For channel forming and U-forming, multiply the result by 2. If forming a channel with a flat bottom, a blank holder is necessary. Multiply the blank holder area in square inches by 0.15 to derive the approximate required tonnage, then add to the bending force derived from the equation. Equation 13-2 is used to determine V-bending forces for air bending in a die such as that used for simple press-brake tooling.

$$F = \frac{KLSt^2}{W} \tag{13-2}$$

where:

F = bending force (lbf [N])
K = die-opening factor: varies from 1.20 for a die opening of 16 times metal thickness to 1.33 for a die opening of 8 times metal thickness
L = length of bent part (in. [mm])
S = ultimate tensile strength (psi [Pa])
t = metal thickness (in. [mm])
W = width of V-channel or U-forming lower die (in. [mm])

The approximate force required to bend or flange a sheet (Figure 13-10) is given by:

$$F = 0.167\frac{SLt^2}{r_1 + r_2 + C} \text{ (theoretical)} \tag{13-3}$$

$$F = 0.333\frac{SLt^2}{r_1 + r_2 + C} \text{ (for wiping dies)}$$

where:

F = bending force (lbf [N])
S = ultimate tensile strength (psi [Pa])

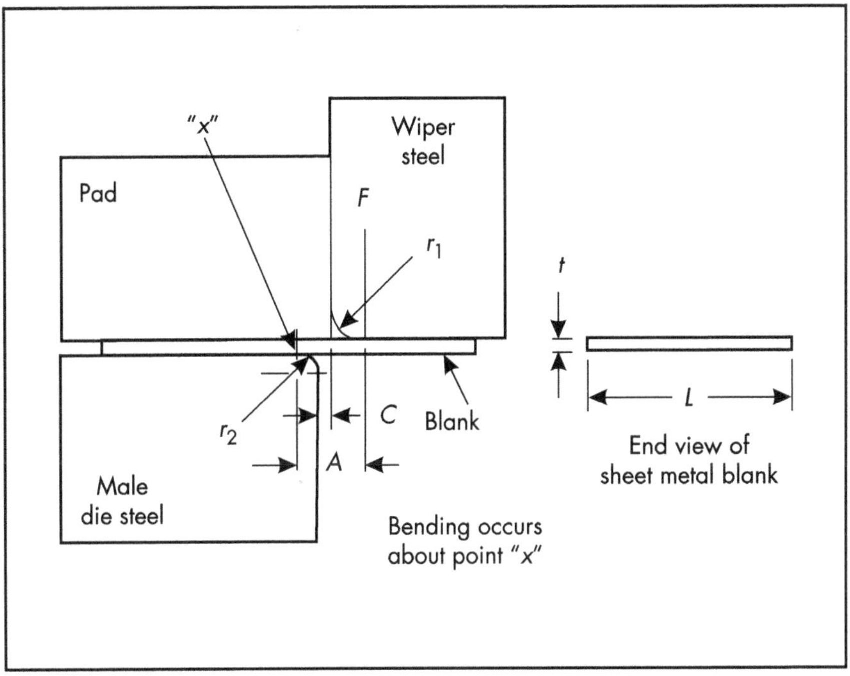

Figure 13-10. Factors entered to determine flanging force requirements.

L = length of bend (in. [mm])
t = sheet metal thickness (in. [mm])
r_1 = punch radius (in. [mm])
r_2 = die radius (in. [mm])
C = die clearance (in. [mm])

The flanging pad must grip the part firmly to insure the part will remain in tight contact with the male die half during flanging. To determine if pad force is sufficient, use the following equation:

$$F_p = \frac{SLt}{3} \tag{13-4}$$

where:

F_p = pad force (lbf [N])
S = ultimate strength of the material (psi [Pa])
L = flange length (in. [mm])
t = material thickness (in. [mm])

REFERENCE

Smith, D. 1990. *Die Design Handbook*. Section 6, Bending of Metals. Dearborn, MI: Society of Manufacturing Engineers.

14

Drawing and Stretching of Metals

DEEP DRAWING OF A CYLINDRICAL CUP

Figure 1-6 illustrated the forces involved in the deep drawing of a metal cup. It is important to note that all of the force required for drawing is transmitted by the draw punch to the bottom of the cup.

The cup-drawing process starts with a flat round blank. The blank is subjected to radial tension and circumferential compression. The metal thickens as it flows toward the draw radius. Deep drawing is unique because of the deformation state of the metal restrained by the blank holder.

Metal Flow

In general, the metal flow in deep cup drawing may be summarized as follows:

- Little or no metal deformation takes place in the blank area that forms the bottom of the cup.
- The metal flow occurring during the forming of the cup wall uniformly increases with cup depth.
- The metal flow at the periphery of the blank involves an increase in metal thickness caused by circumferential compression (see Figure 1-6).

Success Factors

The success of a drawing operation depends upon several factors, including:

- the formability of the material being drawn;
- limiting the drawing-punch force to a lower value than that which will fracture the shell wall; and
- adjustment of the blank-holder force to prevent wrinkles without excessively retarding metal flow.

Figure 14-1 is an example of deep cup drawing in which very little deformation occurs over the bottom of the punch. Nearly all deformation occurs in the metal restrained by the blank holder.

Why and How Deep-drawn Cup-shaped Shells Fail

The maximum force requirement for the drawing process is limited by tensile failure of the material in the sidewall. As this limit is approached, the metal will neck or thin excessively in a localized area near the punch radius.

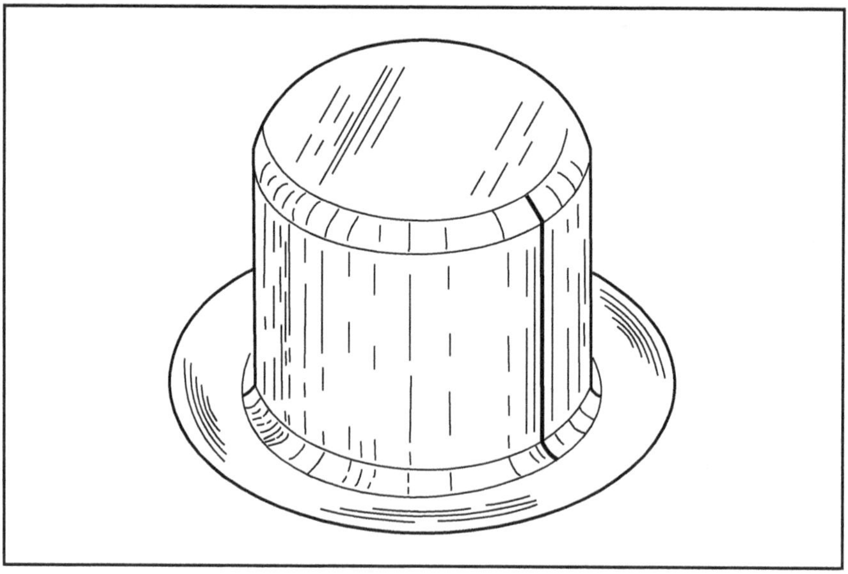

Figure 14-1. Shown is a successfully drawn cup-shaped shell.

Many complex interactions occur during the cup-drawing process. The actual force required depends upon the cross-sectional area of the cup wall and the yield strength of the material as it is worked. Should the process fail, some or all of the following factors may be the root causes:

- The ductility or drawability of the stock may be too low.
- The blank-holder force may be too high.
- Scoring or galling may be present on the die surfaces.
- The blank-holder geometry and draw radius may not provide for metal thickening and smooth flow into the die cavity.
- There may be an incorrect or insufficient amount of drawing lubricant.
- The depth of draw or percentage of blank reduction may be too great.
- One or more redrawing operations may be necessary to obtain the desired depth of draw.

Annealing may be required between redrawing operations, especially when using materials that work harden rapidly.

Draw Radius

The blank-holder draw radius should be approximately four to six times the metal thickness for most applications. It has a large effect on the punch force required to pull the metal into the draw cavity. As the metal passes over the radius, it is bent and then straightened to form the sidewall of the drawn cup. Too small a radius can lead to fracture, because more force is required to pull the metal over a small radius than a larger one. Too small a radius will more severely strain the metal, increasing work hardening. This in turn requires more force to draw the part.

There is little reduction of drawing force achieved by making the draw radius larger than six times metal thickness. Exceeding a draw radius of ten times metal thickness may result in puckering of the metal as it flows over the draw radius. Severe puckering can result in folded metal, which can lock up metal movement and result in fractures.

In cases where all of the metal on the blank holder is to be drawn into the cavity to form a straight-walled shell without a flange, a large radius may result in folded metal as blank-holder control ceases.

Measuring Thickness

Necking failures such as those shown in Figures 14-2 and 14-3 are proceeded by localized thinning that may not be visible in the part. However, the onset of a necking failure can be detected by measuring the metal thickness with an ultrasonic thickness-measuring device.

Application of an ultrasonic thickness gage. The ultrasonic thickness gage is a modern stamping analysis tool that is very useful for on-line process tracking, troubleshooting, and control. It features a portable control box, which provides a thickness readout, and a probe with an ultrasonic transducer. The principle of operation is much like that of the sonic or sonar depth finders used as navigational aids by boaters. In their simplest form, these sonic depth finders measure the time between a sonic pulse being sent out by a transducer attached to the boat's hull and the ar-

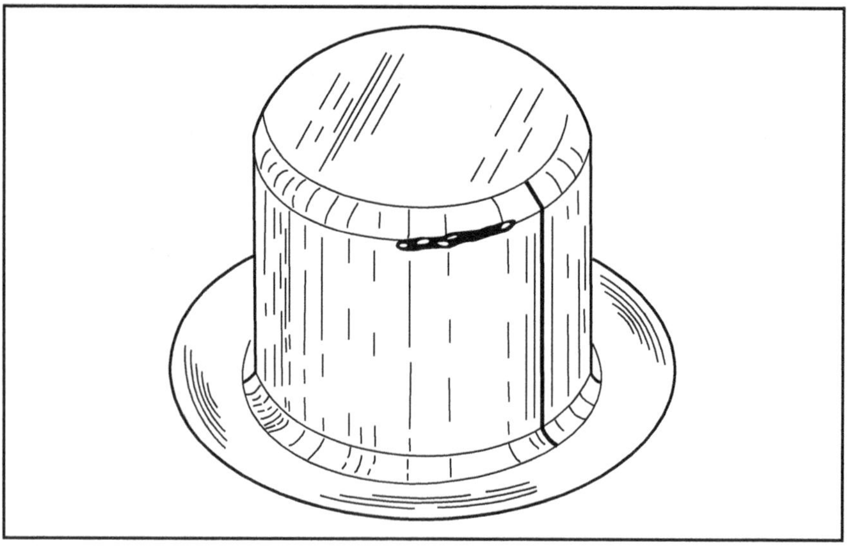

Figure 14-2. Localized thinning (necking) at the punch radius indicates a failure of the drawing process.

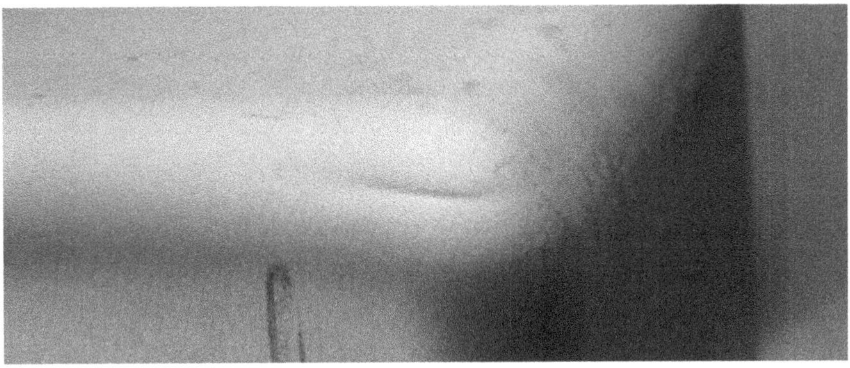

Figure 14-3. Shown is necking failure on the corner of a deep-drawn box-shaped stamping.

rival of the return echo. The speed of sound in water is a known constant. The delay time between the sending of the sonic pulse and the return echo can be easily converted and displayed in units of depth such as feet, meters, or fathoms.

Ultrasonic metal-thickness gages work in a similar way—an ultrasonic pulse is sent by the hand-held transducer, which attaches to the control unit. The return-echo time is very short since the speed of sound in steel is in excess of 16,000 ft/second (4,877 m/second). For stamping thickness measurements, the operating frequency is approximately 15 MHz.

Determining the areas to check. Stampings that are severely drawn or formed (see Figure 14-4) usually have one or more areas where thinning is apt to occur. These areas are candidates for regular thickness checks.

The areas where a necking failure or fracture are apt to occur on a stamping can be predicted with circle grid analysis (CGA) during the development and die-tryout period for new stampings. The failure locations on an existing stamping become well known to pressroom personnel.

Once the likely areas are identified, regular checks should be made with an ultrasonic thickness gage. It is helpful to trend-chart the thinning in each area. This permits corrective action to be taken before a necking failure becomes visible. Causal factors for a pronounced increase in thinning include:

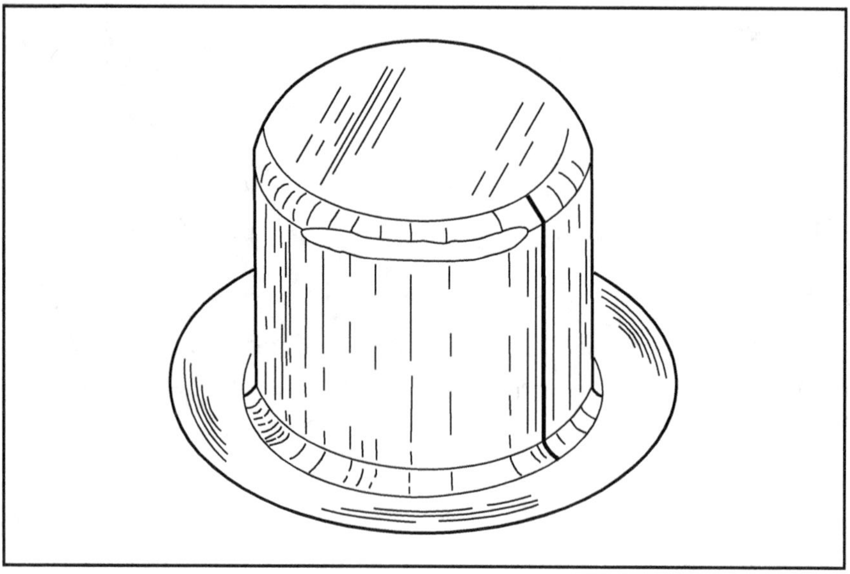

Figure 14-4. Severe failure of this drawn cup is due to a pronounced fracture.

- excessive blank-holder force;
- material problems such as a lack of ductility or drawability;
- material too thin;
- scored die surfaces; and
- material quality issues.

Material Quality

When drawing operations fail, often the material is immediately blamed. The logic is that if the vendor can supply some material that will run, it should be possible to do so consistently. However, the material's formability properties do vary from lot to lot, and even within the same coil.

It is poor economy to specify expensive deep-drawing quality material when good die design and maintenance will permit the use of less costly commercial-grade stock. Formulas and charts giving recommended maximum reductions for various materials may be found in reference books.

If there is proper sidewall clearance in the die, the punch force will not exceed the ultimate tensile strength of the material cross section in the wall. Since some thinning occurs as the ultimate tensile strength is approached, using this figure for force-requirement calculations usually provides a substantial safety factor.

The material's yield strength should normally be used for drawing calculations. The results generally will correspond closely to measured values. It is to be noted that excessive blank-holder forces can cause any cup-drawing operation to fail.

Force and Energy Requirements

In Chapter 11, the force and energy requirements for cutting a low-carbon blank 12 in. (304.8 mm) in diameter and 0.1875-in. (4.763-mm) thick were calculated. The required energy is surprisingly small. Only 1,840.8 ft-lbf (2,496 J) is required. At 60 strokes per minute (SPM), and ignoring frictional losses, 3.347 hp (2.5 kW) will sustain the process.

Several arbitrary assumptions are made. This heavy 12-in. (304.8-mm) diameter, 0.1875-in. (4.763-mm) thick blank is to be drawn into a flanged cup having a diameter of 6 in. (152.4 mm) and a depth of draw of 2 in. (50.8 mm). The yield strength of the material is 40,000 psi (276 MPa). It is assumed that the force and energy required is based on working the cup wall at or near its yield strength. It should be noted that the yield strength is normally specified as a minimum value.

The total cross-sectional area of the wall is 3.534 in.^{2} (2,280 mm^2). Based on the yield strength of 40 ksi (276 kPa), 70.69 tons (64,129 kg) of punch force are required. While the force is much less than required to cut the blank, it must act through a distance of 2-in. (50.8 mm). The work or energy required per part is 141.37 in.-tons or 23,562 ft-lbf (31,946 J).

The energy input per part is 12.8 times greater to draw the part than that required to cut the blank. While the 60 strokes per minute rate may be too high for optimum formability, neglecting frictional losses, 42.84 hp (32 kW) would be required to restore energy to the flywheel.

Conversion of Energy to Heat

In pressworking processes, the materials are worked in a plastic state. High internal friction is present. Nearly all of the energy required for the process is converted to heat.

Studies conducted by the author on drawn automotive-suspension components at Ford Motor Company confirmed that 80–90% of energy often exited the die as latent heat in the stamping. Nearly all of the mechanical energy-to-heat conversion occurred within the stamping itself during deformation. The remainder resulted from surface friction at the part-to-die interface. The most severely strained portions of the part exited the die at a higher temperature than the surface temperature of the corresponding die surfaces. The energy input was determined by calculations of force versus distance. Waveform signature analysis was used to aid in determining the energy input.

The temperature profile of the part and die was measured by a small thermistor probe. The temperature data was used to determine the heat energy in the part and the temperature gradient of the heat energy conducted through the die.

There is close agreement of theoretical versus measured values of mechanical to heat energy. In general, large dies function as effective heat sinks when operated at press speeds of 8–20 SPM.

In the dies studied, ion nitriding of the nodular-iron die surfaces essentially eliminated the metal pickup or galling problems that occurred as the die warmed during operation.

Cooling by chilled water piped through the die may be required for some severe drawing operations, but should be avoided because of cost and complexity. However, water-based pressworking lubricants provide cooling through evaporation, a simple solution in many cases.

STRETCH FORMING

Figure 1-5 illustrated a stretch-forming operation to produce a dome-shaped part. In this figure, the edges of the blank are securely clamped with a *lock bead.* Only the metal in the punch area is deformed. Both the width and length dimension of the metal are stretched. This type of forming is known as *biaxial* stretch.

Stretch forming is a very common operation. The forming of automotive, appliance, and aircraft panels is a widespread application. Typically, 3–10% strain is required to obtain the mechanical properties needed for proper stiffness.

Success Factors

Like deep-cup drawing, stretch forming involving severe deformation is dependent on good material properties, proper lubrication of the punch, and correct die maintenance. To obtain enough stretch to realize good part stiffness, it is important to maintain enough blank-holder force to prevent metal slippage through the lock bead. The blank-holder force required throughout the press stroke is greater than the pressure required to form the lock bead upon initial die closure (Herderich 1990).

Excessive localized thinning or necking, often leading to fracture, is evidence of stretch-forming process failure. Surface roughness of both the die and material should not be excessive. Smooth surfaces may also have high friction. Good lubrication is an important success factor in operations involving severe deformation. Both the die and stock may require an optimum roughness profile to facilitate holding lubricant.

Plane Strain

If the metal is formed by stretching in one direction only, the operation is called *plane strain* or simple stretching and thinning of the metal. When compared to biaxial-stretch forming, this operation allows substantially less elongation to occur before a fracture occurs. (The plane-strain forming method is illustrated in Figure 1-4.)

BENDING AND STRAIGHTENING

Figure 1-3 illustrated a forming method in which the metal is bent and straightened as it passes over the blank-holder radius. Bending and straightening operations permit large deformations with very little thickness change.

A simple type of die employing bending and straightening is used to make U-shaped cross sections with right-angle flanges. Parts of this type are used as stiffeners attached to flat panels in many applications. Examples are automotive body frame rails and cross members assembled by welding.

Bending and straightening operations are employed in conjunction with processes, such as drawing, in the production of complex stampings. A simple example is the stamping of rectangular shells.

DRAWING AND BENDING

The stamping of a rectangular shell involves both cup drawing and simple bending and straightening. True drawing occurs at the corners only. The metal movement at the sides and ends involves bending and unbending. (Figure 1-3 illustrated a section through a portion of a die where bending and straightening occurred as the metal passed over the radius.)

The stresses at the corner of the shell are compressive and result in a thickening of the metal moving toward the die radius. After the metal has been drawn over the radius, the forces are tensile. The metal between the corners is in tension on the sidewall and where restrained by the blank holder. This portion of the operation involves deep-cup drawing (illustrated earlier in Figure 1-6).

Unlike circular shells in which pressure is uniform on all diameters, some areas of rectangular and irregular shells may require differing pressures. The metal at the corners of the blank compresses and will thicken. Both thickening and some wrinkling of the metal at the corners are normal. Figure 14-5 illustrates the appearance of a box-shaped drawn shell made from a rectangular blank. The metal at the corners often increases in thickness by up to 25% or more. An example of a failed rectangular drawn shell is illustrated in Figure 14-6.

Providing Clearance in the Blank Holder

A common error in the construction and maintenance of rectangular shell drawing dies is to machine the blank-holder surfaces

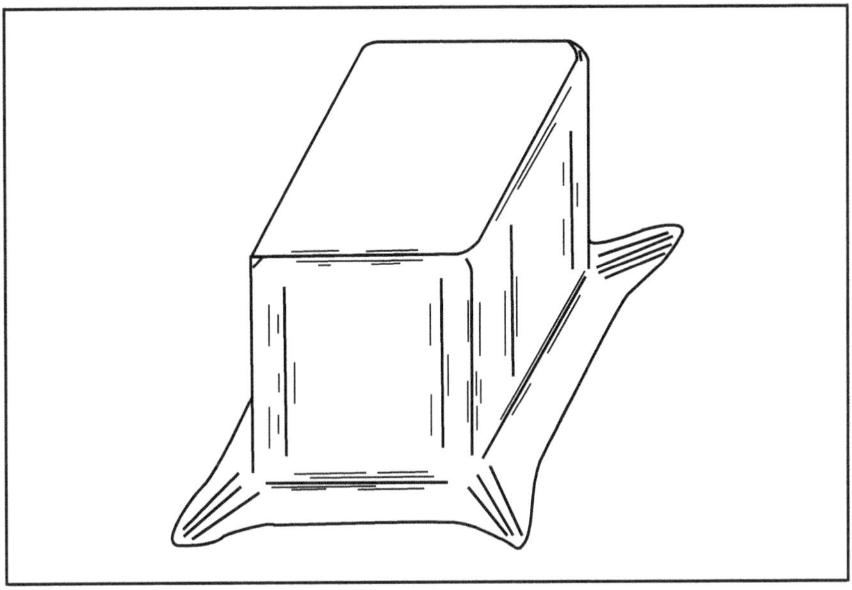

Figure 14-5. A typical rectangular drawn shell shows how metal thickens and tends to wrinkle at the corners.

Figure 14-6. A failed rectangular drawn shell shows the normal location of a fracture due to the metal flow at the corner's locking.

perfectly flat. Clearance provided in the blank holder allows the metal to thicken. This can be accomplished with a pneumatic hand grinder when the die is tried out.

A skilled die-tryout technician will optimize the metal flow by making a series of trial parts and reworking the blank holder as needed. In some cases, it is necessary to increase draw ring and punch radii (with the approval of the product designer).

After the die-tryout work is complete, the die is oil-stoned and polished. If long service life is required in severe applications, ion-nitriding, hard chromium plating, or other processes to reduce surface wear and friction may be beneficial.

Fracture

Minor product changes are often highly beneficial to reduce or eliminate the occurrence of fractures. The corner is the usual location of a fracture in a rectangular drawn shell. The localized thinning or necking that can lead to a fracture is the same failure mode that limits the severity of round deep-cup drawing.

Figures 14-7, 14-8, and 14-9 illustrate fractured drawn parts. Note the very tight coined areas on the blank holders in the corners where the metal locked leading to the fracture. The metal movement's restraining action of the large outboard tabs are also a cause of the fractures.

Parts that emerge from the first operation without fractures, despite being severely worked, may have so much residual stress built up that they fail in subsequent operations. A much more serious difficulty is to have such stampings fail in service.

Figure 14-10 is a photograph of a drawn-shell sidewall. When asked to identify a defect, most observers note the ding or blemish as the fault. However, there is a much more serious problem. Figure 14-11 is a photograph of the same shell sidewall taken with a powerful quartz-halogen lamp inside. The intense light reveals a hairline crack that is a serious defect.

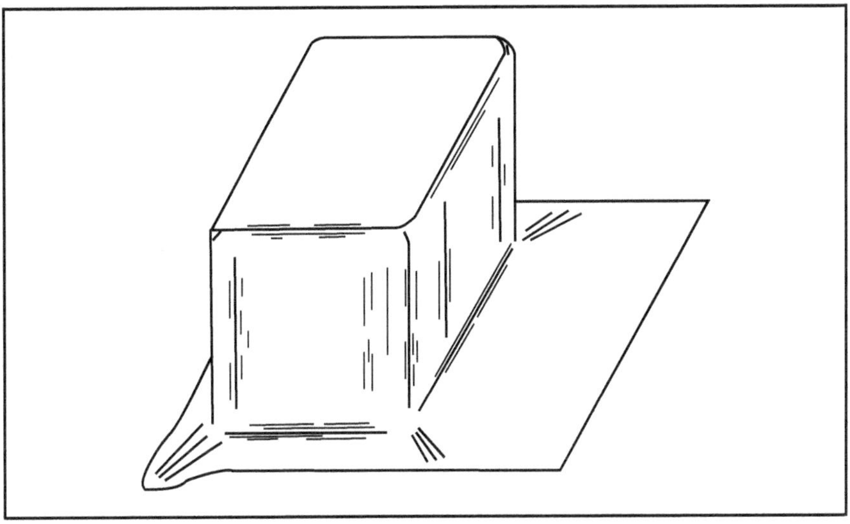

Figure 14-7. Rectangular box draw has a large outboard tab. The tab may severely restrict metal movement into the draw cavity and result in a fracture.

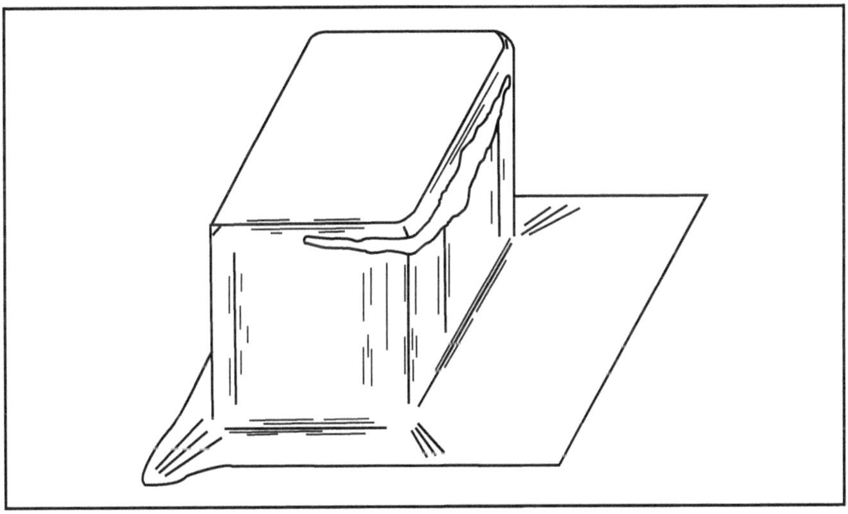

Figure 14-8. Fractured rectangular box draw has a large outboard tab.

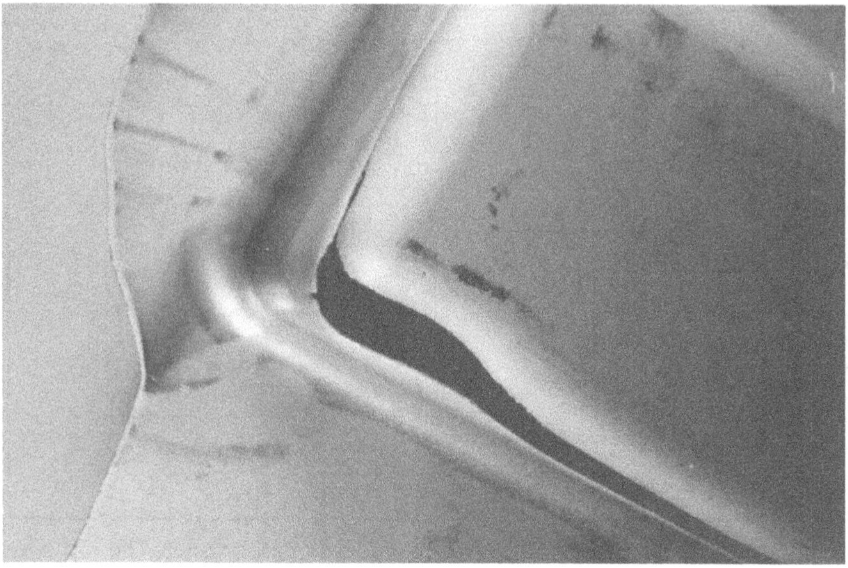

Figure 14-9. A photograph shows the type of failure illustrated in Figure 14-8.

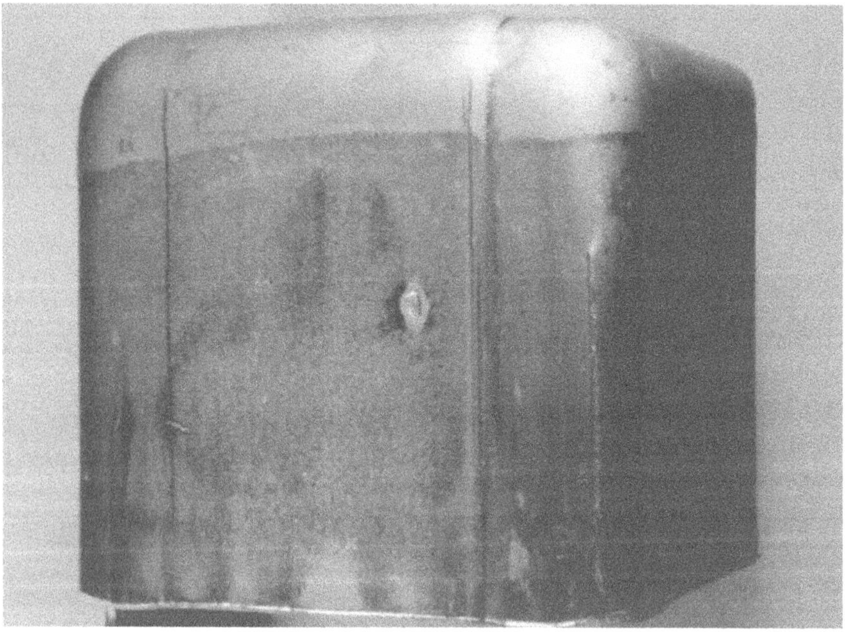

Figure 14-10. Photograph shows a drawn-shell sidewall.

Figure 14-11. A photograph taken with a powerful quartz-halogen lamp inside the rectangular-drawn shell shown in Figure 14-10. The intense light reveals a hairline crack—a serious defect. The crack occurs when a joggle to the left of the crack is formed.

REFERENCE

Herderich, Michael R. 1990. "Experimental Determination of the Blank-holder Forces Needed for Stretch Draw Design." SAE Paper 900281. Warrendale, PA: Society of Automotive Engineers.

15

Stamping Analysis Techniques

Forming flat sheet metal into complex, radically deformed stampings may appear to involve skills and processes that are more of an art than a science. Modern stamping design and development techniques permit the product designer to work with manufacturing and tooling engineers to design parts that can be manufactured with certainty. A complex stamping is shown in Figure 15-1.

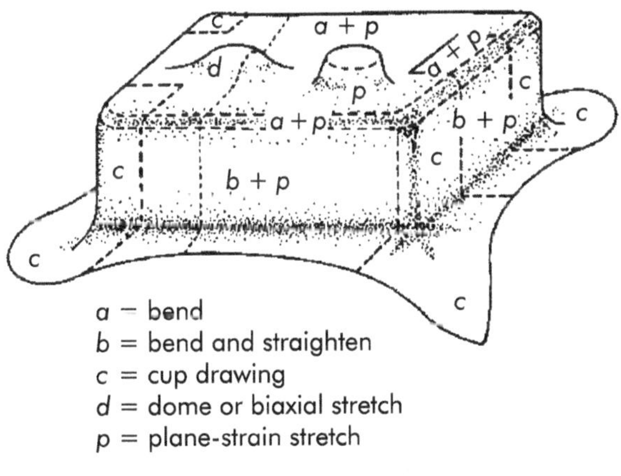

Figure 15-1. A complex stamping is produced by a variety of forming operations, each of which may be analyzed separately. (Courtesy American Iron and Steel Institute)

Stamping designs should be based on data of successful prior designs and formability analytical methods. Uncertainty concerning the manufacturability of complex stampings often results in added expense and delay factors, such as (Committee of Sheet Steel Producers 1984; Wick et al. 1984):

- trial production on temporary tooling to prove process feasibility;
- delays in marketing the product while the process or product design is changed;
- specifying more operations than needed as a safety factor; and
- choosing alternative processes and materials such as molded plastics.

COMPUTERIZED ANALYSIS TECHNIQUES

Easy-to-use computer software programs are available to assure that proposed stamping designs can be manufactured with certainty. Using these programs avoids costly trial-and-error guesswork.

Such software is available to analyze the amount and type of deformation in a stamping design. Computer-aided analysis ties in nicely with CAD design of stampings. The analysis should be applied early in the product-design process. The CAD math data, which describes the part, is used for computerized formability analysis. Computerized analysis falls into several categories, such as:

- simple sectional analysis programs;
- general analysis programs that fully model the part, typically based on finite-element analysis; and
- programs that analyze the stamping, based on the type of deformation occurring in an individual area.

Sectional Analysis Programs

Sectional analysis is good for identifying and troubleshooting a number of simple forming conditions. Computer programs are useful for determining the amount of strain present in a specific area of a stamping so design problems can be anticipated and checked easily.

A moderately priced personal computer has sufficient computational capacity to run sectional analysis programs to determine strain conditions. Estimating the effect of surface friction on metal movement is a useful feature of nearly all computerized formability analysis programs.

General Analysis Programs

To completely model a part using the finite element or finite difference method, general analysis programs are required. Stamping of the whole part is simulated in three dimensions with a single computer program. Many complex interactions occur during the stamping simulation. General analysis programs require a lot of computing power and time to calculate the interaction of many complex variables occurring throughout the forming process (Tharrett 1987).

Simplifying Stamping Analysis

A simplified approach is to break down complete stampings into local regions that can be analyzed individually. Following this approach, a stamping is analyzed as individual zones that interact in a predictable manner as they are formed. Some good programs include an expert systems approach based on a library of successful designs.

CIRCLE GRID ANALYSIS

Measuring Deformation

Another powerful tool to avoid costly guesswork is circle grid analysis (CGA). The CGA technique permits measurement of the deformation that occurs when forming stampings. First, a grid is stenciled on the surface of the blank by dye transfer or electrochemical etching. This grid deforms with the blank and allows accurate calculations of the strain or deformation that occurred during the stamping operation (Committee of Sheet Steel Producers 1984).

Press Shop Applications

Should the part always be found to run well within the safety zone, often a less costly steel or lubricant can be used. If only a few areas on the stamping are close to failure, a blank-holder improvement or minor product change often will ensure the manufacturability of the product.

The CGA system is excellent for training apprentices. By making tooling, lubricant, and material changes and then observing the metal-deformation changes, cause-and-effect patterns can be readily discerned.

Figure 15-2 illustrates a bumper-jack hook. The grid of circles placed on the blank shows different types of deformation required to form the finished stamping. Note that most of the circles are

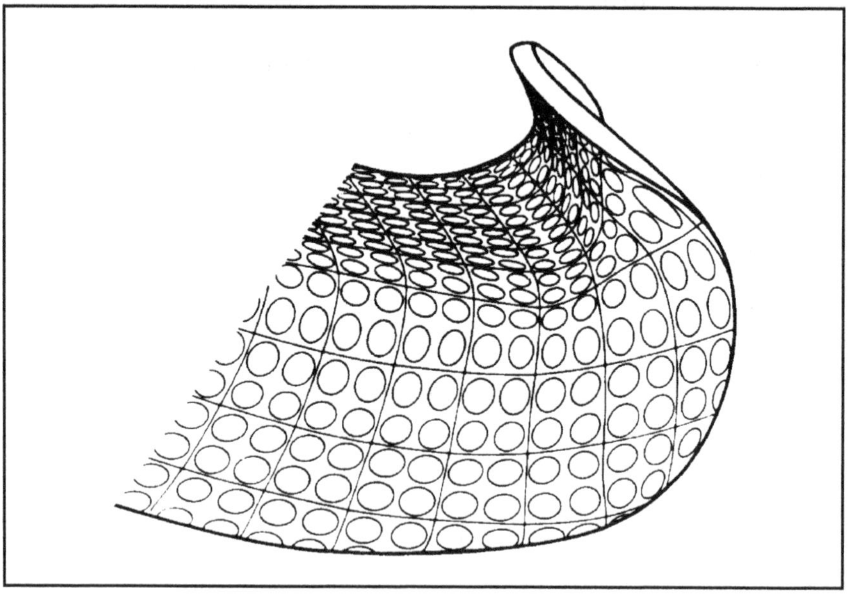

Figure 15-2. A bumper-jack hook formed from a circle-gridded blank. (Courtesy National Steel Corporation)

deformed very little, while a few circles, especially ones near the lip, show pronounced elliptical patterns. If an assumption is made, this stamping may be thought to fracture at the location where the edge is most severely stretched. However, the lip can stretch easily because it is on the edge (Keeler 1969, 1986).

Measuring Forming Severity

The distribution of stretch is useful information by itself. Knowing the location of high stretch concentrations and the direction of the maximum stretch is often sufficient to suggest solutions to forming problems. However, CGA uses a numerical rating system for the deformation of the circles.

The system of rating forming severity is based on measuring the deformation of the circles and plotting the measurements on a graph. Grid measurements are easily made with transparent Mylar™ tape imprinted with a calibrated scale (see Figure 15-3).

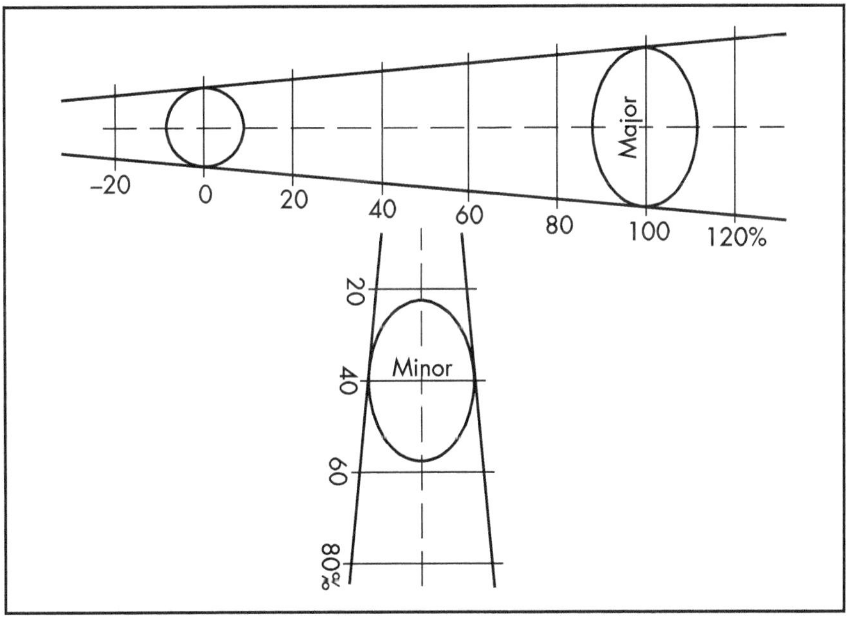

Figure 15-3. Circle deformation is measured with a Mylar™ tape overlay. (Courtesy National Steel Corporation)

The tape is flexible and can be placed around a radius or tucked into a tight corner. The calibration of the tape eliminates any need to calculate stretch. The tape is used to measure the major (length) axis of the ellipse first, and is then rotated to measure the minor (width) axis.

Many combinations of major and minor circle deformations can be found on different stampings. Figure 15-4 illustrates five different types of deformation. Figures 15-4a and 15-4b illustrate large major elongation, while the minor stretch is negative. Circle deformations of this type are observed in the sidewalls of drawn cups and the corner sidewalls of rectangular drawn shells. This combined compression and elongation indicates the metal is subjected to circumferential compression and tensile stretching as it is pulled toward and over the draw radius.

Figure 15-4c is a special case. Here, the minor stretch component is zero. This is called *plane strain*. This stretch condition is found over edge radii or across character lines.

Another special case is shown in Figure 15-4e, where both the major and minor axes of the ellipse are equal—the circle becomes a larger circle. This is called *balanced biaxial stretch*.

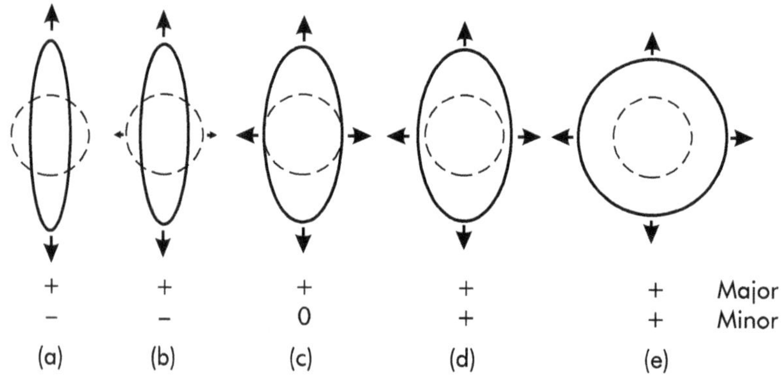

Figure 15-4. Examples of deformed circles are shown. (Courtesy National Steel Corporation)

Plotting the Measurements

Due to the variety of combinations, a method for plotting them on a single graph is necessary. The plotting technique used in Figure 15-5 allows both the major and minor stretch for each circle to be plotted as a single point. The major stretch is plotted on the vertical axis, while the minor stretch is plotted on the horizontal axis. Circles that plot on the left side of the forming-limit diagram have negative minor stretch, while circles that plot on the right have positive minor stretch.

Three of the ellipses from Figure 15-4 are plotted in Figure 15-5. Note that the case of plane stretch (Figure 15-4c) is plotted on the vertical axis.

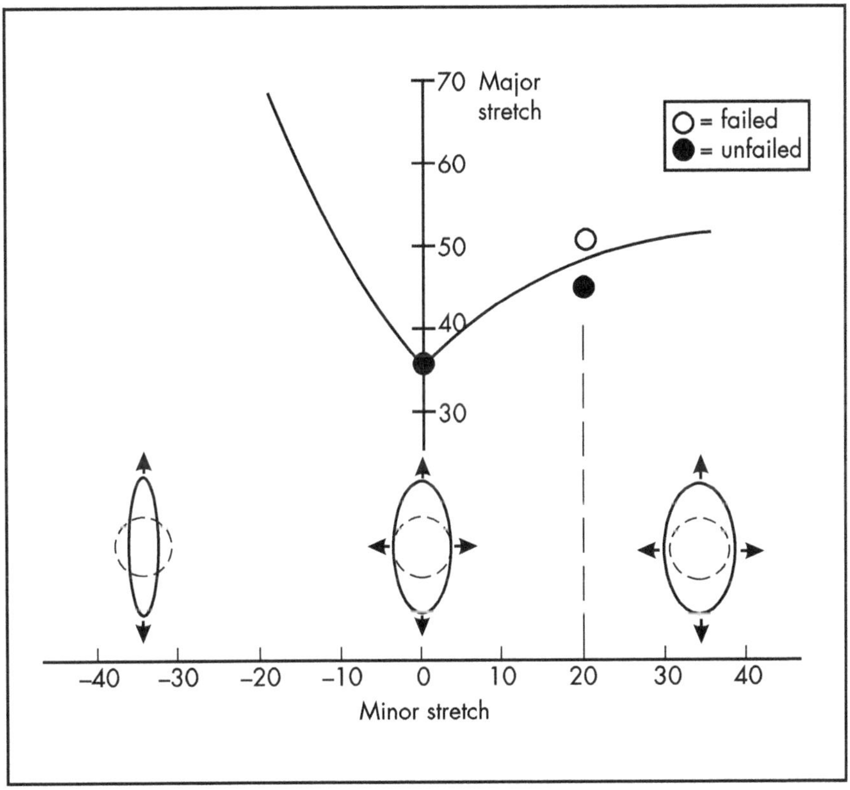

Figure 15-5. Measurements are plotted on the forming-limit diagram (FLD). (Courtesy National Steel Corporation)

Figure 15-6 illustrates an asymmetrical V-shaped curve, which is the forming limit. Circles plotted below this curve show no evidence of necking or fracture, while those above it fail. A graph developed in this way is called a *forming-limit diagram* (FLD). The point where the FLD intersects the major stretch axis is called FLD_o. Here, only plane-strain deformation is occurring. To initially develop this diagram, many samples of failed versus not failed circles from the same material must be plotted.

The shape of the FLD (Figure 15-6) is constant for most low-alloy sheet steel used in the automotive, appliance, agricultural, container, and similar industries. This illustrates how the FLD curve raises or lowers for different steel sheets. The level of the FLD—as specified by FLD_o—is a characteristic of the sheet steel. For example, a thinner sheet of steel would have a lower FLD_o than a thicker sheet of steel. In addition, higher-strength steel would have

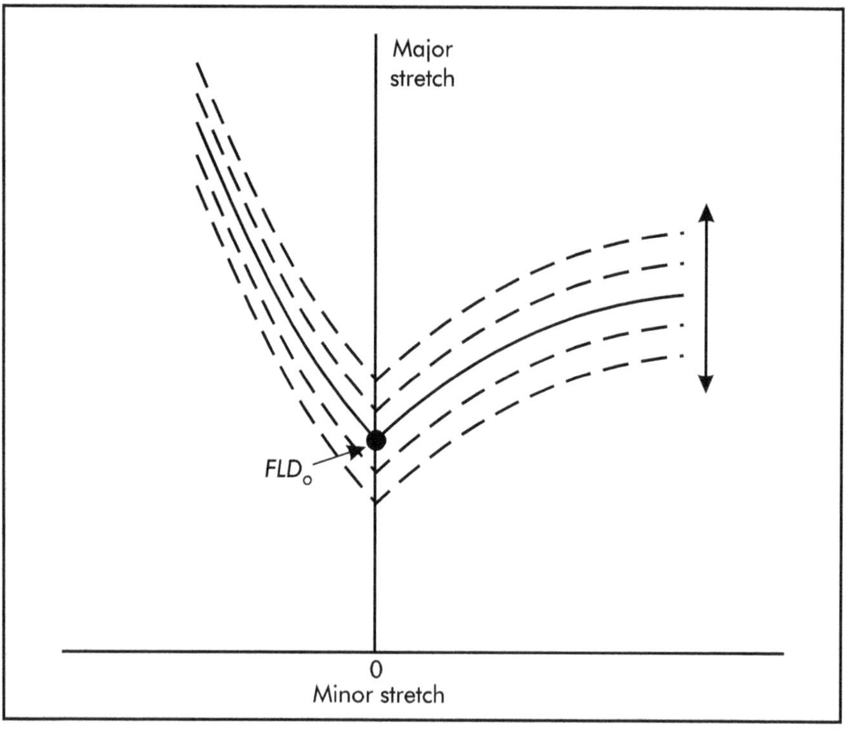

Figure 15-6. Different steels move the FLD_o point. (Courtesy National Steel Corporation)

a lower FLD_0 than lower-strength steel. Thus, the location of the curve can be described by specifying the intersection of the curve's FLD_0 with the minor stretch axis.

Using CGA as a Process-control Tool

CGA is a powerful process-control tool. Data on essential areas of the stamping that are near the forming limit should be checked periodically to determine the effect of die wear on formability. Should a production stamping process start to experience problems, a blank of the material can be quickly gridded and analyzed. The CGA results can be compared with the historical data for the part- and steel-formability specifications.

Making Do Without Etching Equipment

When working in shops that have no etching equipment, the problem areas of the blank can be coated with layout blue and circles lightly scribed with dividers. An overlay pattern can be drawn with simple drafting tools and copied onto overhead transparency stock. Another way to make the measurements is by using a digital dial caliber to measure the amount of deformation before and after drawing and/or forming the blank. The deformation can be read directly from an improvised transparent overlay or accurately calculated using digital calipers and the percentage function on a calculator. These methods are recommended for expediency and not as replacements for the inexpensive electrical etching equipment that uses chemicals and accurate stencils. No shop that does pressworking of sheet metal should be without this simple process-health tracking and troubleshooting tool.

REFERENCES

Committee of Sheet Steel Producers. *Sheet Steel Formability.* August 1, 1984. Washington, DC: American Iron and Steel Institute.

Keeler, S. 1986. *Circle Grid Analysis (CGA)*. Livonia, MI: National Steel Corporation Product Application Center.

——. 1969. *From Stretch to Draw*. Technical Paper MF69-513. Dearborn, MI: Society of Manufacturing Engineers.

Tharrett, M. August 25-27, 1987. *Computer-aided Formability Analysis*. Die and Pressworking Tooling Clinic. Society of Manufacturing Engineers: Dearborn, MI.

Wick, C. et al. 1984. *Tool and Manufacturing Engineers Handbook*—4th edition. Volume 2, *Forming*. Dearborn, MI: Society of Manufacturing Engineers.

16

Die Maintenance Documentation and Tracking

Responsible automobile owners realize that periodic maintenance and good record keeping is essential to assure dependable transportation. Like a car, a stamping die also needs these things if it is expected to produce good parts without abnormal delays.

A die, like an automobile, must be designed correctly. Both dies and cars are assembled from parts that must work together. Dependable, safe performance begins with good design. Careless operation and poor maintenance will cause great difficulty in spite of the designer's best efforts.

New dies may have design problems that must be corrected to assure smooth operation. Dies that do not function correctly are unacceptable.

DIE MAINTENANCE TEAM

The metal-stamping industry has undergone profound changes. Today, pressroom employees and management must work together as a team and be actively involved in solving problems. The greatest asset that an employee can bring to the workplace is his or her knowledge and willingness to contribute to progress.

Figure 16-1 is a cartoon of a foolish pressroom reaction to a die problem. Like a car that frequently breaks down, requiring repair or replacement, dies that fail in service require solutions. Well-

Figure 16-1. A non-team-oriented pressroom reaction to a broken die problem. The cartoon indicates a lack of communication between the pressroom and die repair facility.

managed shops expect employees to accurately communicate when there are tooling problems.

A SYSTEMATIC DIE MAINTENANCE SYSTEM

Simply stated, shops that do not maximize human resources probably will not survive in a free-market economy. The competition in the industry for high quality, dependable delivery, and cost reduction does not permit the luxury of poor communication and inefficient shop organization. Clear, concise communication is an essential factor in efficient die maintenance documentation and tracking. A simple systematic approach to problem identification and communication is required.

Figure 16-2 shows a typical die room response when a die without a work order is dropped off. A work-order system should be part of a closed-repeating cycle to correct problems. This includes eliminating their root causes.

Figure 16-2. Dies sent to the die room without details of the problem to be corrected usually do not get attention.

Using a Die Repair Work Cycle

Figure 16-3 shows a six-step die maintenance work cycle. Maintaining any machine tends to be a repetitive activity. Dies can develop problems that require keen troubleshooting skills. Everyone associated with the process should understand the basic theory of the stamping processes. The operator should understand the basic functioning of the die, press, and auxiliary equipment.

Training is essential at all skill levels. Ongoing training for everyone from the operator to the tooling engineer and even top manufacturing management will increase their skills. Everyone should be able, willing, and encouraged to provide input into solving tooling process problems.

Training operators in basic press and die operation can bring about a prompt resolution of problems. Progressive-die stamping variations are often due to pitch changes that are difficult to predict exactly in the design stage. Moving and redoweling die stations in the die room can correct pitch errors. Left uncorrected,

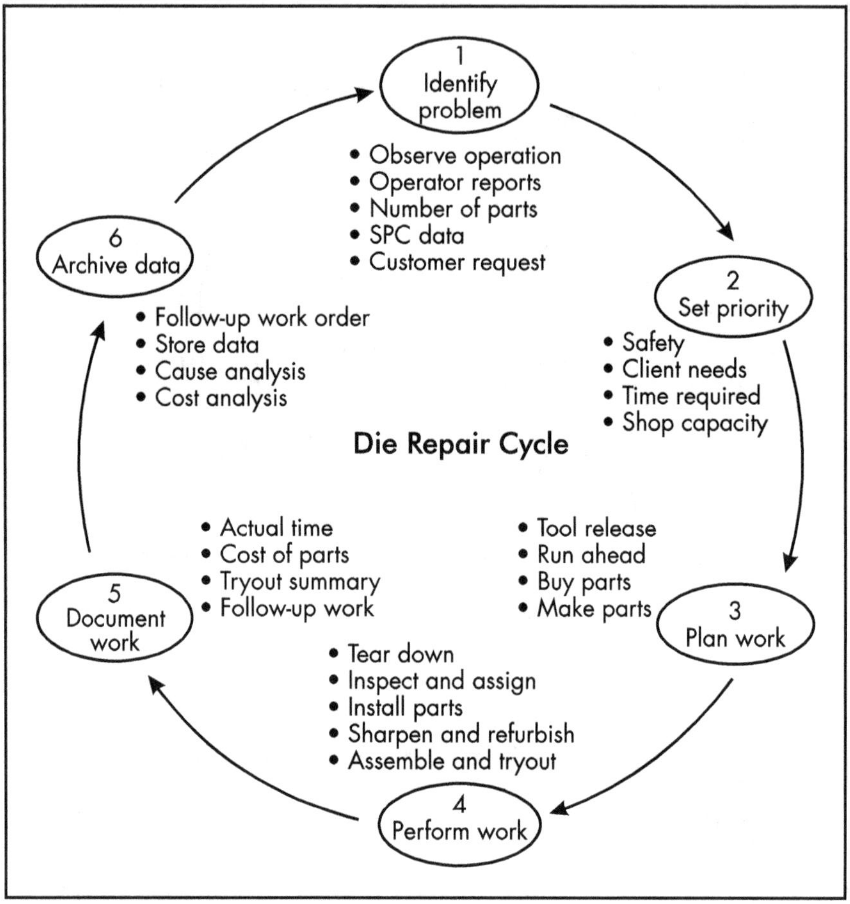

Figure 16-3. Six essential steps in a die-repair work cycle.

the die will always have part-uniformity problems and may not feed smoothly.

Step 1—Identify the Problem

The first step is to identify the problem and originate a work order for the needed correction (see Figure 16-3).

Observing the operation and checking the parts produced is the most common method of finding problems. The operator is in the best position to make long-term observations. An operator's opinions are invaluable to the die setters, tool and die technicians, engineers, and management. Ongoing operator training in the stamping process is essential to insure skillful operation and observation of equipment.

Like maintaining a motor vehicle, dies require periodic shop maintenance based on the number of hits made. The frequency of maintenance, dependant upon on the number of parts produced, requires fine-tuning based on experience with the die or with similar dies. If the die is subjected to misfeeds or multiple hits, it should be checked at once for damage by a tool-and-die technician.

Statistical Process Control (SPC) data can guide the observation process. If the process-control charts indicate that control points are drifting, this means that something has gone wrong with the process. It may be the press, the material, or the die.

Finally, customer quality or functionality issues require immediate attention. Fast, clear communication is the key to determining what has changed and what is needed to correct it. A customer request may involve an engineering change to the part. This usually requires die modification. All die work requires planning and the origination of a formal work order.

Step 2—Set Priorities

After identifying the work required, a work order should be initiated. Some shops use a paperless, computer-entry system for work orders. A clear work order that defines the scope of work is essential for scheduling work.

Problems that pose a safety risk should bc of thc highest priority. Loose die parts, broken cam-return springs, and misalignment are all examples of problems that require immediate correction. Die-component interference can send pieces flying at high velocity throughout the pressroom. Figure 16-4 illustrates this problem.

Customer needs are the next priority after safety. This means, "if we don't fix it soon, we can't make parts for our customer." A die-process failure sets the ultimate priority because it must be corrected or the product cannot be made.

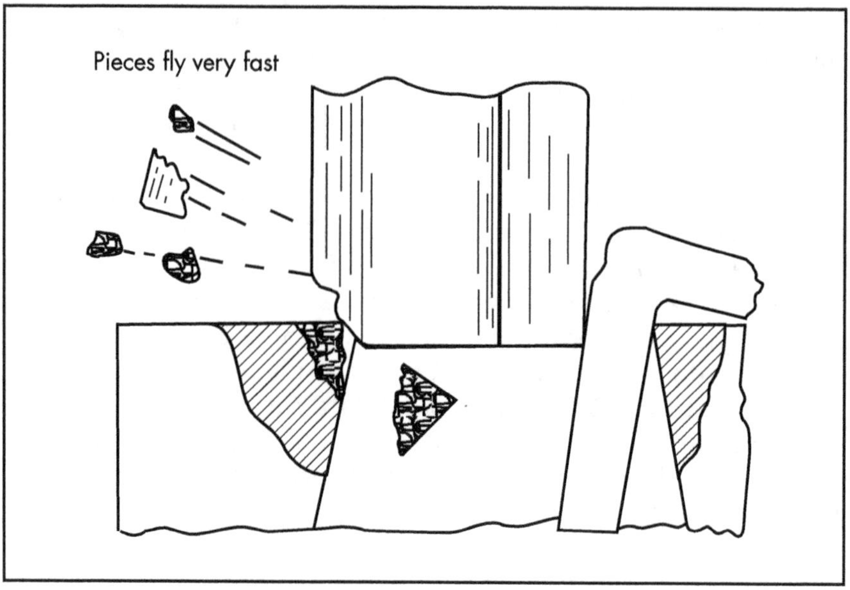

Figure 16-4. Shattered tool steel is a serious safety concern—it can fly at high velocity causing severe injury.

Step 3—Plan Work

Determining required maintenance time is important because time is a finite resource, and production parts must be shipped soon enough to meet customers' requirements. Therefore, it is essential to make an accurate estimate of the time required.

The ability of those in the die room to accomplish work is essential. The die room schedule may require overtime work if there is an emergency involving safety or quality production. In some cases, a repair-capacity shortfall may require the help of a contract die shop.

There are times when an equipment smash-up or breakdown make it necessary to immediately work on a die. This should be a rare exception—never typical.

Getting a tool release from the production schedulers is essential before working on a die. The scheduling department must work with the production control department to make sure sufficient parts are on hand. Stampings must be available to cover the time needed to rework the die. An advance run requires stock, production time, and storage for the stampings.

Planning the work includes the determination of required die parts or details. Some parts, such as springs and punches, are standard items available from vendors on short notice. All special parts should be ordered before starting work. Parts machined by CNC will require cutter-path data. Usually, tool steel requires heat treating, often followed by finish machining. The production of special die details may require from several days to a week or more.

A cooperative team approach to planning is essential. The alternative to planning is confusion, inefficiency, and the failure to achieve a goal of orderly, economical repairs.

Step 4—Perform Work

Step 4 occurs after the halfway point in a work cycle. If the first three steps of clearly identifying the problem, properly prioritizing the work, planning the repair and/or making sure a needed modification are all done properly, the job should go smoothly (see Figure 16-3).

Sample parts and progression strips must be sent with the die to a die maker. He or she can read the progression strips and estimate the required correction needed before opening the die in the die room. After a plan is agreed on, the die is taken to the die room for the required work. A typical sequence of work is as follows:

1. Disassemble and inspect die. The die maker carefully opens and examines the die. The die alignment system, consisting of guide pins and/or heel blocks, is examined for wear. As the die is taken apart, all evidence of wear and interference is noted.
2. Inspect and assign work. A written record of findings is made and the repair plan is finalized. Additional help is assigned as needed, and a plan to defer corrections not made in the allotted time is documented so a follow-up work order can be originated.
3. Sharpen and refurbish as needed. Good die repair skills are essential to repair and refurbish the die. This may involve

more work than initially anticipated in the planning and work-order generation process.

4. Assemble and try out in press. After parts are reworked and new details are installed, the die is tried out. This may be done in a dedicated tryout press in the die room. An alternative is to use the designated primary production press. A record of tryout and actual production results are essential parts of building a die history. This data is essential to find ways to reduce the cost of die maintenance.

Step 5—Document Work

The next step is to document the work done. The following are items that require documentation for archival as die history.

- The actual cost of the work in terms of time expended and the cost of parts is listed in detail and as a total-cost figure. A summary of the die tryout results should list the improvement or change in the die operation and the part produced.
- Follow-up work is often required. Continuous improvement has always been a cornerstone of process-based industries. When the die returns to production, items that can be improved should be noted.

Step 6—Archive Data

Data can be archived in a number of ways. Historically, file card, file folder, or die notebook systems have been used. Increasingly, computerized databases are used to store die history and perform cost analysis. When results of a repair-work cycle are archived, the following steps should be followed.

- Originate a follow-up work order if needed. When a follow-up work order is originated, the cycle is actually restarted to further improve the die and process. The die-repair cycle is a circle, indicating a continuous process. Dies require periodic maintenance that is mirrored in the work cycle.
- The die work history needs to be stored securely. Simple paper-based systems depend on written forms. These systems

are simple to implement and low in cost. Changes in the paper forms are easily made. Duplicate copies of records can be made and stored in a safe location. This insures that duplicate information is available in case the file for a die should be misplaced.

If a computerized die-maintenance system is adopted, the paper work-order and die-history system can work in parallel with it until everyone is comfortable with computerized data storage. Computer records must be backed up and stored off-site. Good backup procedures insure against loss of data.

- Perform root-cause analysis. Achieving continuous improvement proceeds more rapidly if the available resources are directed to solving the most costly problems. First, cause factors must be assigned to the types of process failures that occur. Normally, a limited number of problems cause the majority of difficulties. The causes are placed on Pareto charts. As progress is made toward minimizing the problems, what were minor problems may become the most troublesome. Charting problems helps insure that the available problem-solving resources are directed toward the most costly ones.
- Analyze the cost versus benefit of the work. The cost versus benefit of the maintenance program should be documented. If confusion prevails in the workplace, it is very difficult to assign real cost figures to the problems that occur every day. The work-order system and the six-step die-repair work cycle are good ways to bring order and increased profitability to the stamping shop.

THE REQUEST FOR MAINTENANCE FORM

A die-maintenance-request form should reflect the procedure followed in the die-repair- and cost-accounting process in sequential order. Today it is practical to use paperless data-entry systems that tie in with the plant's mainframe computer. In some cases, a department may have a local area network (LAN) for this function. However, paper forms having two or three carbonless copies remain popular.

Advantages of Paper Forms

There are advantages to using a multi-copy paper form rather than direct computer terminal entry. One advantage of the paper form is that the necessary approvals and signatures can be obtained before inputting the maintenance request into the computer system. The form minus the part(s) used for entry into the maintenance-management computer system remains with the die to provide instant information as to the die repair requirements and status. Familiarity with the paper tracking system is also an important consideration.

Designing the Paper Form

A good form starts with basic information entered into the upper left-hand corner and ends with the results of the correction in the lower right-hand corner. In this way, the form follows the function of the maintenance process. In small shops, reasonably good documentation can be accomplished with manual record keeping and cost accounting. However, if more than 15–20 maintenance technicians are involved, a computerized system is advised to run in parallel with the paperwork-tracking system to assure accurate die history and cost accounting.

Most forms are made from three or more sheets of non-carbon paper. This permits tearing out and sending a copy for computer data entry when the work order is generated. Another copy with the information about the work performed during the repair process will be sent for data entry when the work is complete.

CASE STUDY

The Ford/Mazda Auto Alliance Integrated Stamping and Assembly Plant (Flat Rock, Mich.) depends upon a team approach to solve die problems relating to quality and productivity. Team interaction is key to the success of the system. The stamping shop started with a total of five transfer-press lines, an open-backed inclinable (OBI) press line, and a blanker. Every die was assigned a home line whenever possible.

Auto Alliance solved its throughput-shortfall problem and the need for spare press capacity with a pressroom expansion. This helped to avoid shutting the adjacent assembly plant down in case of an unforeseen press failure. Auto Alliance, like many plants, had problems with press failures due to die-setting errors, as well as metal fatigue under sustained usage.

When the facility was built in the late 1980s, the pressroom was divided into several areas limited to 15–20 maintenance technicians each. This was necessary to cope with the limitations of its paperwork-tracking system. The team size of 15–20 maintenance technicians was considered the upper size limit for a manual paperwork-tracking system to remain workable.

The three-part die-maintenance-request form shown in Figure 16-5 was usually initiated by the transfer press or OBI-line production team leader who sketched a description of the problem in space one (1) on the form. After signing the form in space two (2), the person then took the form to the tool and die team leader for approval.

Team Leader Review

The team leader reviewed the problem with the production team leader and other persons having knowledge of the problem as needed. It was the leader's responsibility to plan the required maintenance. The tool-and-die team leader determined the cause(s) of the problem and identified all dies that required corrective action. The amount of time required to make the correction was also determined.

Tracking and Prioritizing Requests

Usually, only a two-day supply of any given part was produced at a time. Production of more than the usual run of parts might be needed to provide additional release time. The tool-and-die team leader coordinated the required release with the production-planning department. When the plan was complete, the tool-and-die team leader co-signed the form in the blank space (3).

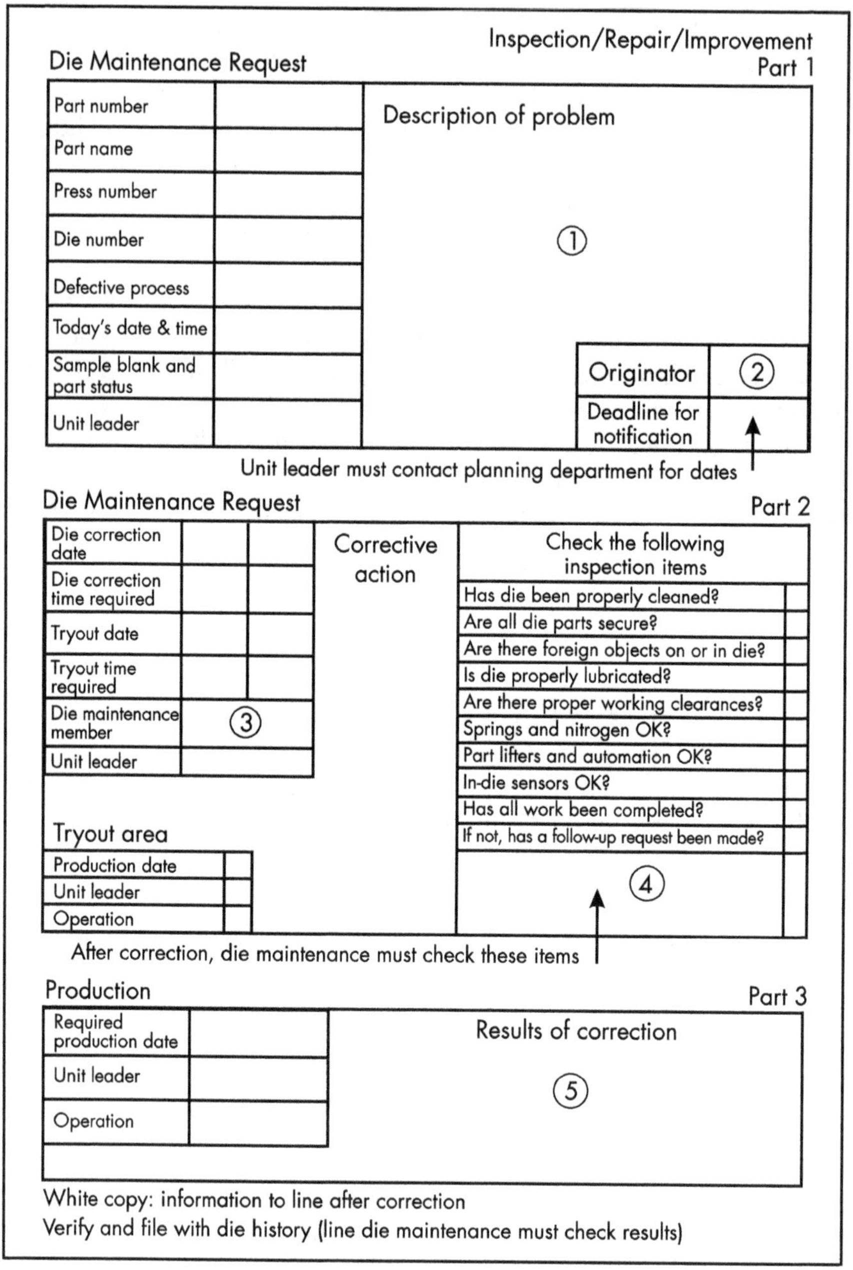
Die Maintenance Request

Inspection/Repair/Improvement
Part 1

Part number
Part name
Press number
Die number
Defective process
Today's date & time
Sample blank and part status
Unit leader

Description of problem
①

Originator ②
Deadline for notification

Unit leader must contact planning department for dates

Die Maintenance Request
Part 2

Die correction date
Die correction time required
Tryout date
Tryout time required
Die maintenance member ③
Unit leader

Corrective action

Check the following inspection items
Has die been properly cleaned?
Are all die parts secure?
Are there foreign objects on or in die?
Is die properly lubricated?
Are there proper working clearances?
Springs and nitrogen OK?
Part lifters and automation OK?
In-die sensors OK?
Has all work been completed?
If not, has a follow-up request been made?
④

Tryout area
Production date
Unit leader
Operation

After correction, die maintenance must check these items

Production
Part 3

Required production date
Unit leader
Operation

Results of correction
⑤

White copy: information to line after correction
Verify and file with die history (line die maintenance must check results)

Figure 16-5. A generic version of a die-maintenance-request form is shown. It can serve as a starting point for developing a maintenance-tracking form. (Courtesy Doug Burke)

One of the three copies was posted on a corkboard at the home line and the other two copies were inserted into a pipe attached to the die for that purpose. The production team leader shipped the required number of sample parts needed for evaluation and try-out to the die room together with the die(s). The parts were identified and stored in a designated area.

At the start and near the end of each shift, both the production-team leader and the tool-and-die team leader met to review problems, including a review of all outstanding maintenance requests posted on the line's corkboard. The outstanding requests were prioritized based upon the extent to which quality and production were affected. There were times when the entire correction needed could not be made due to the size of the job and urgency of other requests. In such cases, the plan was reduced to a partial correction, with a complete correction scheduled for a later date.

Follow-up

The diemaker making the repairs recorded the corrective action on the middle part of the form (4) and filled out the list of inspection items. After the correction was completed, the second copy of the die-maintenance request was filed in the die room records for that die, and the first copy was placed in the pipe attached to the die. When the die was returned to the press line, the first copy was placed in a plastic sleeve attached to the corkboard.

The lower space on the form (5) was filled out by the production-team leader and line-team leader, after the success of the repair was evaluated during a production run. The white copy was then retained in the die repair area as a part of the die's maintenance history.

Conclusion

Simple paper forms can be useful to determine the history of problems with individual dies. However, report generation for cost-accounting purposes requires time-consuming and error-prone manual data tabulation.

Today, most industries rely extensively on computer integration of the manufacturing process for most aspects of their operations. These include inventory control, payroll generation, and government report accountability—just to name several categories of essential data that must be managed and reported in a readily understood form.

The old Ford/Mazda paper-based die-repair tracking system has been abandoned. It has been replaced with a computerized maintenance-management and tracking system. Simply stated, the facility is too large for the paper-based system to work effectively.

CONTINUOUS IMPROVEMENT

A simple paperwork maintenance-tracking system is a good way for a shop relying on verbal orders to improve performance. It can always run in parallel with a computerized maintenance-management and tracking system. Many good computerized maintenance-management systems rely on paper forms for initial data entry and tracking of the repair process.

Tooling design improvements and maintenance cost control are frequently overlooked as areas for continuous improvement. Through effective database development and report generation, opportunities for continuous improvement in die design and maintenance can be identified and corrected.

Solving part quality problems may require a new process and new tooling. Here, a team approach should be used for value-engineering analysis. A cross-functional team including persons from process development, tool engineering, production engineering, and the shop floor can develop a better process and even redesign the part to reduce total cost (Daga and Choubal 2000).

Training is essential at all skill levels. Providing training opportunities to everyone from the operator to the tooling engineer and even top manufacturing management will lead to improved die troubleshooting and good die maintenance practice.

This chapter provides an overview of the topic of die maintenance documentation and tracking. If more detailed information is desired, it is suggested that you contact shops scheduling work with resources similar to your own to obtain advice on maintenance-management tracking procedures.

BIBLIOGRAPHY

Daga, H. and Choubal, S. 2000. *Industry Leadership Through Value Management*, 17th Value Engineering Conference. Bombay, India: Indian Value Engineering Society.

Smith, David. 1991. *Quick Die Change.* Chapter 16—Maintenance Requirements for Quick Die Change Teams. Dearborn, MI: Society of Manufacturing Engineers.

17

Effective Die Maintenance Program Strategy

UNDERSTANDING THE DIE MAINTENANCE FACILITY

Many stamping facilities do not have the die maintenance capabilities to perform repairs involving anything other than simple sharpening and parts-replacement tasks. Perhaps the only intended function of these facilities is to replace perishable details, screws, and dowels, and do minor detail repairs such as sharpening punches and die sections. The only machine tools in such a facility may be a small drill press, lathe, and surface grinder.

However, even though the plant is not equipped to perform anything other than running maintenance, it can and should have a fully functional die maintenance program. A good die maintenance program is a system that maintains dies so that unexpected breakdowns are avoided, all repair costs are accurately accounted for, and die repair cost, per part produced, is minimized.

A Die Maintenance Program versus a Die Maintenance Area

If a systematic die maintenance program—a means of tracking the condition of dies to assure cost-effective, timely maintenance—is not in place, achieving a fully functional system will take time and a lot of patience. Often, developing a program is actually a matter of achieving an orderly method of die maintenance.

It is important to recognize that a die maintenance program and a die room or die repair area are not synonymous. In fact, many production facilities employing metal stamping do not have die repair facilities on site.

In some cases, manufacturing plants are several hundred miles from the die builder that they depend on for major repair and die modification work. The following definitions may be applied:

- A die room or die repair area is a place designated and equipped for working on dies, to accomplish routine maintenance, extensive repairs, and modifications. It is often equipped to build new dies if necessary, although that may not be its primary function. Usually, CNC milling and wire-cut EDM equipment are available to make new die details from readily available cutter-path data. Heat treating of tool steel can be done on site or at a nearby heat-treating contractor.
- A die maintenance program is a means of tracking the condition of dies to assure cost-effective, timely maintenance and the avoidance of production disruptions caused by unplanned downtime. Report generation and "red flagging" dies in need of timely maintenance is a key feature.

Thus, the difference is that the die room is equipped and staffed to build and/or repair dies, while the die maintenance program includes an accounting and data-retrieval system that permits cost-effective management of die maintenance. A die room may be provided at the production facility, stationed at another company location, or die repair may be contracted out to a tool and die shop.

ADDRESSING PRODUCTION-SCHEDULING PROBLEMS

An essential function of any die maintenance activity is proper scheduling and planning of work. Accurate cost accounting, repair cause-code tracking, and a cost-reduction emphasis are indicators that an effective program is in place.

The advantages of low inventory or just-in-time (JIT) manufacturing are obvious and well understood. Eliminating costly safety stocks means that agile, low-inventory manufacturing is ingrained in the plant culture. It is doubtful that any company that has reduced its work-in-process (WIP) would ever want to go back to the older, unwieldy large inventory system of mass manufacturing, with its built-in inefficiencies and high costs.

However, periodic die maintenance must be performed if breakdowns are to be avoided. Therefore, the amount of production release time required must be accurately determined and a plan must be worked out to release the tool from production. If the required release time is greater than the die-run cycle time, a run-ahead plan must be put in place.

Running parts ahead of time often requires obtaining extra stock. The extra stampings must also be stored safely. Clear communication about what is needed and the full cooperation of production scheduling and material control are all essential.

PLANNING DIE MAINTENANCE

To maintain quality production and forecast die repair requirements, there must be some method of planning die maintenance. The planning method can be as simple as a single individual in a small shop, such as the pressroom manager, scheduling dies for repair, to a sophisticated computerized database for prioritizing and scheduling needed repairs. Any good system must take into account key planning factors such as:

- Problems and their root causes must be identified.
- Scheduled inspections and preventive maintenance are important.
- The capacity to do the repair work must be known.
- An accurate estimate is needed of the time required to complete each repair.
- Production scheduling must be considered.
- There must be a means to prioritize backlogged work.

TRACKING THE COST OF MAINTENANCE AND DOWNTIME

Any effective system for tracking the cost of maintenance and downtime depends on a full account of how all maintenance and production time is expended. For example, to determine the true availability and uptime of a press, several factors must first be known, including:

- total available press hours per week, which factors out maintenance time and time not normally scheduled for production; for example, midnight and weekend shifts may not be scheduled as normal production time;
- the amount of time it *should* take to set up the job, run out the production, remove the die, and clean up the production area;
- the amount of time it *actually* took to perform the work listed in the item directly above; and
- an accounting must be taken with downtime cause codes assigned to determine the total difference between the calculated rate and lost time; this information is essential for continuous improvement efforts.

These efforts should result in a surplus of available machine hours that can be utilized. The keys to increasing profitability are increasing available press time and then selling that capacity.

CONTINUOUS IMPROVEMENT

Those who fail to learn from history are condemned to repeat the mistakes of the past. This is also true in good die and process design practices.

Die history is often overlooked as a source of valuable data for accomplishing continuous improvement of the stamping process. Whenever a new die is tried out and placed in service, a record of the results should be entered in the die-service history. By having this continuous improvement data, part design features can be improved and a better die can be built the next time a similar one is constructed.

Die standards are essential to assure that the dies used are suited for their purpose, with safety and durability being prime considerations. They are "living" documents that must be updated frequently. This is to assure that shop-floor and die-maintenance improvements are added so that new die designs incorporate proven beneficial changes, resulting in lower tooling cost without affecting performance. Economical construction and low maintenance costs also must be given careful attention.

Shop-floor Involvement

Achieving continuous improvement in die design should be everyone's responsibility. The die standards activity is normally a manufacturing engineering function. It is important not to isolate this activity from shop-floor improvement activities. As an essential source of continuous improvement information for the engineering staff, this feedback assures up-to-date die standards.

Recognition

A monetary award is not necessarily required as an incentive for pressroom employee participation in the continuous improvement process. The award amount is difficult to compute and may be a topic for controversy. A thank-you letter to the employee, along with posting the employee's picture and details of the value of the improvement on a quality operating systems (QOS) bulletin board is one way to reward the employee. This will encourage other employees to submit their ideas for possible adoption. Close communication among the engineering staff, management personnel at all levels, and shop floor hands-on experts is essential to efficiency and progress. Even if an idea is not adopted, personal thanks and discussion of how to rework the idea into an adaptable form are an essential courtesy (Smith 1993).

Maintenance Work Records

The amount and type of maintenance being performed on each die should be assigned a cause code. In this way, die history can be used to achieve continuous improvement. Printing out the costs for each type of wear or damage for all to see is easily accomplished if the entire history for all dies is maintained in a computerized database.

Computerization and Document Control

It is not possible to know true die maintenance costs unless all costs associated with die maintenance are tracked. This includes routine repair, the cost of damage due to smashed tooling, repair parts, labor, and overhead.

Many stamping facilities improve quality and reduce cost by implementing a computerized die maintenance program. The strategy is to reduce the amount of breakdown repairs and increase the amount of planned maintenance to improve throughput and quality.

The approach usually involves a closed-loop work cycle. The work needed is identified and priorities are set. The work is planned, followed by scheduling and assignment. Progress is reported by each shift. At the completion of the work cycle, work performed is evaluated and entered by die number into the die history file.

Die maintenance can tie in with production and the die-set scheduling system can forecast future die-set schedules—an important planning tool. Reports on die and automation changeover-time performance also can be generated and tracked to evaluate continuous improvement of manufacturing methods.

To further aid in reducing setup times, a changeover-performance system can be implemented. Computer-generated data reports documenting problems that occur during die setting, and any problems that occur upon startup and during production, are essential for team-oriented problem-solving (Krygier 1989).

Manual versus computerized systems. Manual maintenance tracking has a limited capacity for information storage and retrieval. Reports on problem causes and labor hours expended are often needed for management decision-making. These can be very time consuming to generate manually.

Manual systems generally work reasonably well in shops having up to 15 employees in a repair activity. The actual number depends upon how well employees and management cooperate to jointly solve problems. For a manual system to work well in large shops, it may require reducing the maintenance activity into subgroups having no more than 15 employees per group.

Simple manual paper tracking systems may work reasonably well when first implemented. At the start of the program, little die-history data exists to search. However, the paper information will accumulate and become nearly impossible to keep track of.

The accessibility needed for wise resource management and efficient plant administration has resulted in computer-integrated manufacturing (CIM) systems finding widespread application. Employee acceptance generally is excellent when they are involved in the early phases of the CIM-system implementation. The use of a maintenance-management database as part of a CIM system can provide a means to improve quality, reduce cost, and even bring about a needed cultural change.

A work plan can be printed out for the supervisor to use as a management tool. It replaces hand-written reports passed from shift to shift, and clearly avoids misinterpretation. Die-cost history is built and reported.

Choosing software. There are a number of computerized maintenance-management systems on the market. Not all are suited for tracking die maintenance. There are basic requirements for a good system. A team-oriented selection procedure is a good approach for choosing the computerized maintenance-management system (Smith 1991).

Buying software based on price or visual bells and whistles is a poor idea. The goal is to get the new system in use quickly. The people that will enter and make use of the data need training. This may require hiring outside trainers or consultants. A good consulting firm will have a goal-oriented program schedule, and will follow the schedule until all goals are met—without charging more than initially stated.

It is wise to get cost proposals and visit satisfied users. This is helpful to evaluate firsthand how well the program and the vendor support are working.

Using spreadsheets and databases. The computational power of the personal computer and readily available software has increased to rival the capabilities of a mainframe system from the 1980s. Many shops use relational databases and spreadsheets that their employees configure to suit their needs. Some shops will trade favors and provide the matrix they are using. This may suit your needs at very low cost.

SPARE PARTS INVENTORY

Business management books teach many methods for determining optimal spare-parts inventory float levels and economic order points based on mathematical formulas. These formulas are created to address plant and facilities maintenance-management issues.

Most crib-float-level formulas are based on annual consumption versus the cost of keeping parts in inventory. The formula in use may fail to factor in the parts becoming obsolete. Maintaining just enough parts to carry out emergency repairs is a wise alternative to computerized float-level determination based on sporadic consumption.

The textbook approach is apt to establish an arbitrary float level that provides more parts than needed for emergency repairs, but too few parts for a complete refurbishment. Choosing toolroom supply manufacturers and vendors that carry necessary parts in their inventories, and who will ship them quickly, is important. In this way, extensive planned die maintenance can be carried out without maintaining a large die-repair parts inventory.

Emergency Repair Parts

A nitrogen manifold system is a good example of a system that requires a periodic complete overhaul in long-run service. However, it is wise to keep a few repair parts on hand for emergency repairs. Some parts, such as seals, are subject to deterioration when stored for more than a few months.

Emergency repair parts such as a complete spare cylinder, a couple of spare seal kits, O-rings, rupture plugs, charging fittings, and a spare charging console will serve to repair most problems on a temporary basis. Keeping one or two spare port plugs with the same thread size as the cylinders is a good idea. They may be needed in the event that a cylinder needs to be removed to properly balance the draw-ring force in draw-ring-pressure applications.

A few hundred dollars worth of spare parts can serve to repair or correct a pressure imbalance of a nitrogen manifold system costing thousands of dollars. Of course, cylinder wear involving honing or polishing in an emergency or wear attributable to normal extended service requires ordering the needed parts for a complete rebuild that will restore the system to like-new performance. This is no different than overhauling an automobile engine. All parts that are subject to wear are replaced to achieve like-new service.

Parts for Planned Maintenance

A proper rebuilding job on a manifold system requires removal of all cylinders for disassembly, inspection, and replacement of all packing. Any worn or scored piston rods and cylinder bores should be replaced rather than honed smooth since the surfaces subjected to wear are either nitrided or hard chromium plated. The port plugs, charging console, and rupture disks should be removed and the complete manifold carefully washed and oiled with an approved lubricant. All O-rings should be replaced or at least carefully inspected. Likewise, the mating sealing surfaces should be checked for burrs and oil-stoned if needed. Finally, the manifold is carefully reassembled, inspected, and returned to service.

To plan for a manifold system rebuild, a complete set of replacement seals and O-rings must be ordered and on hand in addition to the normal inventory of emergency spare parts. The cylinder manufacturer should be consulted in advance regarding off-the-shelf shipping of any additional cylinders found to be in need of replacement when the manifold is disassembled. If the die can be released for several days, an arrangement for overnight shipping of any needed replacement cylinder rods, cylinder bores, or complete cylinder assemblies can permit a complete rebuild of the manifold assembly within the release time.

Perishable die details that require frequent replacement should be kept in stock and reordered as needed. Expensive die sections that are maintained by sharpening, polishing, or regrinding worn forming-section geometry normally are not stocked.

If a complicated section can be made by CNC milling, or conventional or wire EDM, having a quickly available source of the needed tool steel is wise. For EDM work, pre-hardened tool steel or carbide blocks are commercially available on short notice.

In case of catastrophic damage, emergency repairs are often accomplished by welding or inserting a repair section. Inserting a section can be accomplished by grinding or wire-EDM cutting of a dovetail pocket and making a repair insert.

Reducing Spare-parts Inventory

An essential aid to reducing spare-parts inventory cost is a company policy that permits rapid purchase-order generation for telephone or fax transmission. Some companies have unrealistic approval systems for purchase-order generation that require hand carrying of paperwork for multiple approvals. In addition, spot buying of spare parts may be limited to a low amount such as $500. This leads to a great deal of wasted time generating multiple purchase orders on separate days to accomplish what should be a simple repair task. This waste must be avoided. Internal auditing procedures and a culture that demands honest accounting is a far better solution than allowing a system that must be circumvented to accomplish maintenance.

An important factor in reducing inventory of spare parts is a plan to make or buy the part from a dependable vendor on short notice. This is especially important if the die is owned by someone other than the stamping plant that runs the parts.

Should the Die Owner Pay for Spare Parts Inventory?

Stamping customers who expect low-inventory just-in-time (JIT) production should be informed that a reasonable stock of die parts must be kept on hand to enable fast die repairs. It is fair that the owner of the die, who can source repairs to another stamping shop with little or no notice, pay for and own the necessary spare parts required for normal maintenance.

Of course, if the die were to be outsourced elsewhere, the spare parts would be shipped with the die. It is important not to rely on verbal agreements in this type of arrangement. A written agreement covering the spare-part float level and inventory accountability should be in place.

CALCULATING COSTS

The die repair cost estimation method is essentially the same as that used to build a new die. The same procedure of making a stock list of required new components and determining the required machine and bench time is followed. However, there may be some additional costs to determine that are not normally required for new die construction. These may include in part:

1. *Causal factor analysis*, which involves determining why the die requires repair if not due to normal wear. Future die damage of the type being repaired often can be avoided. For example, to avoid a repetition of die damage, a system of in-the-die sensing equipment can be installed and tied into the press automation and sensing control system.
2. Determine any damage to the die shoe and heel blocks. For example, dowel holes may become elongated and screw holes stripped in the case of mishit damage. The die shoe may be bent. For precision die operations, the shoe may require straightening and welded repairs followed by stress relieving and remachining. If the damage is extensive, replacing the die shoes may be less costly than repairs and require less time.
3. Any special details requiring replacement should be reproduced in advance from the cutter-path information that should

be supplied with any new die. If this information is not available or there is no accurate dimensional information on the die details needing replacement, additional costs will be incurred. A plan to replace them must be worked out and the cost determined.

4. Die repair is an opportunity to correct die design deficiencies. Examples include strengthening weak die sections, making provision for die-grinding shims, and correcting incorrect application of tool steels. In addition, the application of wear-resistant die treatments and coatings as well as upgrading other die materials can be accomplished. Realistically, some of this cost should be charged against the tool cost and not the die repair.

The calculation of die repair costs is carried out the same as when determining the cost of a new die. However, there are also factors that include any required run ahead expenses, excess shipping costs, as well as transportation of the die to an outside repair facility, if that is required. Entry items include:

- cost of production run ahead, if needed;
- cost of premium shipping, if needed;
- cost of shipping the die and sample parts, if required;
- itemized labor costs for bench repair time (hours × shop load);
- itemized machine time costs (hours × machine hour rate);
- itemized cost for standard details replacement;
- itemized cost for custom-machined replacement details, and
- installed cost of any required die protection or die improvements to avoid future repair costs. This can include a credit for the cost-avoidance benefit obtained.

Unanticipated Downtime

Unanticipated downtime due to die damage, poorly completed die repair, or a failure to carry out timely die maintenance tends to escalate exponentially as the downtime delay increases. Largely, lean manufacturing that emphasizes low inventory and JIT production has made freedom from metal-forming downtime a necessity.

The cost of minor in-press die repair such as polishing score, shimming to adjust a flange angle, or replacing a detail may take from 10–30 minutes. Provided there is a slight surplus of press time to accomplish this work, some of the primary cost factors are as follows:

- hourly rate, including shop load of the operator (if he or she is not reassigned to other work during the period of downtime) and the die-repair technician or die maker times the amount of time required to accomplish the repair; and
- cost of press time per hour.

Lengthy Die Repair Time

In a JIT manufacturing environment, long die repair delays can be very costly. The inability to supply parts to an assembly operation often results in large numbers of assembly workers being sent home. The consequence can be very serious. The results can vary from substantial monetary losses and client displeasure to bankruptcy if there are damages specified in a customer's contract.

Run Ahead Production

A run ahead may be required to prepare for release of the tool from production. This is quite likely in the event that parts are being produced on a low inventory or JIT basis. Emergency die repairs may be required to permit a run ahead, which is an additional cost added to the cost of a permanent repair.

Normally, a run ahead requires planning and establishment of a substantial safety stock to permit the die repair to occur. Stocking temporary inventory is a real cost and should be charged to the die or dies involved. A failure to correctly account for this cost will result in the actual cost per piece to be understated.

If the die is set on an infrequent basis and does not require more than several days release time, there may be no need to run ahead to permit repairs. However, if the die requires more time to repair than the production schedule will permit, arrangements must be made for running sufficient extra parts inventory to avoid disruption. Of course, an accurate plan for repair of the die is required to plan the correct amount of release time. Once the die repair time is determined, the run-ahead activity can be planned, which may entail the following preparations:

- Sufficient extra press time to permit the production run ahead must be arranged.
- The press operator(s) and support personnel need to be scheduled. Often, this may entail overtime authorization and the cooperation of the production scheduling activity.
- Sufficient extra material to permit running the job ahead must be ordered. Here, the material-control activity and supplier will need to be involved and will require sufficient advance notice to avoid extra costs for expedited supplier service.
- Low-inventory systems or JIT scheduling often use a pull area for partial storage in standardized containers. Arrangement must be made for the safe and orderly storage of the parts, which may not fit into the pull storage area.
- Excessive part-shipping costs, if any, must be determined. Shipping can be very expensive if air transportation must be substituted for rail or truck transportation. Costs associated with expedited airfreight shipping to avoid shutting down an automotive assembly operation can be very high. For example, shipping a box of stampings weighing 75 lb (34 kg) 60 miles (96.5 km) by helicopter, rather than by truck, typically costs $3,000. Shipping a three-day supply of assembled automotive hoods 400 miles (644 km) by air transport, rather than as part of a unit rail train, has a cost in excess of $250,000.

Cost per Piece for Die Maintenance

When designing stampings, there are a number of cost considerations to take into account for both the stamping design and the processes used to produce it. Initial consideration is given to projected production volumes per month or year for the product application into which the stamping is incorporated. This will determine the class of tooling required to economically make the stamping and hold the part's specified dimensional tolerances.

The big cost items to avoid are excess shipping and any die damage caused by lack of die protection systems. Avoiding future costs through employee training is advised. Correctly planned and carried out die repair will not disrupt a production build schedule. In addition, the repair activity has an opportunity to correct design flaws that will decrease the cost of future die maintenance. This can help reduce the cost per piece for die maintenance.

Often, low-cost tooling is specified to win a contract to produce the stamping. While this may have short-term benefits, it may also result in an ongoing program of upgrading the tooling to improve reliability and hold the customer's dimensional tolerances. In these cases, the tooling upgrade cost is reflected in a higher-than-normal die repair cost per piece. It is far better to agree with the customer on a grade of tooling designed for reliability and low maintenance requirements if this will lower the customer's cost per piece.

Cost Accounting

It is recommended that the cost tracking system treats each job, and the costs associated with the die(s) required to produce the parts, as individual cost centers. Other costs including material, projected press time, press down time, shipping costs, and overhead items must be entered to determine the cost per piece. In this way, the maintenance cost per piece can be reported for all jobs.

The die repair cost can be expected to vary greatly from part to part. Usually, there is a logical reason. For example, the die repair cost per piece for a simple high-volume blank-through part is normally quite low. However, the repair cost for a complicated progressive die performing operations such as in-die tapping or assemblies tends to be high. The alternative is to use secondary operations that are costly and labor intensive, which can result in poorer quality finished parts. The important factor to look for is to be sure that all reasonable means, such as electronic die protection and appropriate wear-resistant die materials, have been incorporated into complicated dies. In short, the value added by elimination of secondary operations must add more value to the part than the combined cost factors incurred, including die maintenance.

Assigning Cause Codes

The cost of poor die maintenance can raise havoc with the total profit margin of any stamping operation and the customers that depend on timely delivery of quality parts. Increasingly, plant activities are treated as cost centers for accounting purposes. Thus, assigning cause codes to costs resulting from poor die maintenance is essential (see Chapter 6). A highly systematic approach to charging departmental cost centers for poor or unnecessary die maintenance is highly recommended to reduce needless rework and repairs.

Establishing the Dieroom as a Cost Center

Having a well-equipped die room at every stamping production facility is not an absolute necessity. There is the alternative of performing minor die repair, such as polishing and replacing perishable details in-house, and sending more extensive work out to another company location or a die-repair vendor shop.

Die repair may be the main activity of a die room, but new die construction also may be a function. It is essential that all costs associated with die room operations be quantified and charged against the correct job. Failure to strictly account for all costs can result in erroneous cost data. For example, if new die repair construction costs are charged against a die that is in the die room for repair, the production job will be incorrectly charged high costs. This can create the illusion that new dies can be built in-house at a lower cost than outsourcing the work to a contract die builder.

Examples of die room line-item costs include the following:

- each employee's base hourly rate;
- fringe benefits such as insurance benefits, pension plan, paid holidays, and vacation time;
- tool-crib spare-die parts inventory, including cost of keeping in stock and expenses attributable to the probability of parts becoming obsolete;
- industrial compensation insurance including employee safety training;

- the cost of floor space and utilities, including taxes, and interest;
- the investment in machinery, including wear and scheduled replacement costs, and
- cost of employee parking space, lunchroom space, lockers, and lavatory.

Putting Numbers on Die Repair Costs

If die repair is done in-house, accurately calculating each of the line items and adding them will provide an approximate cost per hour to repair dies. Here, it is essential to account accurately for all of the time expended and make sure that it is charged correctly.

It is important to have cost codes cover all activities to avoid incorrectly charging time to a job to cover other necessary activities. Non-repair charge codes may include:

- time spent in training and employee involvement meetings;
- medical treatment time and lost time off work due to industrial injuries;
- time expended on cleaning activities, and
- coffee breaks and lavatory time.

The important issue is to charge all time and material used to repair each die correctly so the cost and repair cause codes can be tied together and an assessment of charges made. In this way, the die room is treated as a cost center and each die is a cost center. The repair for each die and repair cause code can easily be accounted for to permit accurate accounting and report generation for action to effect improvements.

Conservative estimates by stamping managers indicate that at least half of all die repair work is due to avoidable damage that occurs in the pressroom. The true total is probably higher. The type of damage observed every day in the toolroom, such as sheared cutting edges, bent dowels, distorted dowel holes, etc., indicates damage due to misfeeds and other common pressroom errors.

It is essential that this data be carefully reviewed with appropriate notes made by the tooling manager. Copies should be sent to all department heads. Action plans should be agreed upon to address corrections. The data should be conspicuously posted for everyone to see with an explanation of the needless impact on the bottom line caused by die damage. Part of the solution may lie in training of proper procedures.

PARETO CHARTING

Pareto's Law, often known as the 80-20 law, was originally an economic theory about the wealth of nations. Through practical application, it has been extended over time to apply to many areas of industrial activity.

It is now a well-accepted fact that for many manufacturing companies, 20% of their problem issues are responsible for 80% of delays and repair costs. The exact ratio is unimportant. The fact remains that a relatively small number of repetitive problems are responsible for the majority of delays and repair costs.

Accurate information on the causes of downtime and repair costs must be available to begin Pareto analysis. This information is collected from each die's history. The examples shown in Figures 17-1, 17-2, and 17-3 chart three major cost factors for a group of dies over a period of four months. Arbitrary cost units are on the left-hand vertical axis. The best indicator would be a unit of money—probably in thousands of dollars per month for many larger plants.

In Figure 17-1, the three highest cost items relating to tooling downtime and repair cost are entered and the severity of the problem is indicated by the height of each bar. The most costly problem is damage or shearing of the cutting edges caused by interference between the die cutting edge. Die shearing can result from a number of factors. The most common are stock misfeeds and improper die opening and closing in the toolroom. Here, operator training, misfeed sensor protection, and better die-handling equipment can be effective in reducing the associated cost. The cost factor dropped from 90 to 30 over a four-month period. The goal of zero is theoretically attainable.

The second most common problem is slug jamming. Like die shearing, slug jamming can have many causes. Proper scrap chutes

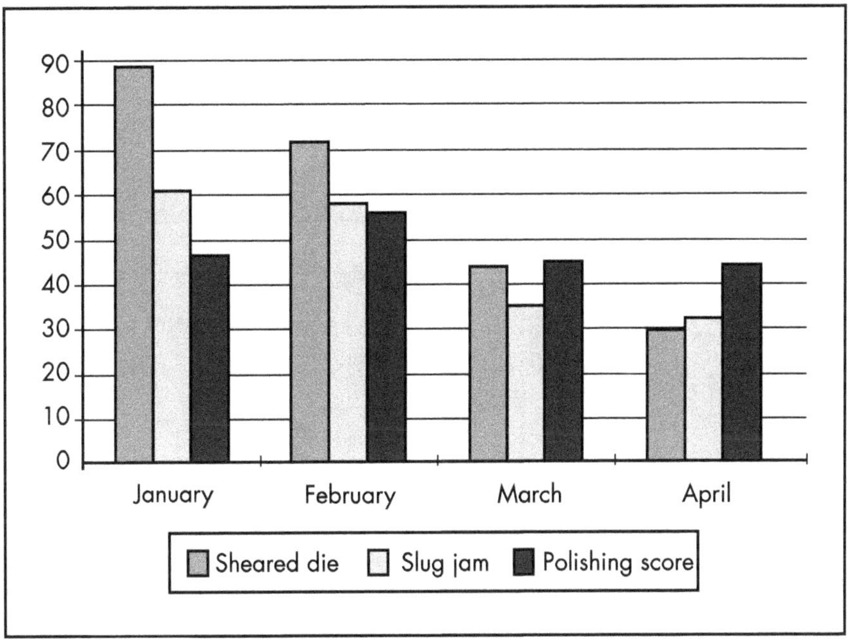

Figure 17-1. Pareto chart illustrates how directing repair and process design resources to several of the worst problems can dramatically reduce the incidence of occurrences over a four-month period.

and conveyors are needed to carry the slugs out of the die. The dies must be built with proper slug clearances and tapered slug holes to assure positive slug discharge. Like die cutting-edge damage, these issues can be addressed by training operators and die makers on how to avoid the problem. Money and additional shop organization may be required to identify the need for better scrap conveying equipment and assure its proper use. The cost factor was cut in half in a four-month period. Here, the goal of zero is also theoretically attainable.

The third item, polishing score or metal pickup, does not improve at all during the four-month period. In fact, applying additional resources to accomplish more polishing actually increases cost. Here, it is obvious that an appropriate solution is not being applied. Score and metal pickup problems may require better tool steels, coatings, and improved lubrication to reduce or eliminate the problem.

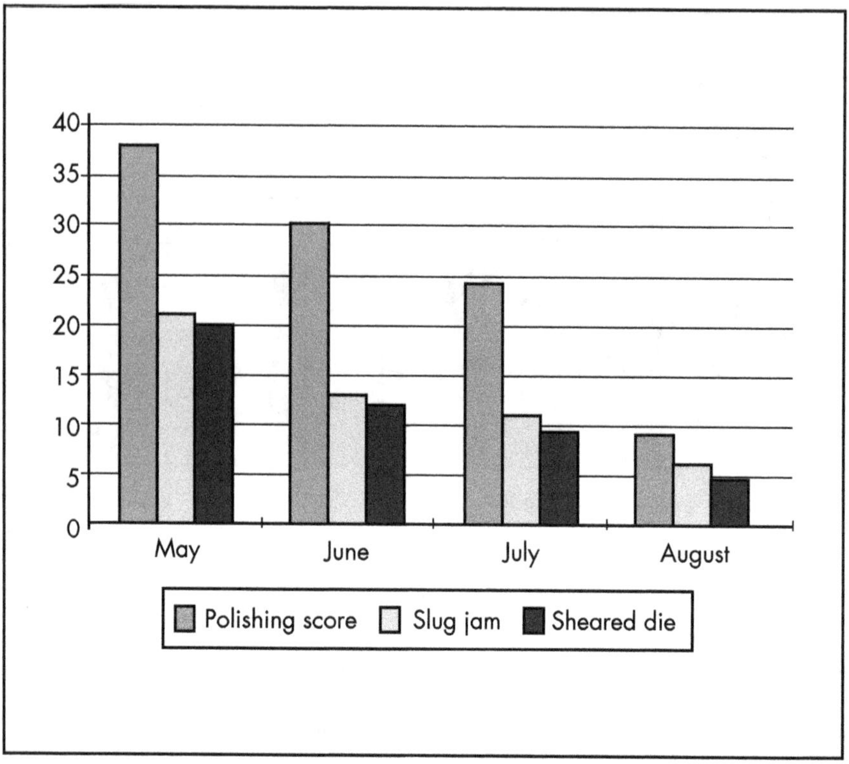

Figure 17-2. Pareto chart shows how directing repair and improving the process can continue to reduce the incidence of occurrences over the next four-month period.

In Figure 17-2, the same three causal factors are still being charted and special emphases are directed toward score problems, slug jams, and sheared dies. Finally, the problem of score is being addressed through improvements such as better die materials, coatings, and lubrication. The total downward cost trend indicates that resources are being appropriately directed.

Polishing score is the only old item left on the chart in Figure 17-3. Two new items have been found to be major cost factors that may have become apparent after addressing the sheared die and slug jamming problems. These are punch wear and feeder error. Feeder errors are a causal factor of sheared dies—in this case one that requires additional analysis and correction. This data can be used for cost justification to upgrade or replace feeders.

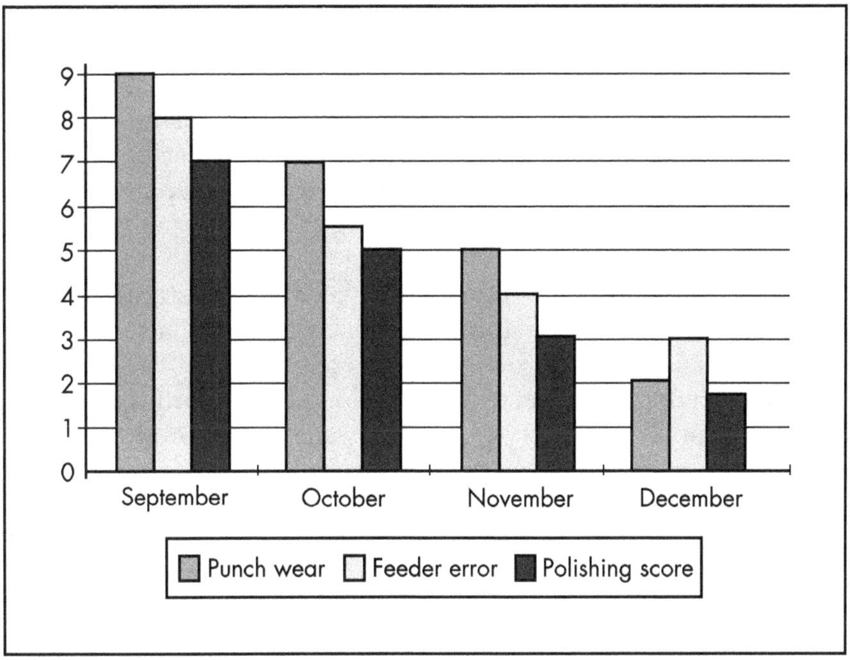

Figure 17-3. Pareto chart shows continued reduction in the incidence of occurrences. Note that the relative cost is much lower than at the start of the charting program. Two new items, punch wear and feeder error, have replaced older items where the cost has been greatly reduced.

DEALING WITH AN OUTSIDE VENDOR

If the die is sent outside the company for repair, it is important to carefully spell out the work to be done. Tryout stock, current part prints, up-to-date die prints, and checking fixtures should be made available to the die repair vendor.

It is essential that the vendor be chosen with care. There must be assurance that the repair work will be done efficiently, correctly, and on schedule.

Die repair is not unlike automobile repair. Once the die is opened up and disassembled, often damage is found that was not apparent from examining defects in the parts and/or progression strip. Of course, this factor makes firm estimates difficult. In any case, latent damage discovered should be corrected if possible. If the vendor is forced to make a patchwork rush repair, perfect results should not be expected. In such cases, proper repairs should be scheduled and carried out as soon as possible.

Mutual trust and cooperation are important factors in a die repair vendor and stamping shop's business relationship. In some cases, very minor problems are settled by an adjustment in the mutually agreed upon cost of the next invoice for repair.

If the vendor does something seriously in error, such as failing to provide any slug clearance through a die shoe, catastrophic damage can result. This damage can deprive the stamping customer of vital parts. A customer will not want to hear why a vendor has failed to deliver the parts. It is possible that the vendor's contract to supply the parts could be lost, along with the vendor possibly assuming liability for the customer's lost production.

The responsibility for lost revenue caused by the die repair vendor's failure to perform is, at best, difficult to assume. A contractual arrangement with the vendor spelling out the liquidated damages for lost revenue is difficult to obtain and even more difficult to enforce. The repair shop is much more apt to be liable under common law for tort damages as a third party if an employee is seriously injured by shattered tool steel.

USED DIES

Accepting a die that has been run at another stamping facility can vary from a profitable business opportunity to a legal nightmare. The key to avoiding problems is to determine why, when, and under what conditions the die will be sent to the plant. It must be certain that the die can be run safely without damaging equipment. In addition, a profit must be made.

Nearly every stamping facility has had to run one or more used dies with which the staff is not familiar. Dies are sent to a different production facility than the one originally intended for a variety of reasons. The reasons for this vary depending on the business specialty and customer base. Important questions must be answered before agreeing to run the die. These range from the design and condition of the die to safety and equipment capability and availability requirements.

Maintaining goodwill with a major customer is important. However, failing to meet or exceed that client's expectations can cost a company money and its reputation as a "can do" shop. Important questions need honest answers.

Who Pays for Refurbishing the Die?

If the die is nearly new and built according to good design by a shop with a reputation for good work, there should be little cost associated with running the die. However, if the die is badly worn and in need of repair, the company outsourcing the tool should agree contractually to pay for any repairs required. This is a reasonable expectation to assure that the tool is capable of running the expected amount and required quality of parts.

It is wise to question the reason why the tool is being sourced to your shop if you do not have experience with the outsourcing company. There should be a formal agreement to permit running it for long enough to cover the costs of adapting the tool to the equipment and to assure a fair profit will be made.

GOOD ENGINEERING PRACTICE

It is essential to make sure that a die can be operated without damage to equipment and in full compliance with government safety regulations. This concern extends to making sure that good engineering practices are followed. These include:

- What are the physical dimensions, shut height, required press stroke, and weight of the die?
- Can the die be safely set with existing die-handling equipment? If not, can the use of die-setting equipment be obtained to safely accomplish the task in the plant layout? Who pays for the use of this equipment?
- Is there a press that has adequate tonnage and torque capacity throughout the stroke to run the die? Overloading presses is illegal and a prime cause of costly press damage.
- Is the upper-die weight correctly identified on the upper-die shoe?

- Does the press to be used have adequate counterbalance capacity at the minimum shop air pressure?
- Can the die be safely fastened in the press with existing die-fastening equipment? If not, what is the cost of adapting the die to be set and fastened safely?
- Are there any die, pressure pad, or cam shock impact problems that can cause the normal die-fastening system to fail? The die owner should be specifically asked this question, especially if the die has a heavy upper pad with substantial travel distance.
- Are all die pads and draw-ring openings safely guarded? This requirement is spelled out in most die standards. However, it is common for a company to fail to follow its own die standards. For example, an unguarded inverted draw die ring can easily result in a foot amputation even though the safety distance to points of operation to which the hands are exposed are adequate.
- Are cam return springs properly enclosed and guarded to contain die parts in the event that the cam return rod retention should fail?
- Does the die have any points of operation that will require respacing light curtains or palm buttons to obtain adequate safety distance?
- Dies are often fastened in the press by clamping to attached parallels. If this is the procedure to be used, are the parallels attached to the die shoes with sufficient high-strength shock-resistant fasteners to exceed the strength of the die-fastening system?
- Are there any processes performed in the die that could cause die parts to shatter in case of a misfeed or double hit? If so, a solid physical guard of thick clear plastic or metal may be required to protect pressroom personnel from shattered tool steel exiting the die at high velocity.

Both OSHA requirements and accepted engineering practice must be met and exceeded where necessary for employee safety.

Die fastening and point-of-operation guarding are especially important items.

Assumed minor issues, such as die fastening and pad or draw-ring guarding, may be overlooked. Dies designed for automated tandem line or transfer-press operations can cause serious injuries when manual (hands in) die operation is practiced. Die fastener failures due to pad impact problems have been known to cause serious injuries. Special shock abatement may be required in such cases (Smith 1989).

Shop Equipment Load Requirements

Press load is a term normally applied to equipment utilization on a press-by-press basis. Well-managed shops normally assign each die to a home or primary process press. Assignment is based on obtaining the most economical utilization of each machine.

To provide for scheduling flexibility, such as accommodating rush orders, it is a good practice to assign a secondary press and in some cases a third machine, all of which are known to satisfactorily run the tooling without difficulty. In each case, die setting and press operating instructions for safe and efficient operation of each die in the alternate presses must be accurately maintained.

A good way to manage equipment utilization is with a computerized database. In some cases, commercial management software is used. However, several major software development firms market integrated spreadsheet and database programs that can easily be configured to provide shop load reports as well as computerized die and pressroom management tracking. Properly configured, such programs can provide real-time answers to queries of operating parameters for dies. The key to making such a database useful is inputting the correct data to generate accurate reports of each press's available capacity. Be sure to enter the press in which the die actually ran to avoid skewing the data to reflect incorrect uptime on the primary press. From correct data, a determination can be made regarding which presses are available to accommodate the die.

Auxiliary Equipment Requirements

Successful operation of the die will require appropriate auxiliary equipment. If the die is fed with coil stock, a decoiler with weight and size capacity at least equal to that used at the plant outsourcing the die will be required if the same size and weight of coils are to be used.

Depending on the stock flatness requirement, a stock straightener with sufficient width and stock thickness capacity to remove coil set and minor coil kinking may be required. Knowledge of the process used at the outsourcing plant is very helpful.

While the author has not conducted a scientific study, it has been observed that nearly all coil-feeding operations have stock straighteners in place. However, during plant audits, personal observation has revealed that the straightening rolls in well over half of the operations are applying no straightening action. This can easily be determined by sighting through the stock straightener. If a daylight opening between the rolls is observed, it is obvious that no straightening action is occurring. In some cases, the straightener is adjusted backwards to impart a reverse coil curvature. This is occasionally needed to provide smoother feeding in progressive dies or to impart a slight curvature needed in the finished part.

Depending on the severity of deformation occurring in the die, a coil straightener is not always required provided the stock feeds smoothly and part consistency is not affected by minor variations in flatness. Based on these observations, it may be agreed with the die outsourcing entity that a coil straightener may not be needed to run the die.

REFERENCES

Krygier, Roman J. Presented August 25, 1989. *"Maintenance Management, a Review of the Program Installed at Body Operations Plants, Ford Motor Company."* Dearborn, MI: Society of Manufacturing Engineers.

Smith, David. 1993. "Visual Indicators in the Workplace." *The Fabricator*, January-February. Rockford, IL: Fabricators and Manufacturers Association International.

—— 1991. *Quick Die Change*. Chapter 27, Selecting Maintenance Software. Dearborn, MI: Society of Manufacturing Engineers.

——1989. "How to Solve Die Impact and Noise Problems with Automotive Pull Rod Shock Absorbers." Technical Paper MF91-259. Dearborn, MI: Society of Manufacturing Engineers.

18

Metallic Springs as Die-pressure Devices

Die-pressure devices and systems should be carefully selected based on the intended service. Factors to help determine the correct spring choices are the required force, deflections, space limitations, stroking rates, and production requirements.

When used within the manufacturer's ratings, steel die springs can provide excellent service life with little or no loss of force. When users experience spring breakage problems, it is usually traceable to a misapplication of the spring.

TYPES OF METAL SPRINGS

Metallic die springs include the following types:

- helical, round-wire metal springs,
- helical, oval-shaped wire-metal springs, and
- dished metal-washer springs, known as Belleville washers or springs.

Nearly all metal springs used in tool and die work are helical compression springs. These have flat ends made by closing the last turn on each end. Often the end is ground flat, especially in the higher-force types, to insure that the spring will be level on a flat surface or a counter-bored hole where it may be placed.

Belleville spring washers are a special type of round, slightly dished, compression spring. They are shaped like a flat washer except that they have a slightly dished contour. Belleville washers find use where large forces are required through short travel distance. Stacking Belleville spring washers together can increase force. Pressworking applications include short stroke in die-spring applications and light-duty die clamps applied by releasing hydraulic pressure. Belleville washers work best in static-pressure applications. Like any spring, Belleville spring washers are subject to failure if cycled repeatedly at or above their rated travel limit.

Most die springs develop force when compressed. However, some die springs develop force when stretched. These are called *extension springs*. A screen-door spring is an example of an extension spring. Extension springs have an eye or loop on each end to permit attachment. Extension springs find many uses in tool and die as well as fixture work.

Other types of springs include spiral springs and flat leaf springs. These types find some use in die applications. A typical use for a leaf or flat spring is to actuate a progressive die, positive-type starting stop.

Compression Springs

Most metal springs used in die construction are compression springs. Compression springs for strippers, pressure pads, and other spring-operated die components can be selected from the ratings given in terms of the amount of force-per-unit of travel. This data can be obtained from manufacturer's catalogs.

Round-wire springs are suited for very light-duty pressure-pad applications because of their low load ratings. They are a good choice for such applications as latch-return springs and for use in progressive-die starting stops.

Compression die springs made from steel wire with an oval or special trapezoidal cross-sectional area are designed for high forces and long service. Winding wire into a helical spring involves bending the metal. The inside of the helix goes into compression while the outside stretches.

Wire with a trapezoidal cross section and smooth, rounded corners is favored by at least one spring manufacturer because the small end of the trapezoidal shape is used to form the inside of the spring helix. This cross-sectional shape produces a better product because the small part of the trapezoidal cross section is easier to compress than the thicker edge of an oval wire having an equal width and cross-sectional area. The result is a spring having less residual tensile stress on the outside of the helix and a more uniform cross-sectional area than would otherwise be the case with oval-wire spring stock.

Ratings

One would suppose that the ISO spring color-coding standard is an industry-wide standard for color-coding springs to identify load or duty rating (see Figure 18-1). Such a system is highly logical in view of the adoption of this standard by both ISO and the North American Automotive Metric Standards Group (NAAMS), which is a working group of the Automotive Steel Partnership. Unfortunately, a simple matter such as adopting an industry standard of uniform identification of the load rating of die springs is not agreed upon by all manufacturers.

Non-standard color-coding of springs includes colors that do not correspond to the actual duty class identified; even springs having two-tone paint schemes. All of this can result in confusion, and may even result in a dangerous die condition if an incorrect duty-class spring fails in an unexpected way and endangers personnel. To further complicate matters, Japan, although a metric-standard country, has a non-ISO standard for die springs.

From a die maker's point of view, there is enough difficulty in designing and maintaining tooling without trying to identify the duty class of a spring die with a variant color-coding scheme. Hopefully, the combined efforts of the North-American-based automakers and ISO will force acceptance of a common spring identification and rating standard.

Helical steel die springs are available in several load ratings or amounts of allowable deflection, expressed as a percentage of the uncompressed or free length and amount of force developed per

Figure 18-1. ISO standard spring color-coding scheme. (Courtesy Danly Die Set Division of Connell Limited Partnership)

incremental unit of deflection. While the quality of steel used to make springs is an important factor in their service life, the amount of allowable deflection is mainly a function of the thickness of the round, oval, or trapezoidal wire used to form the spring.

Springs made of thicker material have substantially lower allowable percentages of total deflection for the same material stress levels. Cycling a spring by repeatedly deflecting it at high stress values will cause the spring to develop fatigue cracks and eventually fail.

Greater wire size or thickness equals greater force developed per unit of deflection. However, the operating stresses developed in the spring material increase with the diameter or thickness of wire used. Therefore, springs wound from thick material develop higher forces per unit of deflection than those made of thinner material.

MATERIALS USED TO MAKE METAL SPRINGS

Plain carbon-steel springs are the least costly and are suitable for low-deflection applications and/or light-duty applications. If deflection is limited, they may last for a long time without failure.

Chrome-vanadium alloy-steel springs, while slightly higher-priced than carbon-steel springs, can last three or more times as long. When selecting die springs for a die design, the best performance with the most reduced downtime can be ensured by using chrome-vanadium-steel springs, and by derating the travel from the maximum deflection recommended by the manufacturer to the deflection recommended for long life.

Other spring materials include stainless steel and a variety of special alloys, including those developed for watch hairsprings and mainsprings. An example of a special spring alloy is Elgiloy™. It is nonmagnetic, very fatigue resistant, and the spring force changes very little over a wide temperature range. Such materials are very useful for instrument springs and applications involving a corrosive environment.

Processing Die Spring Steels

The best die spring steels require careful processing throughout each manufacturing step. This careful processing may include, in part, the following good practices:

- vacuum degassing of the molten metal;
- a continuous casting process carefully controlled to insure uniformity of the rod used to form the spring wire;
- careful process control to draw or roll the wire to the desired shape and size, without atmospheric decarburization, contamination, or unwanted inclusions of oxides or slag;
- use of the best winding and shaping practices to avoid stress concentration or stress risers that may lead to crack formation, propagation, or early failure;
- state-of-the-art heat-treating practices to correctly harden the steel and draw it to the correct spring temper;

- controlled shot peening of the formed and heat-treated spring surface to leave the surface in a desirable state of uniform residual-compressive stress, and
- presetting by compressing to a solid condition to increase set resistance and fatigue life.

SELECTING SPRINGS

For a known spring diameter and length, a spring manufacturer's dimension tables can be referenced to select springs with the desired total force or load capacity. However, if the required diameter and length are not known, a proven seven-step spring selection process may be used to determine the compression percentage, life expectancy, and deflection versus load from the manufacturer's catalog data.

Step One

Step one consists of estimating the level of production required of the die. This should determine the allowable deflection. Short-run dies, in which spring breakage is expected to occur, may use deflections such as the average or maximum deflection. Long-run dies and tooling in constant production should not be deflected more than the long-life percentage.

Step Two

In step two, compressed spring length H and operating travel T from the die print layout are determined. The dimensions may be measured if the die is open on the repair bench. The dimensions are shown in Figure 18-2.

Step Three

The free length C is determined in step three as follows:

- The load classification for the spring is selected. This involves choosing from the light, medium, heavy, or extra-heavy load rating.

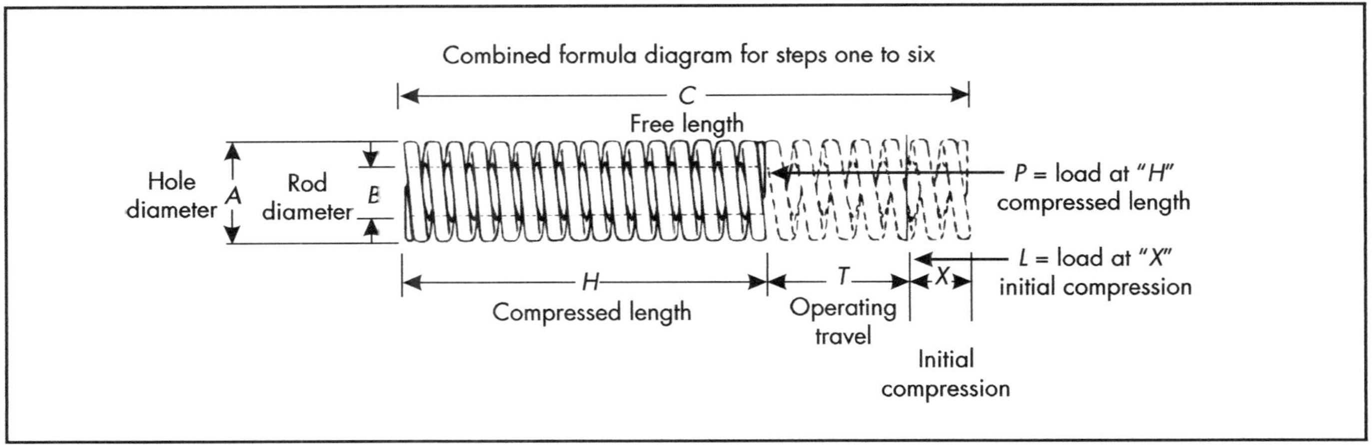

Figure 18-2. Combined formula diagram illustrates the factors needed to determine spring selection in steps one through six. (Courtesy Danly Die Set Division of Connell Limited Partnership)

- Then the figure nearest the compressed length H required by the die design is chosen from the appropriate charts supplied by the spring manufacturer. Take note of the corresponding C dimension, which is the free length of the spring.

Step Four

Nearly all die springs used in pressure pad and cam return applications are precompressed or preloaded to have useful force throughout the working stroke. Step four involves estimating the total initial spring load L required for all springs when the springs are preloaded or compressed X inches or millimeters.

Step Five

In step five, initial compression is determined by:

$$X = C - H - T \tag{18-1}$$

where:

X = initial spring compression or preload (in. [mm])
C = relaxed or free length of the spring (in. [mm])
H = maximum compressed length of the spring during die operation (in. [mm])
T = operating travel when installed in the die (in. [mm])

The X dimension or initial compression produces a calculable force or load L that is determined from the spring manufacturer's data.

The X value or initial spring preload of the total number of springs must be sufficient to provide adequate pressure for stock control upon initial pad or stripper contact as the die closes. The same is true of stripping pressure as the die opens. A safety factor to allow for expected punch-metal pickup or galling is needed to insure dependable operation until scheduled bench die maintenance.

Step Six

In step six, the spring rate for all springs is determined in pounds per 0.10 in. by:

$$R = \frac{L}{10 \times X} \qquad (18\text{-}2)$$

where:

R = total rate for all springs (pounds per 0.10 in. of travel or deflection)
L = load when springs are compressed X in. (lbf)
X = initial compression preload (in.)

European and North American manufacturers generally adhere to the metric system for spring force based on Newtons per millimeter. A Newton is equal to a force of 0.2248 lb. To determine the value R for all springs used under a pad or other metric die application, use Equation 18-3.

$$R = \frac{L}{X} \qquad (18\text{-}3)$$

where:

R = total rate for all springs (N/mm of travel or deflection)
L = load when springs are compressed X mm (N)
X = initial compression preload (mm)

Step Seven

In step seven, the correct spring is selected. First, the free length of the spring C must comply with the length determined in step three.

Next, R, the total spring rate determined in step six, is divided by the total number of springs to be used to get the rate per individual spring. It is often not possible to know this number with certainty since the spring diameter is not yet determined.

Placement of the springs around the die details, under die pressure pads, and in other limited die-space applications must be determined. The required spring diameter and allowable deflection (depending on the duty class of spring needed) are determining factors in making the selection.

Once the number of springs and spring rate are determined, refer to the manufacturer's catalog to choose springs having the desired rate. If the number of springs is not known, divide *R* from step six by the rate of the spring selected to determine the correct number of springs.

Table 18-1 lists data for the maximum allowable deflection recommended for four different ISO die-spring classifications.

Table 18-1. Allowable spring deflection versus relative spring life

ISO Light Load Color-code Green Allowable Deflection			ISO Medium Load Color-code Blue Allowable Deflection		
Long life	Average life	Maximum deflection	Long life	Average life	Maximum deflection
25%	30%	40%	25%	30%	37.5%
ISO Heavy Load Color-code Red Allowable Deflection			**ISO Extra-heavy Load Color-code Yellow Allowable Deflection**		
Long life	Average life	Maximum deflection	Long life	Average life	Maximum deflection
20%	25%	30%	17%	20%	25%

CONSIDERATIONS WHEN REPAIRING DIES

When repairing dies that do not have enough spring force or that have experienced excessive spring breakage, the following systematic process can help pinpoint the problem. In determining the length of a spring, higher spring forces require selecting larger diameter and higher load class springs.

For best economy and space savings, light- and medium-load springs can be chosen. If a heavy-load spring is used, it should have a free length equal to six times the travel. If an extra-heavy-load spring is required, a free length equal to eight times the travel should be used. If ratios lower than these are used because of height limitations, the number of springs required will need to be substantially increased. In such cases, self-contained nitrogen gas springs should be considered as an alternative.

The required pad and cam return forces should be carefully calculated in the die design process. Springs are often the best choice from a cost and reliability standpoint. However, if extremely high forces are required, nitrogen and hydraulic systems should be specified.

Nitrogen Cylinders and Hydraulic Pressure Systems

In cases where high initial compression is required, high-pressure nitrogen cylinders or hydraulic pressure systems may be required. Both nitrogen and hydraulic die-pressure systems have the advantage of providing high forces upon the initiation of travel. In other words, initial spring compression, which uses up available spring travel, is not needed when nitrogen or hydraulic die pressure systems are used.

In the event that the die fails to have enough pad or cam return force, self-contained nitrogen cylinders are available that are size-for-size compatible with many popular die springs. Replacing some or all of the die springs with self-contained nitrogen cylinders can serve to increase the initial contact force and total system force.

Replacing springs with self-contained nitrogen cylinders is not a simple substitution process. In many cases, provision must be made for a hard wear surface for the nitrogen cylinder rod to contact. This can involve substantial modification of the die, especially if the cylinder rod end is in line with a pilot hole used to counterbore a spring pocket. Depending on die geometry, the hole may need to be fitted with a hardened insert or a wear surface of air-hardening weld overlayment.

Spring Mounting and Care

The location and mounting of springs in pockets, around pilots or bolts, in tubes, or by other methods are determined by the space available, service requirements, and whether or not the spring will malfunction because of slug interference, misalignment, or other causes.

Springs must be well supported in a hole, over a rod, or by other means to guide them adequately under stress. Lack of support can result in crushing, twisting, binding, or surface wear (Smith 1990).

When springs are set in holes, the bottom of the hole should have a flat bottom to provide a flat seat and eliminate the possibility of crushing or deforming the ends. The edges of the holes should have a small chamfer to prevent interference with the movement of springs.

If the unguided length of the spring is greater than the diameter, a center guide rod may be used. The guide rod also serves to retain the broken pieces should the spring fail.

Tubular steel spring cages or cans placed around the spring as illustrated in Figure 18-3 are a good alternative for a rod to guide the spring. The can or cage serves several important purposes. It helps prevent the entry of dirt and debris into the spring pocket. Dirt caused by flaking zinc is especially a problem with dies used to work galvanized steel. Another important function of placing spring cages or cans around springs in bored pockets is to retain any broken pieces of failed springs. This is especially important if there is a possibility of a spring fragment flying and causing personal injury. Another consideration in retaining pieces is to prevent them from causing severe interference if pads bottom out.

Good-quality spring cans have a hard surface treatment for wear resistance. They are available in a variety of outside diameters and lengths. The hole *H* is sized to accommodate a shaft or rod if desired.

ANALYSIS OF SPRING FAILURES

Properly selected and used, metal die springs provide long trouble-free service. If die springs fail frequently, there is a rea-

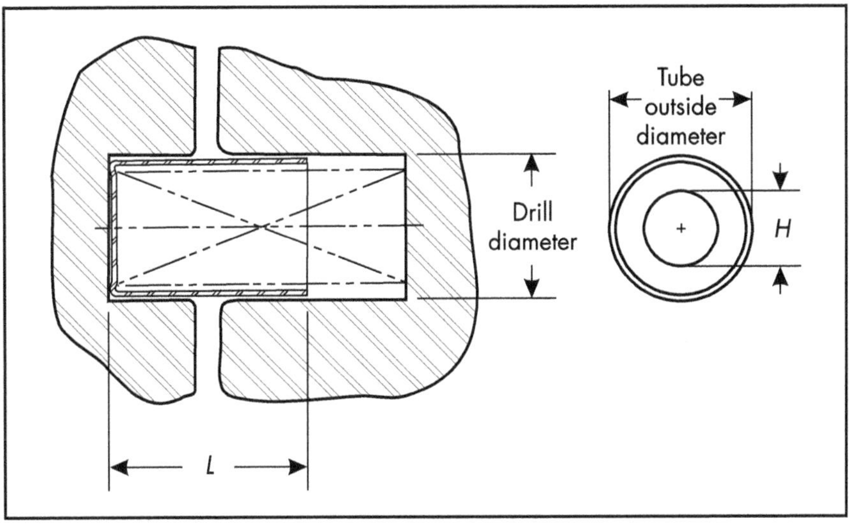

Figure 18-3. A spring cage or can is used to keep debris out of spring pockets and contain any broken spring pieces in the event of spring failure. (Courtesy Danly Die Set Division of Connell Limited Partnership)

son, and normally an alternative is available to reduce or eliminate the failure rate.

In conducting both public and in-plant training, the writer has asked the class attendees if they have had problems with die springs breaking. The answers range from hardly ever to yes—broken springs are a serious downtime, cost, and safety concern.

The next logical question was to ask how many attendees had problems with the valve springs in their automobile engines breaking. With the exception of a very few persons who have exceeded the mechanical endurance of valve springs in racing engines, the usual answer was that virtually no one had a problem with automotive-valve springs breaking.

Die engineering is solidly based on mechanical engineering principles. Any mechanical failure has a cause and in most cases a straightforward solution. The automotive-valve spring comparison leads to a sensible conclusion. Since valve springs and die springs are made of similar high-quality steel, then die springs that fail must be excessively stressed.

A source of confusion is the tendency of a few manufacturers of die springs and die nitrogen cylinders to use negative comparisons of competing products in their advertising literature. In the author's opinion, there is a best application for all types of die pressure systems.

Excessive Deflection

Table 18-1 illustrates how die springs are rated for maximum deflections based on the required life expectancy. If die springs are deflected beyond the long-life rating, it is assumed that spring failure is likely to occur and should be expected. If high deflections are required by the design, all springs so deflected in the die should be replaced based on the number of strokes completed before failures start to occur.

While automobile designs vary, their engines operate at approximately 2,000 revolutions per minute at highway cruising speeds. Therefore, a four-cylinder engine would undergo 1,000 spring compressions per valve during each minute of operation. Thus, automotive die springs routinely withstand well over 100,000,000 compression cycles during the conservatively rated nominal life of the engine. High-speed pressworking is accomplished at speeds of 300 to over 2,000 strokes per minute (SPM). A typical speed for an electrical connector die is 1,200 SPM. Such dies will complete over a million hits in a typical 16-hour, two-shift operation. In such costly precision tooling, spring breakage could result in catastrophic damage.

Most spring failures result from excessive deflections. This causes stress cracking that leads to rapid failure. A partial listing of bad shop practices includes:

- replacement of springs with a higher load class resulting in deflections in excess of the die design criteria, leading to stress-cracking failures;
- using the wrong load class spring due to a color-coding error;
- failure to specify that spring suppliers follow the widely accepted ISO standard color-coding system (include this requirement in your die construction standards and contractually insist that all vendors follow it);
- neglecting to specify that tooling construction sources use ISO standard springs and that the deflections be specified

for the desired life expectancy—if there is doubt, a copy of the spring supplier's invoice should be supplied; and

- shortening die springs with abrasive cutoff wheels or cutting torches—this does not provide a flat surface on the end of the spring, resulting in lateral bowing.

WINDING SPRINGS IN-HOUSE

Many tool and die makers are taught how to wind springs as part of their apprenticeship training. The usual material for springs made in the toolroom is music spring wire. This material is also commonly known as piano wire, although the term music spring wire is the correct term for the commercial product used for nearly all springs that are wound in-house. Winding spring wire onto an arbor in a lathe is one way to make springs in the toolroom. However, this practice is discouraged for safety reasons. For example, a finger or other body part may become entangled in a loop of the wire as it is fed into the lathe. Should this occur, serious injury such as amputation may result.

Commercial spring-winding machines are used in a few large toolrooms that need a variety of special springs on short notice. This in-house ability is especially handy if prototype or jig and fixture work requires the development of special springs. Some spring winders are hand cranked and can be operated by a single individual. This type greatly reduces the possibility of injury. Appropriate safety equipment such as approved safety glasses should always be worn to avoid injury when working with springs.

In general, the use of springs that are catalog items will help insure that the spring will meet the engineering specifications of the manufacturer. However, for prototype and instrument work, having the knowledge and equipment available to wind a special spring quickly is valuable.

CONCLUSION

The use of coiled metal die springs is one of the most widespread die-pressure system applications. Readily available engineering data predicts that metal fatigue will not cause failure problems if springs are carefully manufactured and not over-deflected. In general, the design of dies with total deflections (including initial compression for preloading) below the manufacturer's recommendations for long life will result in long trouble-free service.

There is an obvious need for North American manufacturers to adopt the ISO color-code designations as their internal standards for duty class, which is also the standard of the North American Automotive Metric Standards Group. Maintaining tooling to meet low inventory reliability requirements makes standardized procedures a necessity. If your company adheres to ISO spring standards, any tooling built in non-ISO countries should be built to ISO standards to avoid maintainability problems.

Designing dies with metal spring deflections greater than those specified for long life is advised only for tooling designed with redundant springs and an absolutely foolproof means to contain broken springs and spring attachments such as cam return rods. This is advised to avoid the potential for personal injury and unscheduled downtime. In general, spring deflections greater than those specified for long life are apt to fail in service. This is almost a certainty if the maximum deflection rating is chosen. Here, a preventive maintenance program to replace all springs as the end of their useful life approaches can save time and money, and avoid unplanned breakdowns.

REFERENCE

Smith, David. 1990. *Die Design Handbook*. Section 22, Die Sets and Components. Dearborn, MI: Society of Manufacturing Engineers.

19

Gas and Hydraulic Die-pressure Systems

Metallic die springs have two major limitations. They occupy much more space in the die than gas and hydraulic die-pressure systems, and deflection must occur to develop force. Compressed air, nitrogen, and hydraulic die-pressure systems have the advantage of providing high forces upon the initiation of travel—there is no need for initial compression or preload.

The die maker is often required to redesign the die-pressure system. The required pad and cam return forces are part of the die design process. Springs are often the best choice from a cost and reliability standpoint. Gas and hydraulic systems should be specified if high forces are required. The amount of force required usually determines the type of pressure system to specify. New die tryout and die maintenance functions require knowledge of how to maintain and modify die-pressure systems.

PNEUMATIC CYLINDERS

Except for automation applications, pneumatic cylinders find little use in new die designs. Pneumatic, nitrogen, and hydraulic cylinders satisfy the requirement for adjustable forces. A wide range of mounting styles and configurations are available. Commonly available piston sizes vary from 1–10 in. (25.4–254 mm) in diameter. Stroke lengths range from 0.5 in. (12.7 mm) to over 10 in. (254 mm), with special types available by special order.

Types

The initial force available from a pneumatic cylinder is equal to the piston area times the applied shop air pressure. Typical shop air pressures vary from 75–100 psi (517–690 kPa). To obtain higher pressures upon die closure, one-way check valves may be placed on the inlet line if approved by the cylinder manufacturer. Most commercial valves for this purpose are of the self-bleeding type. This design exhausts the pressure in the cylinder upon removal of the supply air source. This feature is important if a pad or cam must have the pressure removed for maintenance.

Many types of die air springs have hollow pistons that act as surge tanks. The amount of pressure increase as a function of piston travel is available from manufacturer's data sheets. Some types require use with an external surge tank recommended by the manufacturer. Piping diameters must be large enough to avoid restriction. Operations that require carefully balanced pressures under dynamic conditions may require hoses or piping of equal length with individual ports in the surge tank to insure even pressure.

Safety Precautions

Die makers repairing older dies should be alert to some dangerous cylinder designs. Some designs have attempted to provide very high pressures at the end of the stroke by means of a high compression ratio. The pressures involved can exceed the working limits of the components. A rare catastrophic failure mode that can occur in such cases is diesel ignition of the lubricating oil vapor. Compressed air is an active oxidizing agent. That factor combined with the heat generated when the air spring is compressed can lead to rapid seal deterioration and very dangerous diesel detonations. Nitrogen gas does not support combustion. If the required air pressure exceeds an input value of 75–100 psi (517–690 kPa), compressed air can become reactive.

Exhausting all pressurized air from the system, including every cylinder and surge tank, is a necessary precaution. Some dies have more than one air-spring pressure system. For example, the upper die may nave a pneumatic spring pad as well as a separate cam-return air-cylinder system. In addition, there may be pneumatic part-lifter systems and stock-positioning devices. Many individual pneumatic systems for die automation and parts transfer are used in tandem- and transfer-press production systems. Draining is a part of the power lockout procedure, before performing any work on the die or the press. Failure to do so exposes the person(s) working on the die to injury in the event that a pneumatic system actuates.

Another energy source that power lockout requires dealing with is gravity acting on objects that may fall. In the case of large die repair, many trim, cam flange, and similar dies may have large cams that operate on an angle. These may be returned by compressed air-actuated cylinders. Such slides may weigh 300 lb (136 kg) or more. It is necessary to safely block these in the open position. In some cases, keeper pins for securing the slide-lock open position assure the die repair technician's safety. Cams, like any other part where unexpected movement can cause injury, require power lockout and tagout. Before cycling the press, all lockout devices are restored to normal operation.

Some designs of older pneumatic die springs have large hollow piston rods retained on the piston by a light-duty steel snap ring. These rings can fail due to rusting caused by moisture normally present in plant air supplies. If the die is disassembled to inspect the pneumatic springs under pressure, there is a danger that the piston rod may suddenly break free of the cylinder. Such incidents have resulted in severe injury to the face and upper torso. The piston rods have left pronounced indentations in the metal roofing of high-bay press rooms that are 60 ft (18.29 m) or more in height. The recommended procedure is to inspect such systems by disassembly rather than a pressure test. Any cylinders found having inadequate means to retain a piston or piston rod are repaired. If the manufacturers of air springs of this type are still in business, they should replace the two-piece piston and rod assembly with a safe, updated version. If this is not possible, designs not subject to this defect should be substituted.

On systems known to employ intrinsically safe cylinder designs, each cylinder should be carefully inspected for visually apparent damage. Make sure that each cylinder collapses easily before further work, such as disassembly of a system component, is attempted. Next, check systems employing self-relieving check valves at each cylinder for proper action upon removal of the air supply.

Maintenance

The key to any good maintenance program is to avoid unexpected downtime. Breakdown maintenance has no place in a well-managed pressroom. A few common-sense practices include:

- Supply pneumatic systems with clean dry air via an efficient air dryer on the compressor system with filters having moisture traps at the point of use.
- Use air-line lubricators and a minimum amount of an effective, environmentally and hygienically safe cylinder oil.
- Lubricate new hoses that connect the press air valves to the die by flushing them with the approved cylinder lubricant. This will assure no delay in lubricant reaching the die cylinder system from the air-line lubricator normally installed on the press.
- Establish and maintain a pneumatic system-rebuild program based on the number of hits and elapsed time since the last complete overhaul. The rebuild schedule should provide for the best economy and lowest occurrence of breakdowns.
- Use good-quality hose and piping. Use hard copper or steel tubing with quality fittings for fixed runs.
- Where repeated flexing is required, use wire-braid hydraulic hose.
- Never grind or machine a cylinder to modify it in a way not approved by the manufacturer.
- Use an oil-mist collecting device on the outlet of four-way valves that cycle die automation. Make sure the oil-mist collector does not cause significant back pressure.
- Always control the movement of air automation cylinders by using adjustable restrictors with check valves in the reverse flow path. The way to install them is to restrict the air exhausted from double-acting cylinders.

NITROGEN DIE-PRESSURE SYSTEMS

Five to 20 times the force available in an air spring of the same size can be obtained from nitrogen die cylinders. The early designs of these systems were largely based on hydraulic cylinder systems. Much of the hose, packing, and fittings used are available as generic hydraulic parts from hydraulic suppliers.

Reusable hose fittings are often supplied by cylinder manufacturers. However, to save time and assure properly attached fittings, the crimp-on style is recommended. In well-organized shops, one manually operated crimping fixture and appropriate dies will suffice to make new system hoses and replacements for worn or damaged hoses. Crimp-on fittings are especially advised for use with high-pressure Teflon®-lined hose that has an exposed, woven stainless-steel braid. Reusable fittings are especially difficult to install on this type of hydraulic hose.

Piped Nitrogen Cylinders

Currently produced high-pressure nitrogen systems use a maximum charging pressure ranging from 1,500–2,000 psi (10.3–13.8 MPa). The forces available in high-pressure systems at full-rated pressure range from 0.5–6 tons (4.4–53 kN). The smaller sizes are sometimes used as cam return cylinders. A variety of mounting styles is available. These piped systems are used with an external surge tank. The tank, cylinders, charging console, hose, and all fittings are usually ordered as a package from the manufacturer.

In die repair work, older systems have a peak charging pressure ranging from 500–650 psi (3.4–4.5 MPa). Low-pressure systems are not used in new designs. The newer high-pressure systems rate the cylinders based on the amount of force developed at the maximum charging pressure. The older systems have ratings based on cylinder bore diameters in inches.

Filling

Figure 19-1 shows the use of a snap-on nitrogen fill fitting. This type of hydraulic fitting is widely used to fill die high-pressure

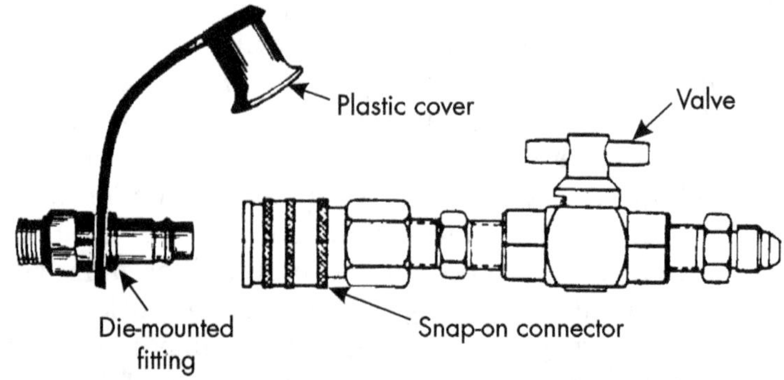

Figure 19-1. A snap-on nitrogen fill fitting is used for charging high-pressure nitrogen systems.

nitrogen systems. The fill fitting is installed on the die, while the filling assembly, which is shown with a shut-off valve, is installed on the hose from the regulator. These fittings can be hand-connected at full system pressure of 2,000 psi (13,789 kPa). There is no pressure drop across the fitting. The author is not aware of any other fill fitting that has these advantages.

When setting dies with a piped or manifold nitrogen system, it is important to be sure that the system is charged to the correct pressure. This information should be marked on the die. It is convenient to place a small mark of paint or fingernail polish on the gage located on the charging console.

Where possible, charging the nitrogen system(s) should be done before the die is set. Some dies must be charged in the press to avoid cocking the die if the nitrogen system is located off center. This also may be required for double-action press dies in some presses.

The four-bottle nitrogen charging cart illustrated in Figure 19-2 was developed specifically for quick die change applications. A four-position valve (Figure 19-3) permits rapid sequencing between the tanks. To make the most efficient use of the available nitro-

Figure 19-2. A four-tank nitrogen-charging cart was developed specifically for quick-die-change applications.

gen, filling is started with the tank having the least pressure and sequenced to the next lowest until the desired pressure is reached. The regulator is permanently shock-mounted to the cart. To permit rapid exchange of tanks without the use of tools, a special handwheel-type connector (Figure 19-4) is used. All fittings and piping are generic hydraulic components rated well in excess of maximum tank pressure.

Figure 19-3. A four-position valve permits rapid sequencing between the tanks. The regulator is permanently shock-mounted to the cart.

When constructing and maintaining high-pressure equipment of this type, it is essential that it is properly engineered and safe procedures are developed for its use. All personnel must be instructed in correct safe procedures. Any problems with the use of the equipment must be corrected before proceeding with further use.

Wear and Leaks

As the system wears, leaks can develop. One manufacturer's accepted figure for nominal leakage is 25 psi (172 kPa) per week.

Figure 19-4. A special handwheel-type connector is used to permit rapid exchange of tanks without the use of tools.

Whenever the packing in one cylinder starts to fail, it is best to rebuild the entire system.

As an aid to the designer, overlays of the system components are available from the manufacturer. CAD users can store these as macros.

NITROGEN MANIFOLD SYSTEMS

Manifold systems have the advantage of few external fittings to leak. The manifold itself can serve as both a surge tank and a structural component such as a die shoe. A nitrogen manifold may vary in size from a thickness of less than 2 in. (50.8 mm) to 10 in. (254 mm) or more.

The design and fabrication of a nitrogen manifold plate itself is straightforward die engineering and machine work. Nitrogen manifolds are machined from solid steel plate. Some die shops machine their own manifolds and buy commercial cylinders. It is important to test the plate using ultrasonic flaw-detection equipment to insure that there are no voids or slag inclusions prior to any machining. Failure to perform this test can result in nitrogen leakage through the manifold body.

Figure 19-5 illustrates the use of a nitrogen manifold system. This type of die construction is popular for drawing automotive body panels such as roofs, fenders, hoods, deck lids, and inner doors. In a tandem line there is an advantage in that a part turnover is not needed to get the drawn panel into position to load onto the trim die. This often permits higher line speeds without the cost of automation.

A disadvantage of using the die design shown in Figure 19-5 in conventional double-action presses is accelerated press wear because of tonnage curve violations. Another problem is reverse shock loading of the gear train. This is due to the clearance being suddenly

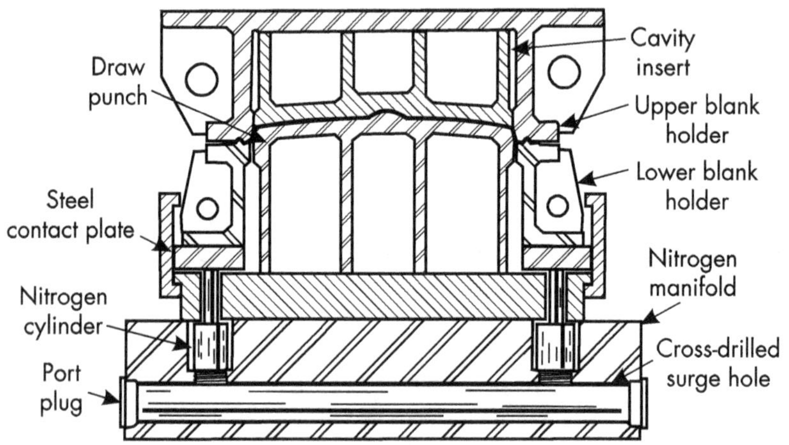

Figure 19-5. Inverted automobile roof draw die shows the use of a nitrogen manifold to provide draw-ring pressure in a single-action press.

taken up by the high nitrogen pressure at the start of the upstroke. This is especially troublesome if the die is driven by the blank holder slide of a double- or triple-action drawing press.

The die shown in Figure 19-5 is correctly termed a *stretch-form die* if enough blank-holder pressure is used to prevent metal movement on the blank holder. Stretch-form dies normally use a lock bead to prevent movement (Smith 1990).

A nitrogen manifold, which also functions as the lower-die shoe, has 24 six-ton (53.4-kN) nitrogen cylinders. These provide a total of 144 tons (1,281 kN) of pressure at 1,500 psi (10,342 kPa) nitrogen pressure. This force is applied to the draw ring. The cylinder travel is 4 in. (102 mm).

SELF-CONTAINED NITROGEN CYLINDERS

Figure 19-6 illustrates a cross-section of a self-contained nitrogen cylinder. This type of cylinder features a seal material tightly contained between U-shaped packing cups. Up to 2,000 psi (13.8 MPa) of nitrogen gas is charged into the cylinder through the fill

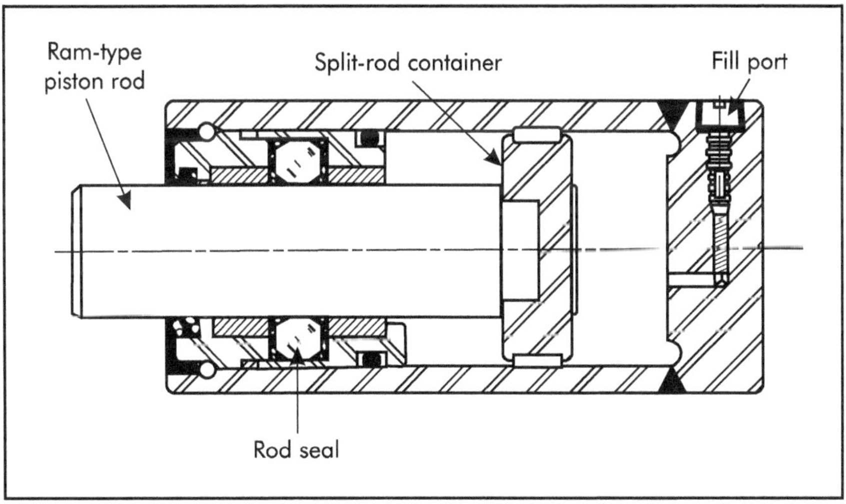

Figure 19-6. Cross-section of an older successful design of a self-contained nitrogen cylinder is shown. Essentially all current designs are based on improvements to this design. (Courtesy Dadco, Inc.)

valve. The effective piston diameter is the actual rod diameter. The split-rod retainer serves only to guide and retain the rod—the gas pressure is equal on both sides of the container.

Popular available sizes of self-contained nitrogen cylinders range from 0.75–5 tons (6.7–44.5 kN) in a wide range of stroke lengths. Many mounting styles are available including types without mounting brackets, which are placed in a drilled or cored hole.

The leakage rates are extremely low. The normal failure mode is a blown seal after the packing material becomes worn away. This type of nitrogen cylinder is favored by many die designers because of its simplicity of installation and low initial cost per ton when compared to air springs and other types of nitrogen cylinder systems. The popularity of the cylinder style and competition within the die-component industry has resulted in many improvements over the basic design shown in Figure 19-6.

Inverted Draw-die Applications

The cost-to-tonnage ratio of the larger sizes of self-contained cylinders compared to manifold systems favors the use of self-contained cylinders in large, inverted draw-die applications. Figure 19-7 illustrates the use of 5.5-ton (4,990-kg) self-contained nitrogen-gas cylinders to provide draw-ring pressure in an inverted draw die. In such applications, it is wise to provide cored inspection windows in the lower die shoe to permit a periodic visual check for failed cylinders. Failure will be evidenced by a collapsed cylinder. Normally, several extra cylinders are specified to compensate for the loss of any one cylinder and to avoid interrupting a production run.

Cam Return Applications

Self-contained cylinders are increasingly used for heavy cam-return applications because of the large forces available and its simplicity of installation. Figure 19-8 illustrates a self-contained nitrogen-gas cylinder used to return an aerial cam.

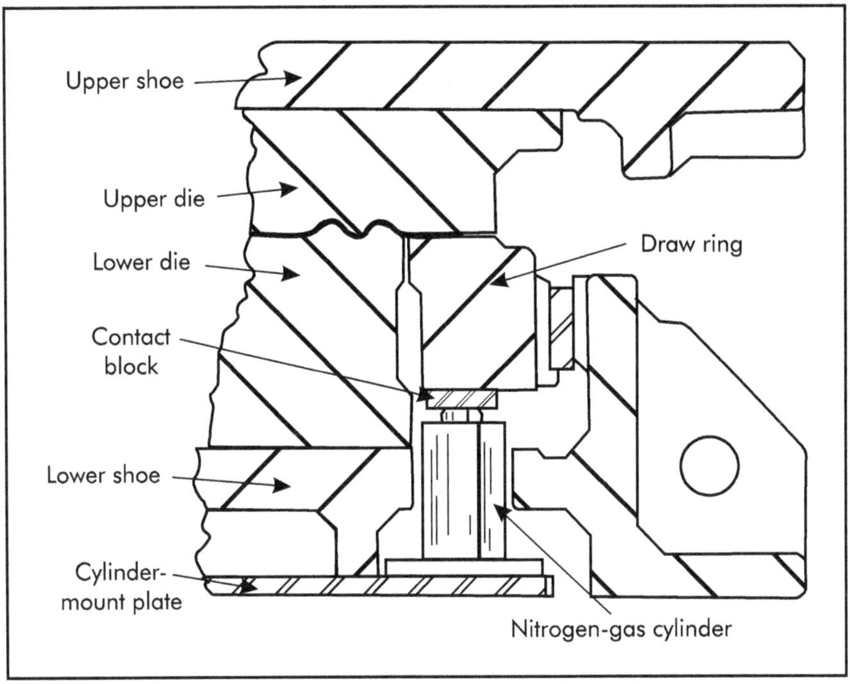

Figure 19-7. A 5.5-ton (4,990-kg) self-contained nitrogen-gas cylinder is used to provide draw-ring pressure in an inverted draw die.

Floating Draw-die Punch

Figure 19-9 illustrates self-contained nitrogen-gas cylinders used to float a draw-die punch in a double-action press. For large automotive body panel dies with punches weighing up to 8 tons (7.26 metric tons), six 3-ton (27-kN) presses are used. Four cylinders are sufficient. The two extra cylinders permit the production run to continue should any one cylinder fail. Cylinders generally last beyond 500,000 strokes before seal replacements are needed in this application.

Cylinders without mounts are often used. They are placed in holes bored in the blank holder adapter plate or bull ring. The punch adapter plate should be inlaid with an air-hardening weld where the cylinder rod makes contact. The use of hardened-steel contact

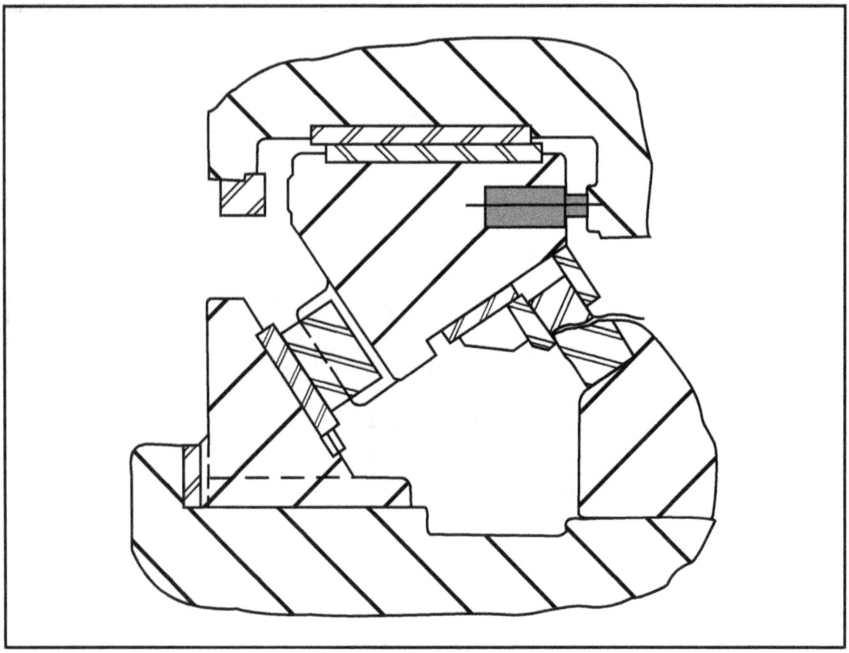

Figure 19-8. A self-contained nitrogen-gas cylinder is used to return an aerial cam. (Courtesy Dadco, Inc.)

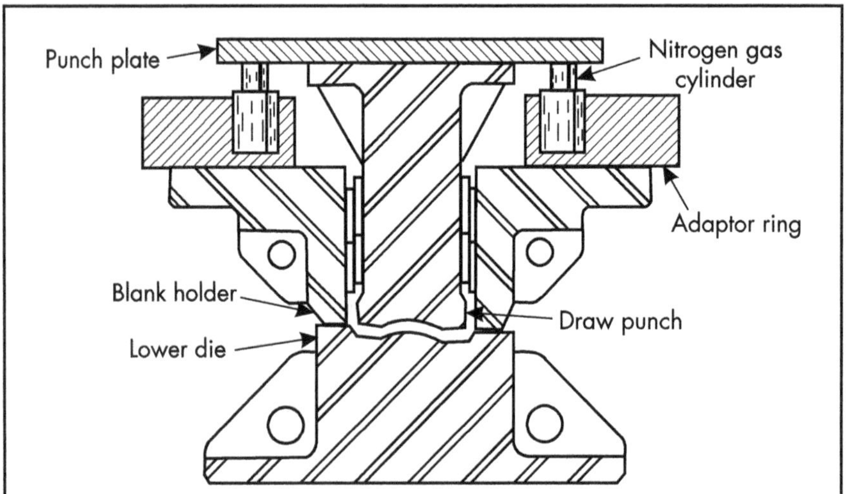

Figure 19-9. Self-contained nitrogen-gas cylinders are used to float a double-action draw-die punch.

blocks is not advised because they are apt to come loose and enter the die.

In addition to the obvious advantage of eliminating a die-setter task when setting the die, the floating-punch draw die is not affected by lateral movement of the press slides caused by worn or incorrectly adjusted presses. The chances of a punch or blank-holder wear plate loosening and falling into the die cavity are also practically eliminated.

Filling

Self-contained nitrogen cylinders are filled from portable nitrogen tanks as shown in Figure 19-10. The tank, regulator, and other equipment are available from suppliers of industrial gases and nitrogen cylinders.

Before charging a cylinder as shown in Figure 19-10, it is essential to make certain that is has been assembled in accordance with the manufacturer's instructions. Extra care should be exercised to make sure that any required retaining rings are in place

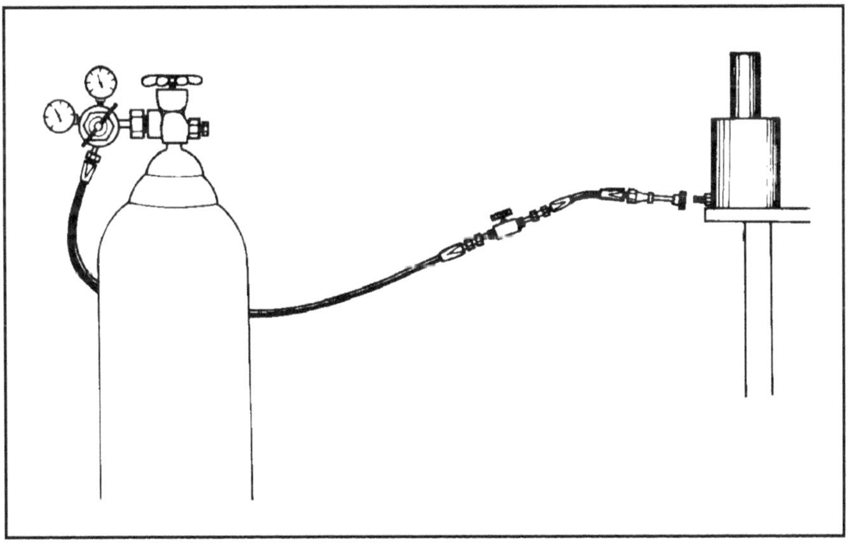

Figure 19-10. Filling of a self-contained nitrogen-gas spring is shown. (Courtesy Dadco, Inc.)

before applying pressure. The cylinder rod should be pulled into place with a T-handle screwed into threads in the rod end, and the position of any required snap in the cylinder body and rod busing should be correctly determined before applying pressure.

External Pressure Adjustment

In cases where the ability to adjust the pressure of a self-contained cylinder is desired, it can be converted to a piped system with miniature stainless-steel-braid hydraulic hose and fittings. Figure 19-11 illustrates this system used with an external fill valve to permit pressure adjustment.

If it is desired to operate different parts of a system or even individual cylinders at different pressures, a modular control panel may be used, as illustrated in Figure 19-12. This unit permits monitoring and adjusting each part of the system individually.

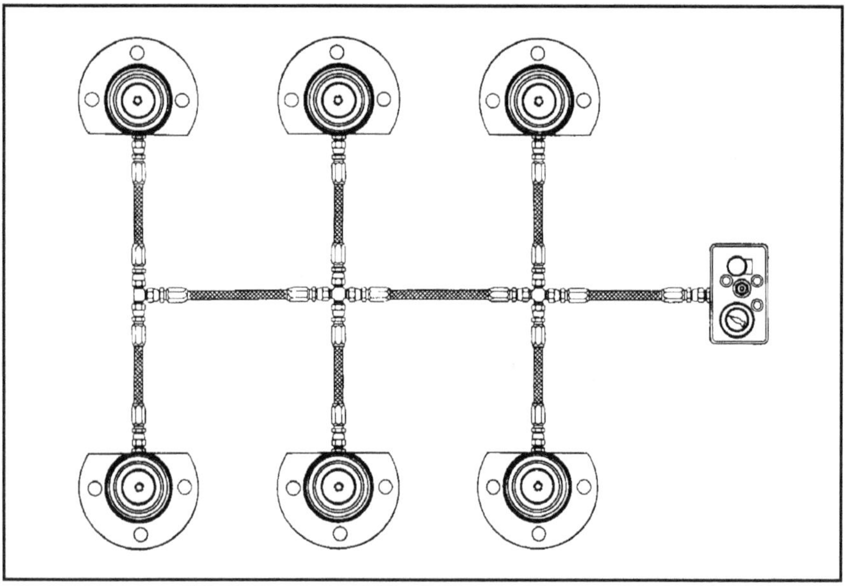

Figure 19-11. High-pressure miniature hydraulic hoses are used with an external fill valve to permit pressure adjustment. (Courtesy Dadco, Inc.)

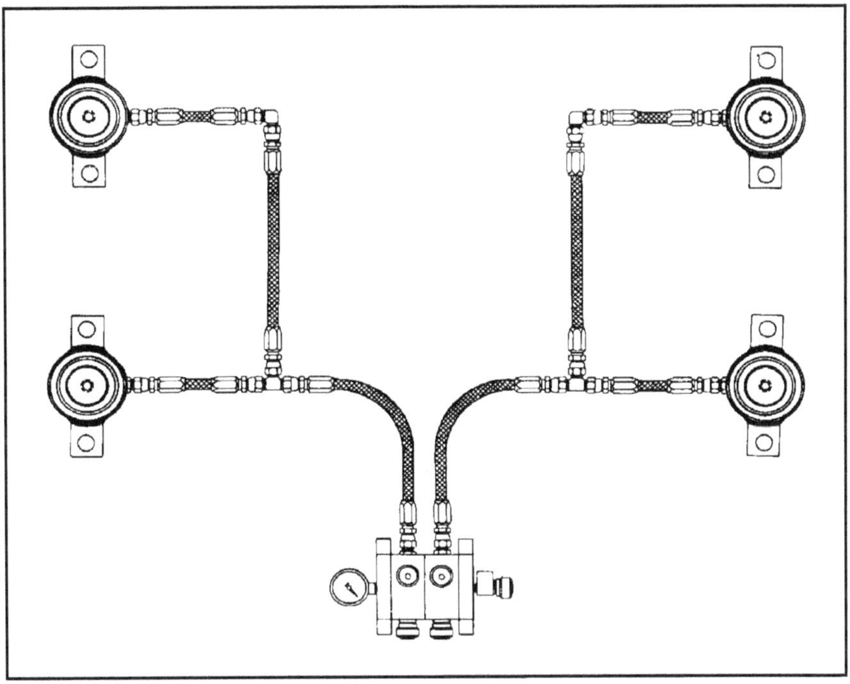

Figure 19-12. A modular control panel allows checking and adjusting nitrogen pressure in a system split into two separately adjustable halves. (Courtesy Dadco, Inc.)

HYDRAULIC DIE-PRESSURE SYSTEMS

The use of hydraulic cylinders for die-pressure systems has a long history in the pressworking industry. Die hydraulic systems predate the use of nitrogen pressure systems. Unfortunately, many tool engineers will specify these once popular systems only when other systems will not work for an application. The principal reason for specifying them is to obtain controllable cylinder-refilling delay to avoid crushing a newly formed flange when the die opens.

Advantages

There are many important advantages to the use of hydraulic die-pressure systems. These include:

- the ability to obtain more pad-pressure force than is obtainable in a given space or volume with any other die-pressure system;
- adjustable cylinder refilling delay to permit part removal without distortion;
- controllable linear force upon die closure;
- no bottom of stroke pressure buildup in draw-ring applications;
- the avoidance of back loading of a press gear train with stored gas pressure on the press upstroke; and
- long system life and low maintenance if used properly.

Limitations

Most system limitations are traceable to a lack of understanding of how hydraulic die systems should work and what is required to obtain good results. The main concern is housekeeping problems in case of a leak. Leaking oil, unlike air or nitrogen, makes a mess and demands attention. Limitations include the following:

- There is a need to maintain a leak-free system to avoid housekeeping problems.
- Hydraulic system design may be unfamiliar to some tool designers.
- Die makers may require training in system maintenance.
- The systems are designed with a maximum press ram velocity requirement that is determined by oil passage and piping sizes. If this is exceeded by excessively long press strokes or high stroking rates, system failure may occur.
- Problems may result if the hydraulic fluid used is of the incorrect type and viscosity.
- There is a need to use fire-retardant fluids, which may require a different type of cylinder packing for fluid compatibility.

In general, hydraulic die systems tend to be underutilized because they are not understood by those who must maintain them. However, for long-run jobs requiring precisely controlled high forces, they are often superior to other systems.

REFERENCE

Smith, D. 1990. *Die Design Handbook*. Section 13, Dies for Large and Irregular Shapes. Dearborn, MI: Society of Manufacturing Engineers.

20

Press Selection and Maintenance

Success in any activity requires that a number of tasks be done correctly. In metal stamping, everything from low-quality parts to accelerated die wear and damage is the result of poor press selection, die setting, and machine maintenance. No responsible machine shop manager should neglect machine tool lubrication, alignment, or repair.

The same logic must be applied to press selection and maintenance. A power press is a machine used to supply force to a die that is used to blank, form, or shape metal or nonmetallic material. Thus, a press is a component of a manufacturing system that combines the press, die, material, and feeding method to produce a part. The designer of the manufacturing system must also provide proper point-of-operation guards to safeguard pressroom personnel. Each of the components of this manufacturing system is important. A foundation for understanding the system is acquiring a working knowledge of the press.

PRESS TERMINOLOGY

The following terms are used to describe some of the principle characteristics and specifications of power presses. Many of the terms apply to both mechanical and hydraulic gap-frame and straight-side presses.

The bolster size is used to determine the maximum size die shoe that can be accommodated in the machine. The minimum and maximum amount of vertical open space between the ram and bolster must be known to determine if a die will fit the press. Of course, the force capacity is very important. In the United States it is illegal to overload a stamping press.

Shut Height

The space available between the press bed or bolster and the slide or ram is called the *shut height*. It is always measured with the press shut or at bottom dead center. It may be specified as the vertical space between the ram and either the top of the bed or bolster as illustrated in Figure 20-1.

Figure 20-1 illustrates two examples of shut-height measurement. Measurement *A* is taken from the ram to the bed, and measurement *B* from the ram to the bolster. Since the bolster is needed to add stiffness to the press bed, measurement *B* is the one that

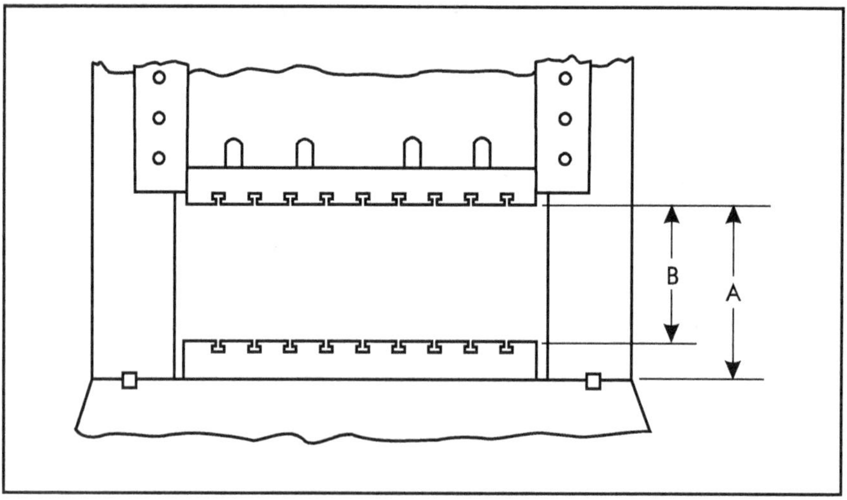

Figure 20-1. Example of shut-height measurement taken from the ram to the bed (A), and from the ram to the bolster (B).

should be used when determining the allowable maximum die-shut height.

When a die must be put in an existing press, the distance from the top of the bolster to the bottom of the ram determines the range of closed heights of the dies that will fit into the press. This distance should be measured with the screw adjustment at maximum and minimum values.

At times, more shut height than that which can be accommodated with the press bolster in place is needed. Some shops have removed the bolster and fastened the die directly to the press bed. This is a poor practice. The bolster is needed to stiffen the bed and spread the load evenly.

Bed and Bolster

The *bolster* adds stiffness to the press bed. It has tapped holes, or preferably T-slots, to permit the die to be fastened in the press. T-slots permit dies to be changed quickly and fastened in the press more securely than tapped holes.

Gears

In mechanical press drives, *gears* are used extensively to permit increased flywheel speed and provide greater press torque capacity. Gears are expensive press components. Proper lubrication is an absolute necessity. Where pairs of gears must work together, such as twin-end drives, it is essential that the load be shared equally.

Pitman

The link between the crank and slide is generally termed a *pitman*. When the press is driven by eccentrics, the link between the eccentric and connection is called an *eccentric strap*. The flexible attachment of the pitman or eccentric strap to the slide is called the *connection*. A means to accomplish shut height adjustment by an adjustable screw mechanism is nearly always a part of the connection system.

Press Slide Connections

A mechanical press connection is the point of attachment of the pitman or eccentric strap to the slide. Ball- and socket-type bearings are frequently used in smaller machines. A connection bearing of the type shown in Figure 20-2 has both a bronze-lined, saddle-type bearing and a wrist pin to transmit force to the slide.

Connection Strength

The connection is designed to transmit large compressive forces to the slide. If subjected to an extreme overload, the ball-and-socket type may be damaged by a crack or deformation of the socket. Large overloads may extrude the bronze bearing material out of saddle-type bearings and the wrist pin may be bent or broken.

Slide Adjustment

The connection may incorporate a screw adjustment mechanism. Larger machines have an electrically or pneumatically powered adjustment screw drive. Normally, a mechanical brake in the motor automatically engages to hold the adjustment in place. In the case of presses having multiple connections, a single motor is used in conjunction with shafts, bevel gears, and flexible couplings to drive all adjustment screws in synchronism. This is illustrated in Figure 20-2.

TYPES OF PRESSES

There are over 300,000 presses in use in the United States. Mechanical and hydraulic presses fall into two predominant types: gap-frame and straight-side presses. The frame types used in mechanical presses are similar to those used in many hydraulic presses. Most power presses are actuated by electrical, motor-driven mechanical, or hydraulic energy. However, gravity-drop hammer and foot-powered kick presses are also used.

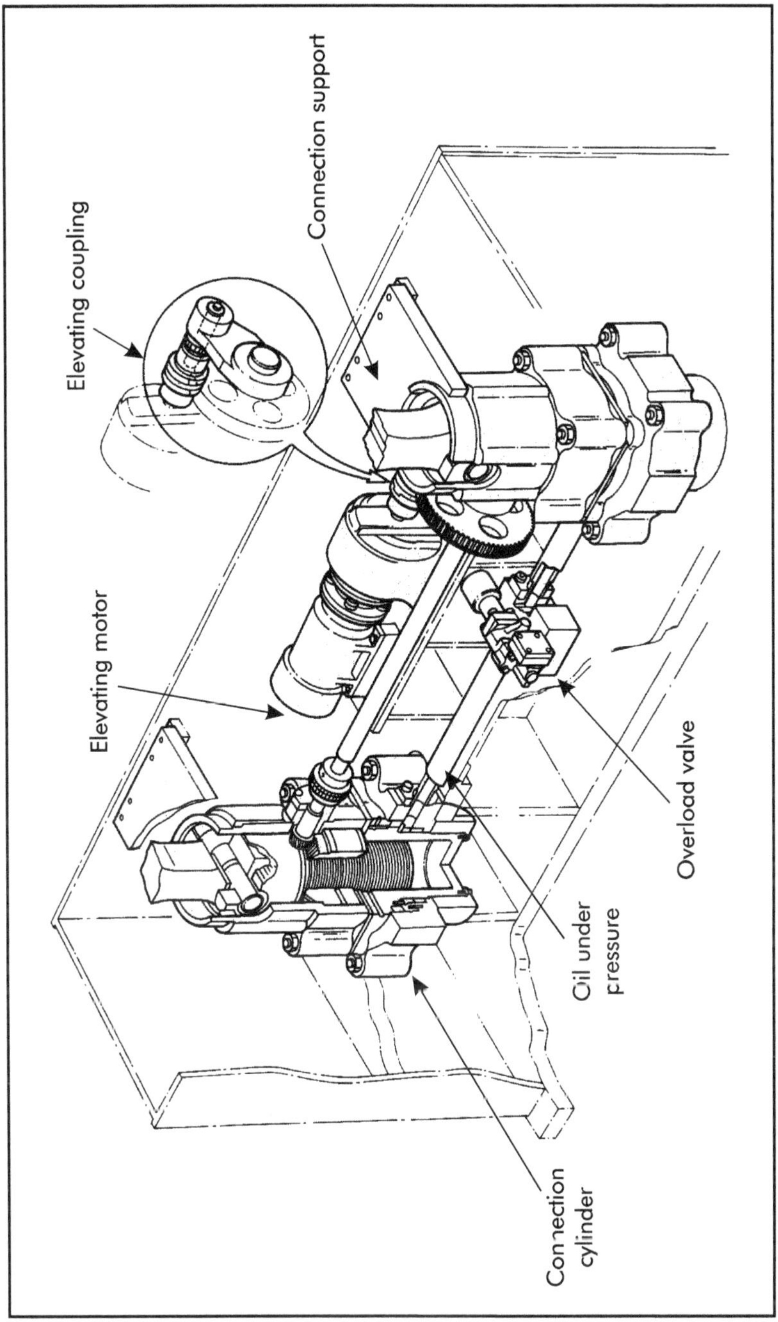

Figure 20-2. Precharged hydraulic cylinders under each connection provide fast-acting, easily reset overload protection. (Courtesy Verson Corporation)

The drive systems, clutches, brakes, counter-balance systems, die cushions, and electrical, hydraulic, and pneumatic features are similar in both the gap-frame and straight-side types. The type and size of press selected is mainly determined by the work to be done.

Gap-frame Presses

Gap-frame or *C-frame presses* derive their names from the C-shaped throat opening. In addition to the familiar open-back design, this style of machine has a long history of use for portable hole punching. The press brake is also a type of gap-frame press.

Gap- or C-frame presses have many useful features, including excellent accessibility from the front and sides for die setting and operation. The machines also cost substantially less than straight-side presses. The open back is available for feeding stock as well as ejection of parts and scrap.

Another advantage of a gap-frame press is that the machine is easier to set up than a straight-side press. The die setter has much greater access to locate and bolt the die in place. Gap-frame presses generally have less height than straight-side presses of comparable tonnage. This is a valuable consideration when overhead clearance is limited.

Gap-frame presses with force capacities up to 250 tons (2,224 kN) and larger are less costly than straight-side presses having the same force capacity. In the 35–60-ton (311–534-kN) force range, a gap-frame press costs approximately half that of a straight-side press.

The main disadvantage of gap-frame presses is an unavoidable angular misalignment that occurs under load. Limiting the amount of angular misalignment requires very robust construction—this adds to the weight and cost of the machine.

Gap-frame presses have more deflection than straight-side presses for a given load. The deflection has both a vertical and angular component. The angular deflection or misalignment that occurs is due to the spreading of the throat opening as tonnage is developed. In many applications, this angular misalignment under load may not be objectionable. This style of press is popular for short-run work, where high accuracy of die alignment or close part tolerances are not necessarily controlling factors. For low-tonnage high-speed work, precision gap presses are widely used. Here, the work is done before the bottom of the stroke. The light loading avoids angular deflection problems.

However, straight-side presses are generally recommended for any application where angular machine deflection would cause unacceptable part quality and accelerated die wear. The lower cost of gap-frame construction machines may be poor economy if accelerated tooling wear and quality problems result.

Open-back Inclinable Type

Figure 20-3 illustrates three types of gap-frame presses. Two of the types are a style of machine known as the *open-back inclinable* (OBI) press. The press frame is secured in a cradle, which permits the machine to be inclined backward. This is done to facilitate gravity loading as well as part and scrap discharge out of the open back of the press.

The frame of most older OBI presses is made of cast construction. The most commonly used materials are gray cast iron or steel. Figure 20-3a illustrates two pretensioned tie-rods across the open front of the machine. Lugs are cast into the frame of the machine to accept the tie-rods, which are installed as an option to reduce angular deflection.

The development of timed air blow-off devices and a variety of small conveyors have lessened the demand for the OBI-style press. Today, many OBI presses are operated in the vertical position. While the OBI style is not obsolete (see Figure 20-3b), many press builders supply OBI presses only on special order.

Open-back Stationary Type

The open-back stationary gap-frame press shown in Figure 20-3c is more compact and often a more robust machine than the older OBI style that it has largely replaced. The OBS press has a heavy box-like structure. The press shown in Figure 20-3c features a guided plunger connection and an eccentric drive.

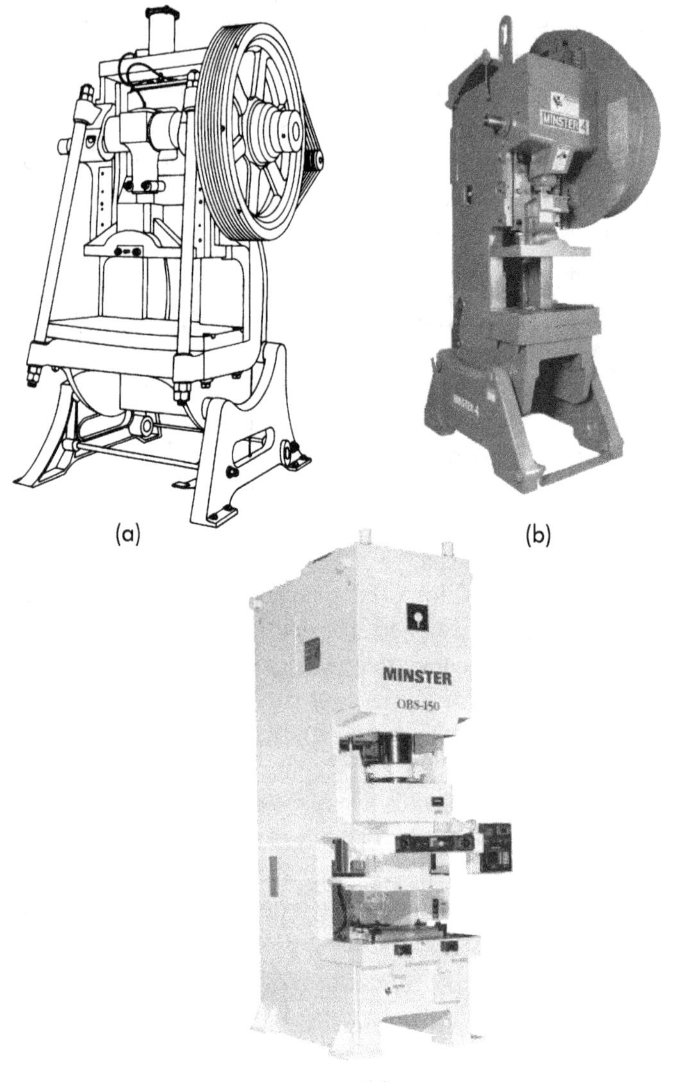

(a) (b) (c)

Figure 20-3. (a) Shown is an older-style, unguarded open-back inclinable (OBI), gap-frame press with tie rods added to reduce angular deflection; (b) a modern gap-frame-OBI mechanical press; (c) a modern open-back stationary (OBS) press, featuring eccentric drive and a guided plunger connection. (Courtesy Minster Machine Company)

Measuring Gap-frame Press Stiffness

An older accepted American standard that has been used by builders of gap-frame presses to measure stiffness is 0.0015 in. per in. (0.380 mm per mm) of throat depth. This is measured from the centerline of the connection to the back of the throat opening. The measurement includes both vertical and angular deflection. Angular deflection is by far the greatest concern because it results in misalignment between the punch and die.

ANSI standard B5.52 M specifies the allowable vertical and angular deflection in machines built to metric standards. A practical measurement procedure is to place a jack capable of full-rated tonnage at the centerline of the slide and bed. The vertical deflection is then measured between the centerline of the slide and bed. It should not exceed 0.002 in. per in. (0.05 mm per mm) of distance from the centerline to the back of the throat opening. Some press builders design for a lower value of deflection. Since this measurement is intended to pick up an angular value, measuring the difference from front to back across the slide face will give similar results. For many applications, angular misalignment under load is harmful. If the job cannot be run in a straight-side press, some reduction in angular deflection can be achieved by installing tie-rods across the open front of the press.

Adding Tie-rods to Gap-frame Presses

Some older gap-frame presses have existing lugs for tie-rod installation on the front of the machine. This is illustrated in Figure 20-3a. In other cases, the manufacturer may supply lugs, which can be welded in place. Provided the rods are properly prestressed, a significant reduction in deflection results from adding tie-rods.

The best method of installing tie-rods is to use them in conjunction with tubular steel spacers around the rods. The spacers help ensure that the tie-rods are not over-stressed, and serve to further stiffen the machine.

At best, adding prestressed tie-rods to a gap-frame press will result in reduced total angular misalignment. The physically limiting factor is that the cross-sectional area of the tie-rod and spacer is small compared to that of the press frame. Also, adding tie-rods to the front of the press limits access to the die opening, and can make the point of operation more difficult to guard properly.

The spacer should have 1.5 times the area of the tie-rod. In addition to adding stiffness, the spacers will reduce the alternating load in the tie-rod thread. The tie-rod area should be sufficient to support half the force capacity of the machine. A conservative nominal prestress in the tie-rod is approximately 14,000 psi (96,524 kPa).

Straight-side Presses

Straight-side presses derive their name from the vertical columns or uprights on either side of the machine. The columns together with the bed and crown form a strong housing for the crankshaft, slide, and other mechanical components.

The housing or frame of most straight-side presses is held together in compression by prestressed tie rods. Some straight-side presses have solid frames. Generally, a solid-frame straight-side press is less expensive than one having tie-rods. However, tie-rod presses are easier to ship disassembled and have a better ability to withstand overloads.

A major advantage of the straight-side press compared to the gap-frame machine is freedom from angular misalignment under load. Maintaining true vertical motion throughout the press stroke minimizes tool wear and achieves accurate part tolerances.

Components

Figure 20-4 illustrates some of the principle mechanical components of a straight-side press with double-end drive gears and two connections. The bed is the base of the machine. The columns support the crown. It also has gibs or gibbing attached to guide the slide. The crankshaft end bearings may be contained in the columns or crown.

The crown serves many functions depending upon machine design. Typically, the clutch, brake, motor, and flywheel mount on the crown of the press. The gears shown in Figure 20-4 may be open, having only a safety guard designed to contain the gear in

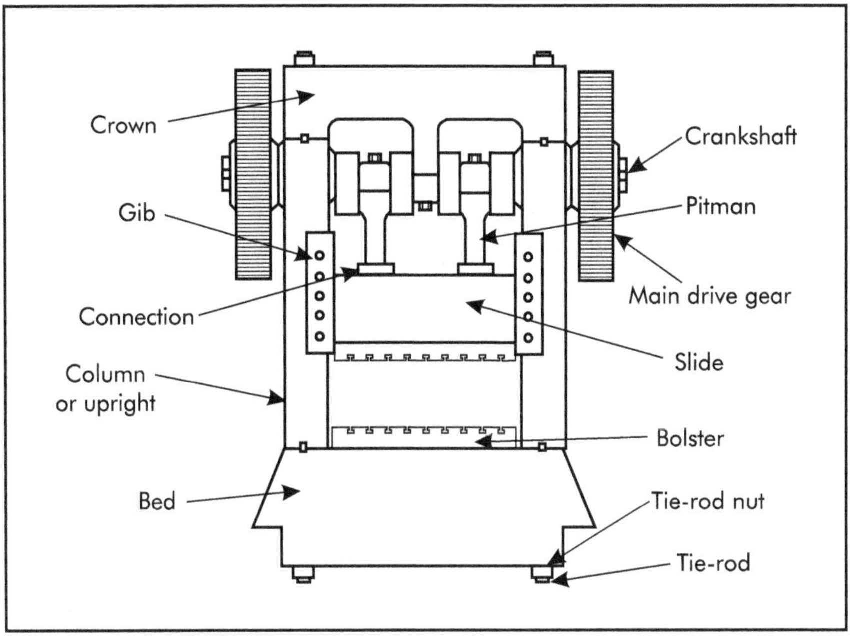

Figure 20-4. A straight-side mechanical press with double-end drive gears and two connections (Smith 1994).

case it should fall off due to a failure such as a broken crankshaft. In modern designs, the gears are fully enclosed and run in a bath of lubricant. They are in separate enclosures to permit using a heavier viscosity lubricant than that used for other machine parts such as the bearings. The latter are often supplied from a recirculating lubricant system. The separate gear housing and lubricant bath system serves to lessen noise and insure long gear life.

Figure 20-5 illustrates the placement of pneumatic piping tanks and controls for a typical straight-side press. The system is typical of a good pneumatic arrangement for a press equipped with an air-actuated friction clutch and die cushions.

Production of Precision Stampings

Many high-volume, close-tolerance stampings are made in straight-side presses. These include electrical connectors, snap-top

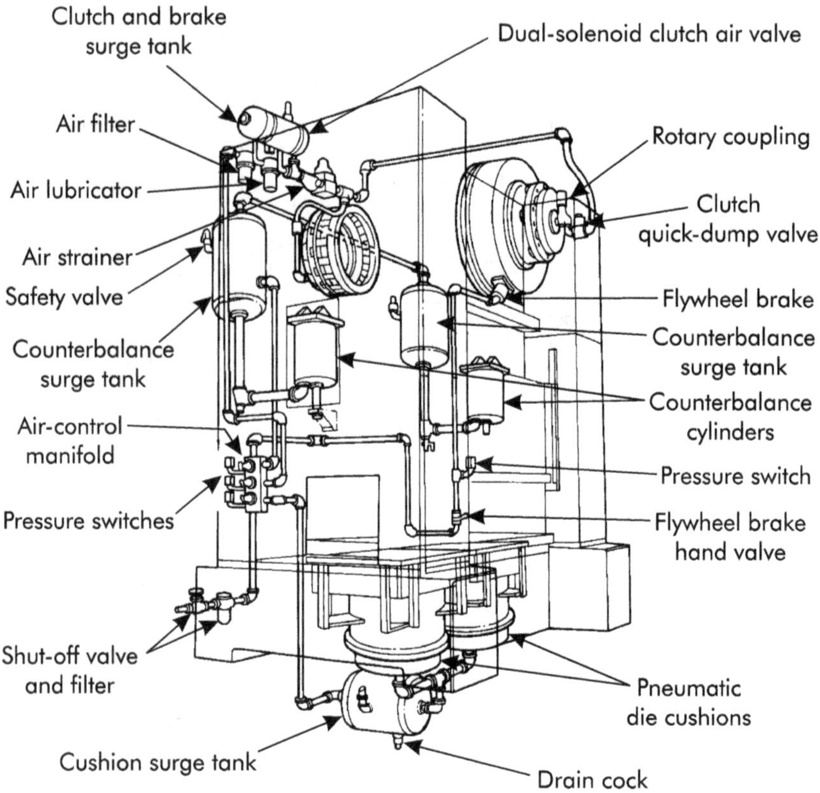

Figure 20-5. Pneumatic piping, tanks, and controls installed on a straight-side press. (Courtesy Verson Corporation)

beverage cans, spin-on oil filter cartridge bases, and refrigeration compressor housings. Tiny computer connectors are stamped at press speeds of 1,800 strokes per minute (SPM) or more. Often two to eight or more parts are completed per hit. Precision stampings are also produced at low speeds. For example, large refrigeration compressor housings may be stamped at press speeds of approximately 12 SPM. The housing consists of two mating halves, which must fit together precisely to properly align the internal parts.

OVERLOAD PROTECTION

Hydraulic

Hydraulic presses limit overloading by restricting the maximum pressure supplied to the actuating cylinder(s). Overload protection for mechanical presses can be provided by placing a hydraulic overload cylinder in series with the force delivered to each connection as shown in Figure 20-2.

When a preset maximum limit is exceeded, an overload valve dumps the precharged oil from the overload cylinders and trips a limit switch, stopping the press. The cause of the overload condition is corrected. Then, the overload system is recharged by actuating a key-locked switch. Such systems can accommodate maximum overload errors of approximately 0.75 in. (19 mm).

Mechanical

Other types of press overload devices include shear collars, Belleville washers, shear pins, and stretch links. These mechanical devices are simple and low in cost. Shear collar and Belleville washer overload devices are set under the connection in place of the hydraulic overload cylinders shown in Figure 20-2.

There are two major problems with mechanical overload protection. The failure point of stretching or fracture for overload devices is uncertain and subject to change with repeated cycling. In addition, failure of these devices may not be immediately detected. This may result in a slide out-of-level condition that can score the press gibbing, damage the die, and result in poor quality stampings. The machine must be shut down and the failed part replaced. This results in unscheduled maintenance and production delays that are normally unacceptable.

Tie-rods

Most, but not all straight-side presses employ tie-rod construction. The rods hold the press housing in compression. They provide a means to move large presses in sections. Should the press become stuck on bottom, they can be heated to relieve the prestress. The rods also limit press overloading.

As long as the press columns are maintained in a preloaded condition by the tie-rods, the deflection in the die space occurs at a linear rate as a function of increasing tonnage. However, once an overload condition exceeds the tie-rod preload, the crown lifts off the press column. Once crown lifting occurs, press stiffness is greatly decreased, limiting overloading.

MECHANICAL PRESS DRIVES

Both gap-frame and straight-side presses are built with direct and geared drive systems. In non-geared or direct-drive presses as they are also known, the flywheel is mounted on the end of the crankshaft. The flywheel is motor driven by means of a belt drive. Directly driven presses are capable of much higher operating speeds than geared types. Speeds range from under 100 to over 1,800 SPM for short-stroke high-speed operations.

Direct-drive Mechanical Press Applications

Non-geared presses find widespread application in blanking, high-speed production, and shallow forming operations. They have several major advantages over all other press types. First, the design is simple (see Figure 20-6). There are few bearings and no gears to wear out. In addition, frictional losses are lower than that of mechanically geared and hydraulic presses. The high operating speeds provide much greater productivity than that of geared presses. The direct-drive press is very popular for precision progressive die and high-speed perforating operations.

Two main factors limit the application of the direct-driven press. First, the full-rated force of the machine is only available very close to the bottom of the stroke, typically 0.060 in. (1.52 mm) from bottom dead center. A second disadvantage is that the ability to deliver rated forces is substantially reduced if the press is operated at less than full speed.

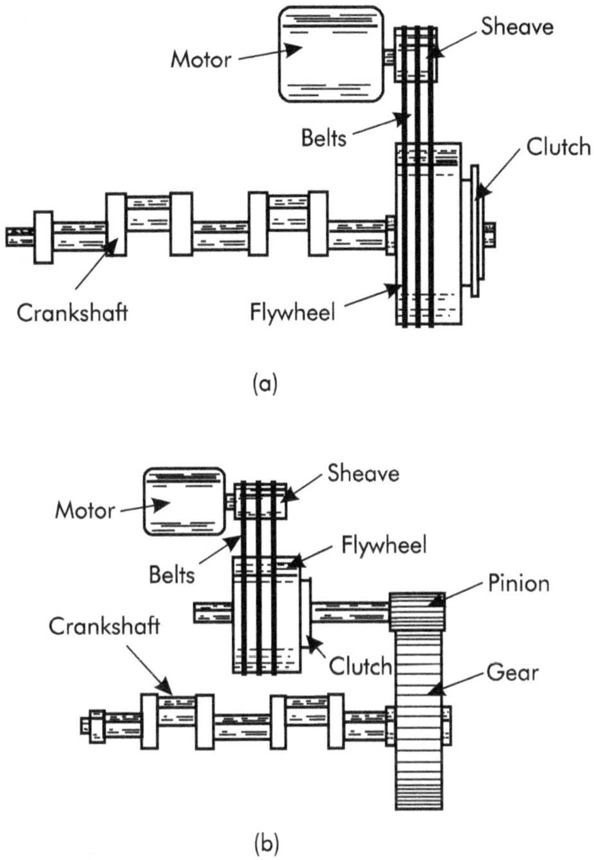

Figure 20-6. (a) A non-geared drive is shown with a double-throw crankshaft directly driven by the flywheel through a frictional clutch. (b) A single-geared drive is shown with a double-throw crankshaft driven by a single-end gear and frictional clutch.

Single-gear Reduction Presses

In single-gear reduction presses, the flywheel is mounted on the back shaft and the power is then transmitted through a pinion to a main gear mounted on the crankshaft. Some single-gear reduction presses have main gears mounted on both ends of the crankshaft, which is mounted on a pinion. Single-gear reduction presses typically operate in the speed range of 16–200 SPM.

Because single-gear reduction presses utilize gear reduction, with the flywheel on the high-speed back shaft, more flywheel energy can be provided for a given flywheel weight than in a non-geared press. This greater amount of flywheel energy provides greater torque capacity, thus making the single-geared press better suited for drawing and heavy forming operations than a direct-drive press. A single-end drive, single-gear reduction press system is illustrated in Figure 20-6b.

Angular Misalignment Due to Crankshaft Twist

Figure 20-7 illustrates how an angular misalignment proportional to the torque transmitted through a crankshaft with two throws occurs in a single-end drive press. The result is that the side of the ram nearest the driven end of the crankshaft will reach bottom dead center before the other end. The amount of ram tipping will be approximately proportional to the force delivered by the machine if the ram is uniformly loaded.

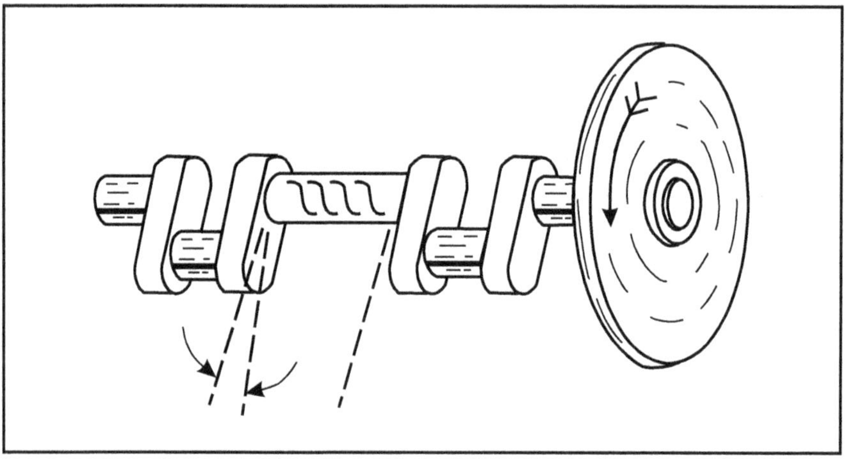

Figure 20-7. Angular misalignment proportional to the torque transmitted through a crankshaft with two throws occurs in single-end drives. This factor results in a ram left-to-right tipping alignment error.

The error will be made worse if the largest load is placed on the side of the press opposite the driven end of the crankshaft. Presses having driving gears on each end of the crankshaft are often specified for heavy presswork. However, single-end drive presses can be used for precision high-speed presswork by using a very large and rigid crankshaft in comparison to the machine force capacity. High-speed press crankshafts have short stroke lengths. They have very low amounts of torsional twist. Because the shaft diameter is large compared to the stroke length, they are termed *eccentric-type crankshafts*.

Twin-end Drive Presses

Figure 20-8 illustrates a single-gear reduction twin-end drive on a press that has a crankshaft with two throws. By driving the crankshaft equally on both ends, there is more accurate left-to right ram-to-bed alignment under load than is the case of the single-end drive system. It is important that the machining and timing of the gears, keyways, and crankshaft be accomplished in a precise manner to avoid binding and ensure smooth operation.

Double-gear Reduction Presses

Presses having two gear reductions from the flywheel to the crankshaft are termed *double-gear reduction presses*. These machines normally achieve a speed range of 8–30 SPM. They are used for difficult applications such as heavy deep drawing, cold forging, and forming large parts such as truck frame rails. Large transfer presses also frequently employ double-gear reduction. Figure 20-8 illustrates one type of clutch and gearing arrangement.

Single-connection Presses

Straight-side presses with single connections often are built to provide very high force capacities in a machine having a relatively small bed size. Figure 20-8 illustrates a high tonnage, single-action, straight-side, eccentric-type mechanical press.

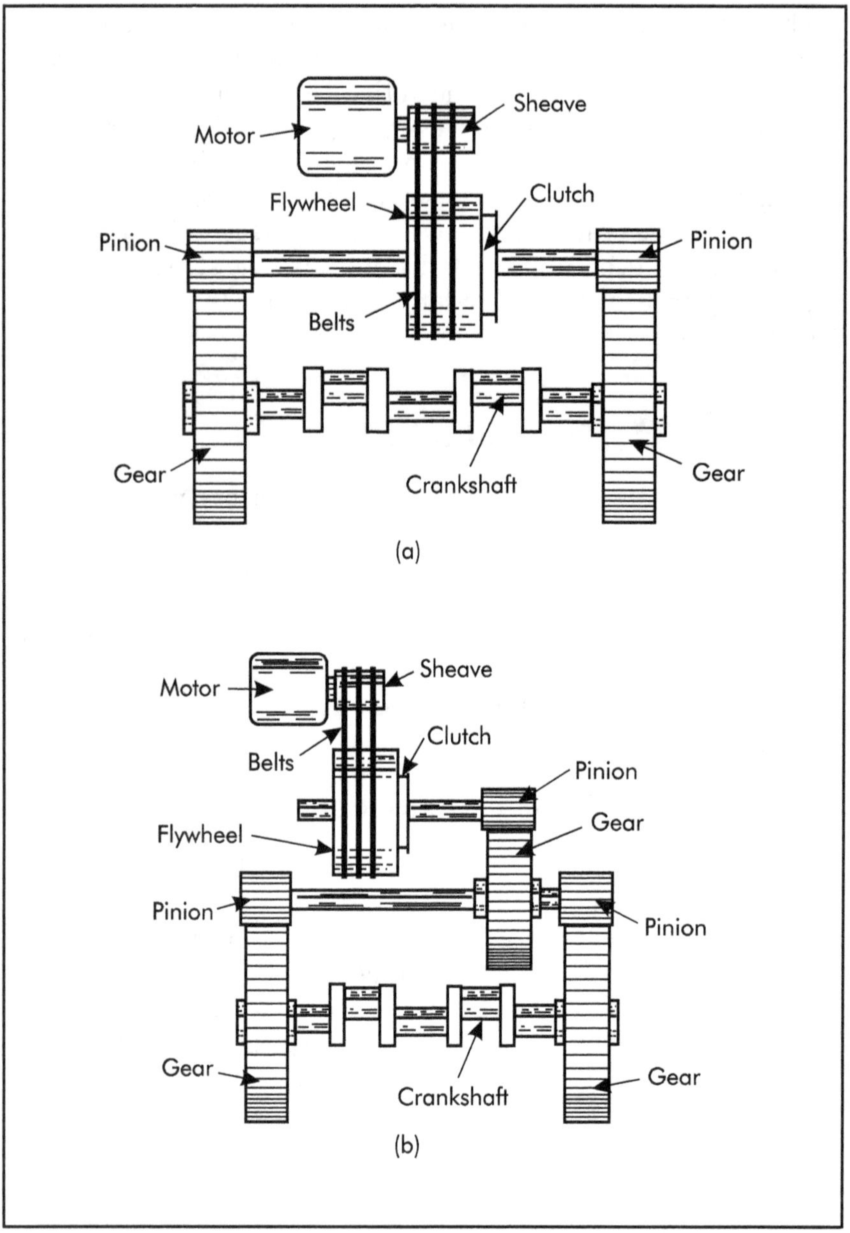

Figure 20-8. (a) Providing a driving gear on either end of the crankshaft avoids the angular misalignment under load illustrated in Figure 20-7. (b) Shown is a typical style of clutch and gearing arrangement used on double-gear reduction presses.

The gear train is of the double-reduction type. A large gear on either side drives the eccentric. This type of machine is very useful for heavy forming, as well as warm and cold forging work. The double-geared eccentric is capable of transmitting a great amount of torque. Presses of this type are capable of developing full tonnage relatively high in the press stroke. This factor makes them very useful for closed-die forging work.

The heavy construction together with the narrow bed and slide result in low deflection under load. A press of this type is ideal for blanking work involving thick high-strength materials. Very stiff machines, with bed sizes no larger than necessary, are subject to much less snap-through energy release than presses with bed sizes that are wider than necessary for the application.

To avoid ram-tipping problems, the load must be carefully centered under the connection. While this is always very important, it is especially necessary in single-point presses. Keeping the load centered minimizes the pressure on the gibbing and lessens die wear.

Two- and Four-point Presses

Double- or quadruple-gear-driven eccentrics normally rotate in opposite directions to aid in slide guiding and avoid lateral thrust. This feature is found in most two- and four-point presses, which are top-driven by eccentrics.

The timing of the gearing from the left-to-right side of the press is critical. Couplings are provided on the shaft, driven by the clutch, which are either adjustable or may be fitted with offset keys.

Both crankshaft and eccentric-driven straight-side presses with two connection points are in widespread use. Ram parallelism from left to right is mainly controlled by the connections, while front-to-back alignment is provided by the gibs. The press shown in Figure 20-9 has two connections that are driven by gear-actuated eccentrics.

Two-point presses are guided by the correct adjustment of the pitman straps from left to right. Some front-to-back guiding of the slide is provided by the saddle bearing and wrist-pin-type connection. However, the majority of the front-to-back guiding is provided by the gibbing. Again, it is important to center the load, especially from front to back.

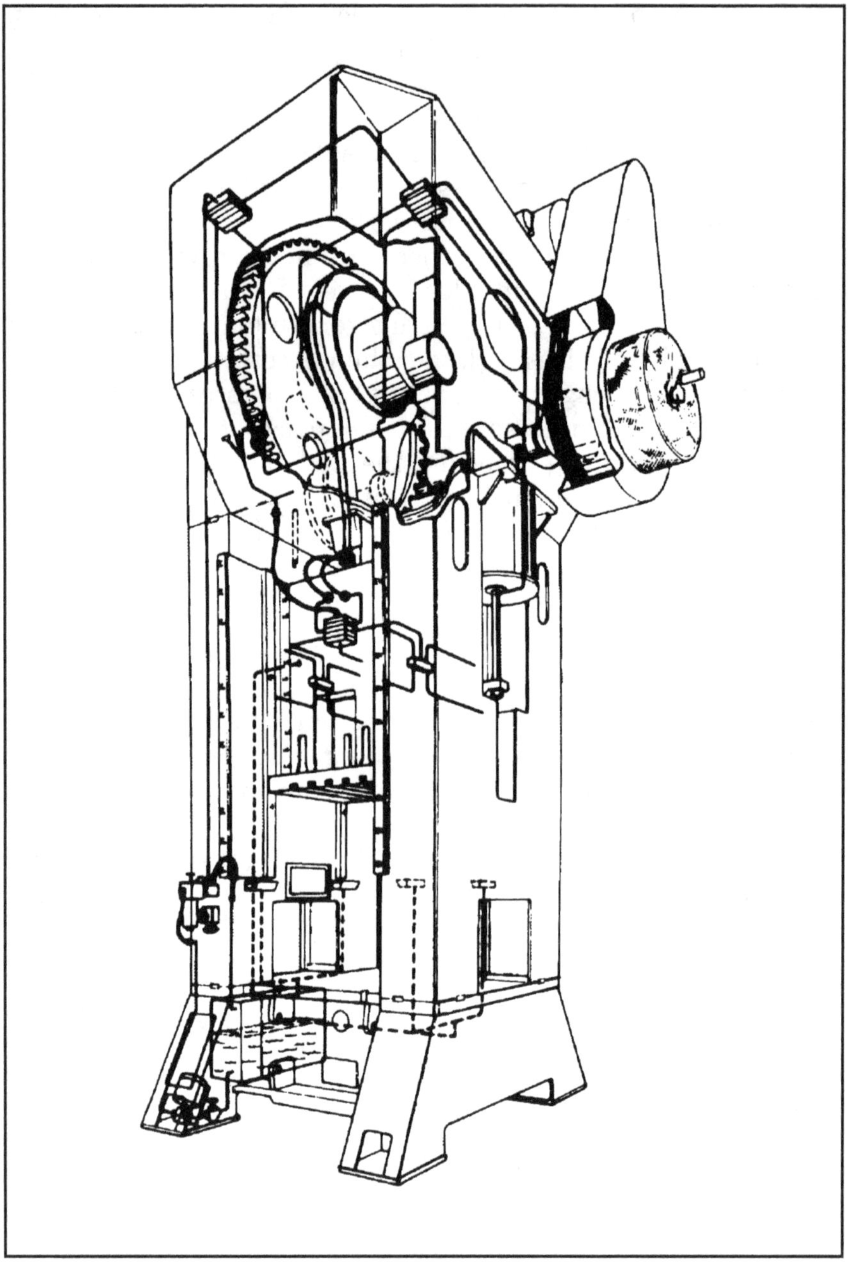

Figure 20-9. A single-action, straight-side, eccentric-type mechanical press is shown with a cascade lubrication system and large reduction gears in the press crown. (Courtesy Verson Corporation)

Figure 20-10 illustrates a two-point eccentric-driven straight-side press. Typical electrical control components needed to power and control the machine are shown. Note that the rotary cam switches are separately driven. This adds an extra measure of safety in addition to chain-breakage monitors.

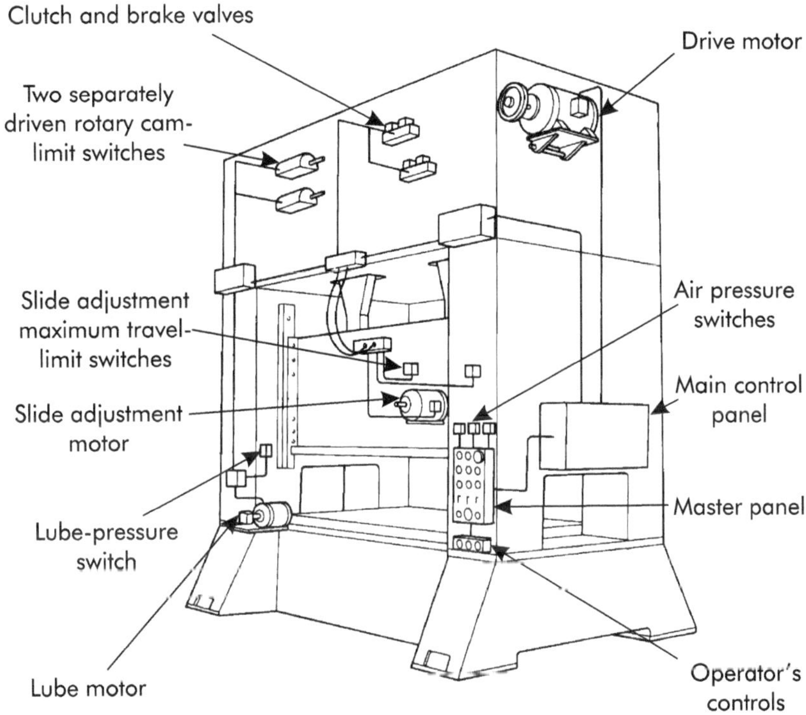

Figure 20-10. Two-point, eccentric-driven, straight-side press is shown with typical electrical control components. (Courtesy Verson Corporation)

SPEED REDUCTION AND TORQUE CAPACITY

The motor furnishes energy to the flywheel. Once the flywheel is up to speed and not being cycled, the motor need only supply enough energy to make up for frictional losses. The flywheel stores the energy until some is used to perform work.

The press must take the energy of the flywheel and transmit it through the clutch, gears (if a geared press), crankshaft, connection, and slide to perform the required work. The energy stored in the flywheel increases as the square of the flywheel rotational speed. Thus, depending upon the speed adjustment setting, the available flywheel energy can vary greatly for presses with variable speed drives.

When the flywheel speed is reduced, the flywheel energy decreases by the square of the speed reduction. Likewise, if the speed is increased, the flywheel energy is increased by the square of the speed increase. This may cause problems if the press is slowed for work such as deep drawing, which requires more energy than cutting and bending.

There must be enough time between successive strokes for the motor to restore flywheel speed. Otherwise, the press may stall. Belt slippage and motor overheating can result from inadequate flywheel energy. If the variable-speed drive cannot be increased, a press with sufficient energy to accomplish the work will be required.

MECHANICAL VERSUS HYDRAULIC PRESSES

Mechanical presses are built with force capacities to 6,000 tons (53.4 MN) or more. Force capacities of 50,000 tons (445 MN) or more are available in hydraulic presses. The very large hydraulic machines are used in hot- and cold-forging applications as well as various rubber-pad and fluid-cell-forming processes.

Both single- and double-action hydraulic presses are used for forming large parts for the automotive and appliance industries.

Features

Deep drawing and forming applications often require large forces very high in the press stroke, all the way to the bottom of the stroke. Some mechanical presses do not develop enough force high enough in the downward stroke to permit severe drawing and forming applications.

In most hydraulic presses, full force is available throughout the stroke—a very important characteristic. Figure 20-11 illustrates why the rated force capacity of a mechanical press is available only near the bottom of the stroke.

Another advantage is that the stroke may be adjusted to match the job requirements. Only enough stroke length to provide part clearance is required. Limiting the actual stroke will permit faster cycling rates and reduce energy consumption. The ram speed also can be adjusted to a constant value that is best for the material requirements.

Overload Protection

The force that a hydraulic press can exert is limited to the pressure applied to the total piston area. The applied pressure is lim-

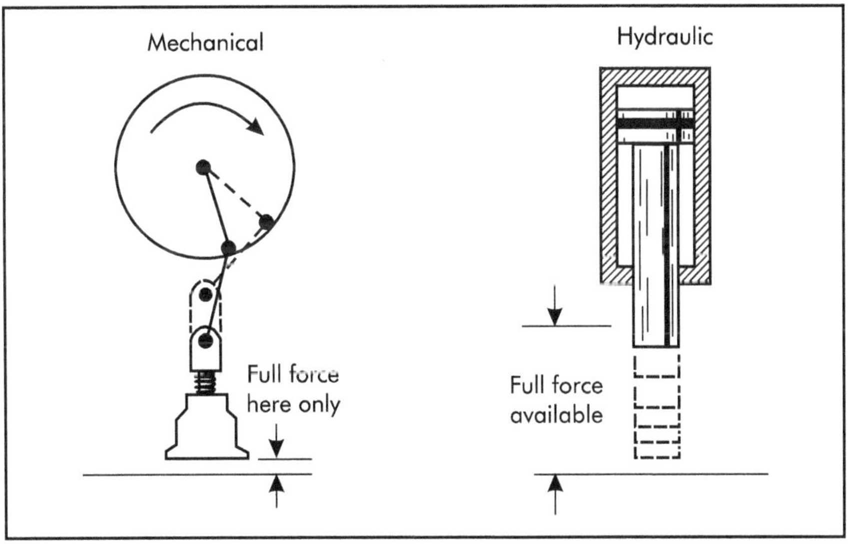

Figure 20-11. The rated capacity of a mechanical press is available only at the bottom of the stroke. The full force of a simple hydraulic press can be delivered at any point in the stroke.

ited by one or more relief valves. A mechanical press usually can exert several times the rated maximum force in case of an accidental overload. This extreme overload often results in severe press and die damage. Mechanical presses can become stuck on bottom due to large overloads caused by part-ejection failures or die-setting errors.

Hydraulic presses may incorporate tooling-safety features. The full force can be set to occur only at die closure. Should a foreign object be encountered high in the stroke, the ram can be programmed to retract quickly to avoid tooling damage.

Force

When choosing between a mechanical and hydraulic press for an application, a number of items should be considered. The force required to do the same job is equal for each type of press. The same engineering formulas are used.

There is always a possibility that an existing job operated in a mechanical press requires 20–30% more force than the rated machine capacity. The overloading problem may go unnoticed, although excessive machine wear will result. If the job is placed in a hydraulic press of the same rated capacity, there will not be enough force to do the job. Always make an accurate determination of true operating forces to avoid this problem.

Mechanical presses have the full-rated force available only very near the bottom of the stroke. A chart giving distance from bottom of stroke versus available force is called a force or tonnage curve. The force curves for six different mechanical presses are shown in Figure 20-12.

In some cases, a sharper coined impression may be obtained at a rapid forming rate. Jewelry and medallion work make use of both high-force hydraulic presses and drop hammers. Each process has its own advantages. Often die pressures in excess of 250,000 psi (1,724 MPa) occur in hydraulic press medallion work.

Mechanical presses with high force capacities are physically much larger than their hydraulic counterparts. Few mechanical presses have been built with force capacities of 6,000 tons (53.4 MN) or more. Higher tonnage and/or compact construction are practical in

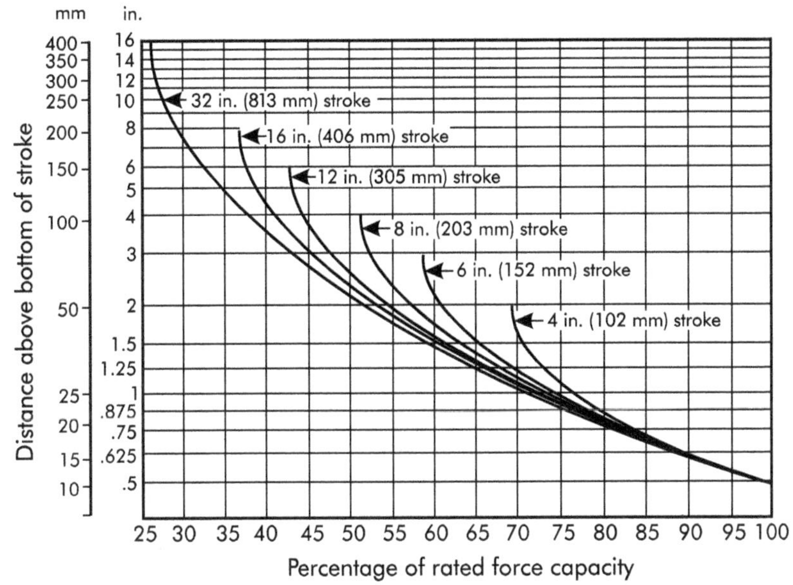

Figure 20-12. Force curves for six different mechanical presses. (Courtesy Danly Machine Corporation)

hydraulic presses. Hydraulic presses for cold forging are built with up to 50,000 tons (445 MN) or greater force capacity. Some hydraulic fluid cell presses have force capacities over 150,000 tons (1,335 MN).

The pressure at which the press delivers full tonnage is important. The most common range is from 1,000–3,000 psi (6,894–20,684 kPa). Some machines operate at substantially higher pressures. There is no set rule on the best peak operating pressure for a press design. Obviously, higher pressures permit the use of more compact cylinders and smaller volumes of fluid. However, the pumps, valves, seals, and piping are more costly because they must be designed to operate at higher pressure.

The force of a hydraulic press can be programmed in the same way that the movements of the press are preset. In simple presses, the relief-valve system that functions to provide overload protection may also serve to set the pressure adjustment. This allows the press to be set to exert a maximum force that is less than press capacity. Usually there is a practical lower limit, typically about 20% of press capacity. At extremely low percentages of force capacity, a stick-slip phenomenon known as *stiction* in the cylinder rod and piston packing can cause jerky erratic action.

Press Construction

The bed size, stroke length, speed, and tonnage of a hydraulic press are not necessarily independent of each other. Press construction depends upon the amount of total force required and the size of dies to be used. Figure 20-13 illustrates how bed size does not directly relate to press-force capacity. Shown are presses that use the tie-rods for ram-guiding jobs that do not produce lateral loads. Figure 20-13a illustrates a large-bed-size press with low-force capacity. Note the small cylinder size. Some uses for these presses include cutting and pressing soft materials such as fabric, and wood or plastic laminating applications. The press design illustrated in Figure 20-13b is suitable for high-force applications such as heavy forging.

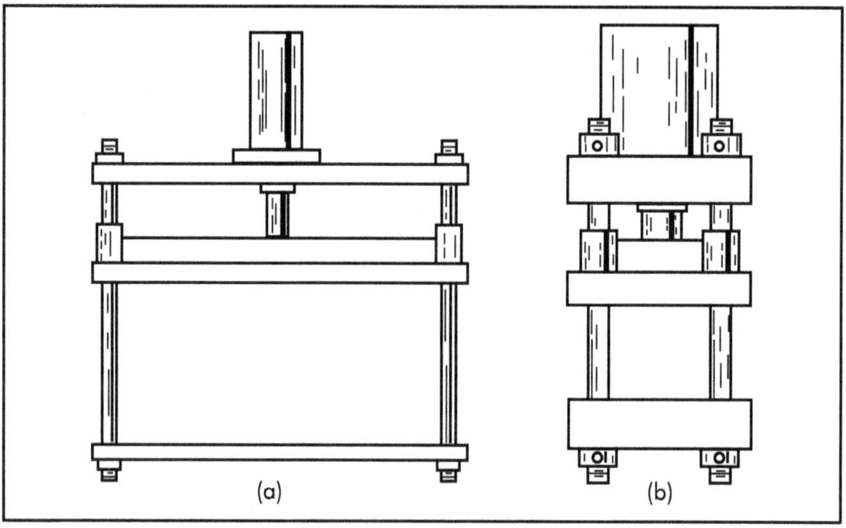

Figure 20-13. (a) Cutting soft materials and laminating work may require a low-force-capacity machine with a large bed area such as the one illustrated. (b) Heavy work, such as coining and cold forging, requires a compact machine capable of delivering high forces over a small bed area.

Like mechanical presses, hydraulic presses are available in many types of construction. Factors to consider when deciding between a hydraulic and mechanical press are stroke length, actual force requirements, and the required production rate.

Press Speeds

Most press users are accustomed to describing press speeds in terms of strokes per minute. Speed is easily determined with a mechanical press. It is part of the machine specifications.

The number of strokes per minute made by a hydraulic press is determined by calculating a separate time for each phase of the ram stroke. First, the rapid advance time is calculated. Next, the pressing time or work stroke is determined. If a dwell is used, that time is also added. The return stroke time is added. Finally, the hydraulic-valve reaction-delay time is added to determine the total cycle time. These factors are calculated to determine theoretical production rates when evaluating a new process. In the case of jobs that are in operation, measuring the cycle rate with a stopwatch is sufficient.

The forming speed and impact at the bottom of the stroke may produce different results in mechanical presses versus their hydraulic counterparts. Each material and operation to form it has an optimal forming rate. For example, drop hammers and some mechanical presses seem to do a better job on soft jewelry pieces and jobs where coining is required.

In deep drawing, controllable hydraulic press velocity and full force throughout the stroke may produce different results. Often parts that cannot be formed on a mechanical press with existing tooling can be formed in a hydraulic press. Hydraulic presses can be provided with controllable force throughout the press stroke and variable blank-holder pressure distribution.

Ram Tipping Controls

Some two-piston presses similar to the design shown in Figure 20-14 feature a system of linear-position transducers and servo valves to vary the force to each piston to maintain the ram level

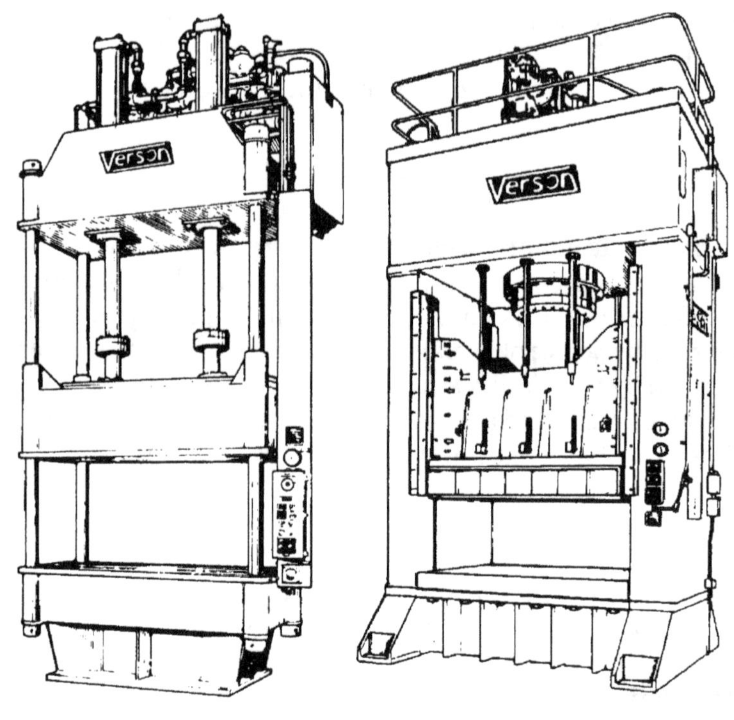

Figure 20-14. (a) A two-cylinder, four-post hydraulic press is suited for light- to medium-duty work that does not involve lateral (side) loads. (b) A straight-side hydraulic press is designed for applications requiring close alignment and high forces. (Courtesy Verson Corporation)

with the bed. How well such a system works depends on the accuracy of the position sensors and reaction speed of the servo valving. Some manufacturers have marketed lightweight dual-ram presses featuring servo leveling with an implied claim that they are a direct replacement for two-point mechanical presses in critical work such as progressive-die operation. Taking any advertised claim at face value can result in a highly unsatisfactory tooling operation. The ability of a sensing system to acquire data and redirect oil through servo valving faster than the several milliseconds during which a snap-through release occurs is extremely difficult to achieve. In turn, the inability to react quickly enough to prevent ram tipping may be harmful to the press, tooling, and process.

The press shown in Figure 20-14a is not necessarily available with servo anti-tipping controls. Nor would it be a needed feature for the recommended press application. If heavy punching work and limited ram tipping under lateral or snap-through loads were required, a press such as that shown in Figure 29-14b would be a recommended choice.

Straight-side presses, such as the one illustrated in Figure 20-14b, are much better able to withstand off-center loading and snap-through energy release than the type shown in Figure 20-14a. Quality features to look for in a press designed for severe work when ram tipping is to be minimized are a single-piston design together with a large ram or slide with long guiding and eight-point gibbing. However, loading should be carefully balanced and cutting dies timed to minimize snap-through shock to the best extent possible in any pressworking operation.

UPGRADING EXISTING PRESSES

Older presses often can be upgraded for smoother, more reliable operation and reduced tooling wear. In some cases, rebuilding a damaged press can pay rapid dividends in reduced tooling repair costs and better part quality.

Electrical controls, which may no longer meet safety requirements, can be replaced. Usually, the most satisfactory way to retrofit the press is to install a new control package specially designed for the application. Such systems are available from several suppliers.

DIE CUSHIONS

When a single-action press is used for drawing operations, the manner in which the blank-holder pressure is applied to control the flow of the metal blank is important. The application of pressure to a blank holder is one of the features of a double-action press. Single-action presses lack this feature and therefore require supplementary blank-holding equipment.

Dies are sometimes built with a blank holder using compression springs, air cylinders, or high-pressure nitrogen cylinders to supply the holding pressure. This greatly increases the cost of the die. A press cushion can serve every die with this requirement, lowering the cost of tooling.

Pneumatic Die Cushions

A pneumatic die cushion is supplied with shop air pressure. The design normally uses either one or two pistons and cylinders. The recommended capacity of a die cushion is about 15–20% of the rated press tonnage. The size of the press-bed opening limits the size, type, and capacity of the cushion.

Figure 20-15 shows a pneumatic die cushion. This illustration shows an inverted-type cushion in which the downward movement of the blank holder, through pressure pins, forces the cylinder

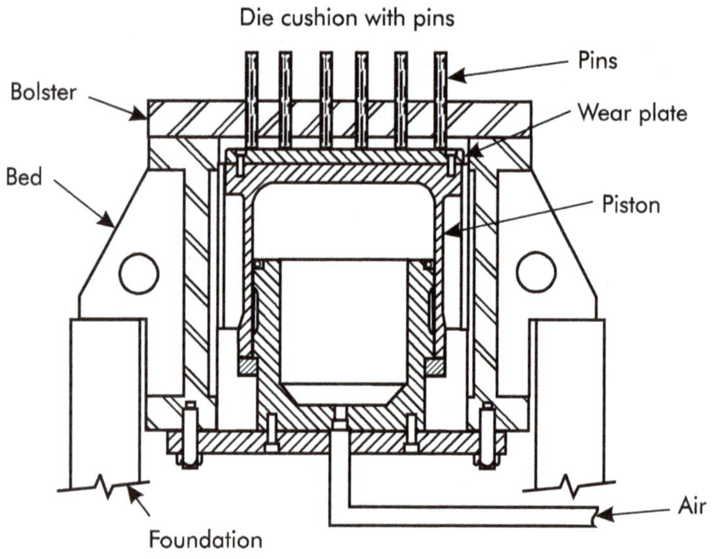

Figure 20-15. Sectional view through a press bed and bolster illustrates a pneumatic die cushion.

against a cushion of air inside the cylinder, and moves the air back into the surge tank (not shown). The external components, such as the surge tank, regulator, and pressure gage, are essentially identical in function to a press counterbalance system. On the upstroke, the air in the surge tank returns to the cylinder. Other designs function without surge tanks.

It is very important to load the cushion evenly to avoid premature wear and cushion failure. The die designer should incorporate equalizing pins in the lower shoe if required to accomplish loading equalization. Often, these can be actuated by the upper heel blocks. In some cases, special pin drivers made of structural tubing may be attached to the slide to actuate equalizing pins. Operator safety must be considered when performing this operation because additional pinch points are created.

To avoid press and die problems, it is of extreme importance that the correct length of pins be used. In drawing and other critical operations, a pin that is 0.060 in. (1.52 mm) longer than the others can easily cause a wrinkled or fractured part. When problems are encountered, and the pressure setting is found correct, the pins should be carefully measured with a micrometer or vernier caliper to determine if they are of the same length. If the pins are the same lengths, they can be reinserted into the bolster pin holes, and checked with a dial indicator as shown in Figure 20-16. If a variation in length is found, it is probably caused by unevenly worn depressions in the top of the piston wear plate or the cushion is damaged. The wear plate can be repaired by regrinding.

The bolster pin holes will become larger with use. If this occurs, hardened bushings can be fabricated and used to repair the worn holes. Generally, this is a permanent solution. The die also can be a source of difficulty. With extended use, the pins can wear uneven depressions in the underside of draw rings, pads, and subplates.

Hydraulic Die Cushions

Hydraulic die cushions have the advantage of taking up less space than air cushions. They also can be equipped with fixed or servo-actuated relief valves. Programmable controllers and servo valves are used to control resistance throughout their travel.

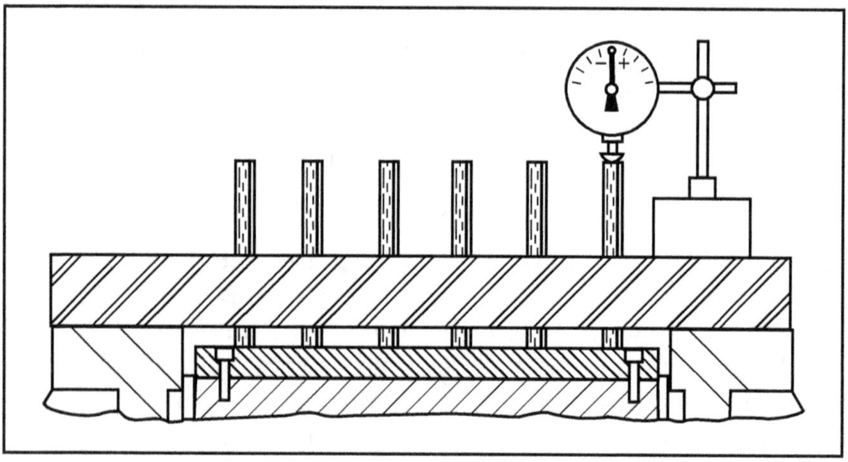

Figure 20-16. Troubleshooting a die cushion system with a dial indicator.

CENTERING THE LOAD IN THE PRESS

Maintaining correct die clearances is a basic requirement for producing high-quality stampings. If the load is not centered in the press, critical die clearances will be changed.

Off-center loading actually results in an out-of-level condition much like that caused by setting a die on a slug. This is because the press must deflect to develop tonnage. If the load is centered, the deflection is uniform and the slide remains parallel to the bolster. If the load is not centered, the greatest deflection occurs in the side of the press with the greatest load, resulting in an out-of-parallel condition. This condition can cause rapid press wear as well as die damage and part quality problems.

In the case of cutting dies, too little clearance between the punch and die will result in excessive cutting pressures, which will break down the cutting edges quickly. Too much clearance will result in excessive burrs. Ram tipping will also result. This will cause rapid wear of the gib liners and other important press parts.

Figure 20-17a illustrates an exaggerated view of the press deflection that is a normal result of developing the tonnage. Just as a spring must change shape or deflect to develop pressure, a press must deflect to develop tonnage. The illustration shows a small die centered in a large straight-side press. Not all of the press tonnage

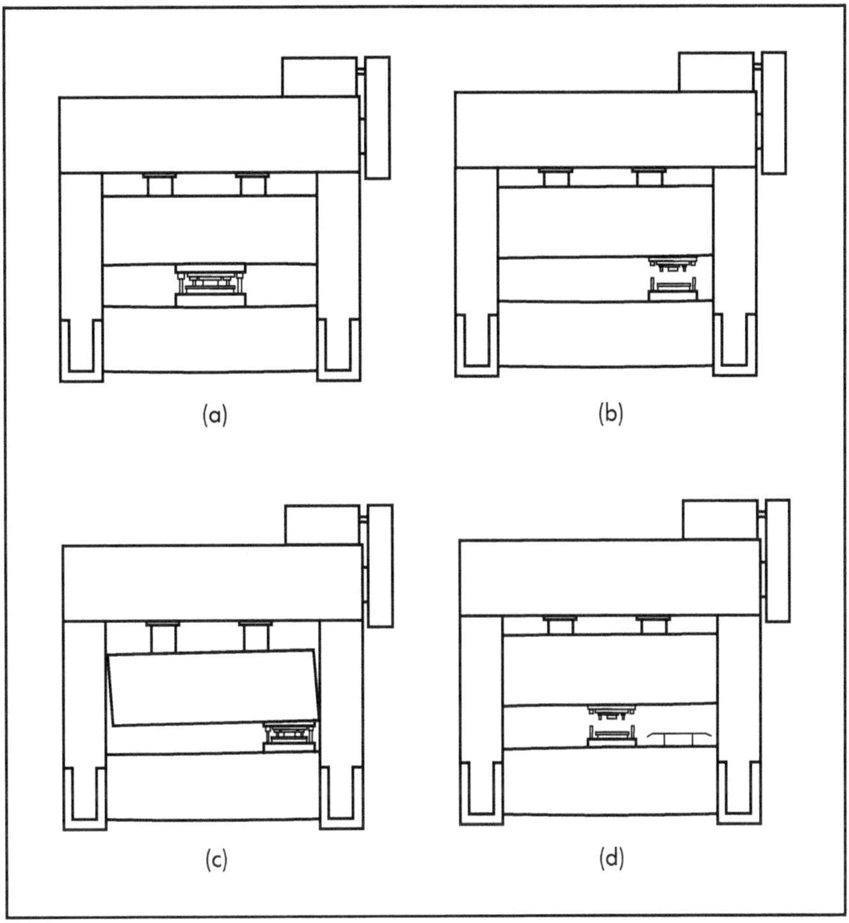

Figure 20-17. (a) Illustrated is an exaggerated view of the press deflection that is a normal result of developing the tonnage needed to do presswork. (b) The die is offset to one side of the press for operator or setup convenience. (c) This creates unequal loading and results in an out-of-parallel condition when the press deflects. (d) To correct off-center loading of a press, a simple production aid or stock guide may be all that is required.

capacity is safely usable. As a rule, at least 70% of the press bed should be occupied with a centrally placed die shoe if full press tonnage is to be developed. Developing full tonnage with a small die can result in damage to the press slide and bed due to excessive localized deflection.

Offsetting Dies

The most common reason for offsetting dies from the center of the press is for the loading convenience of the operator. Offsetting the die may satisfy ergonomic considerations. The operator does not have to reach or bend as far to place and remove parts from the die. Productivity and operator comfort go hand-in-hand.

Another reason dies may be offset in the press is the lack of a proper coil stock guide. Figure 20-17b illustrates a die offset to one side of the press to move closer to the coil feeder.

Formulating a good alternative to offsetting dies usually requires a careful analysis of all factors. Common-sense solutions may include asking the following questions:

- Is a smaller press available so the operator won't need to reach as far to place the blank?
- Can a production aid be constructed to permit the part to slide into the correct location by gravity or automatic movement?
- Is it possible to operate two dies in the same press to correctly balance the loading?
- Is the placement of nitrogen cylinders on one side of the press to balance the load a practical solution?
- Is there a good reason for offset loading such as moving a progressive die to one side to balance the force?

Figure 20-17d illustrates a simple stock guide used to correct the problem of out-of-location stock.

Unbalanced Loading

Figure 20-17c illustrates the result of unequal loading. The result is an out-of-parallel condition when press deflection occurs. The short-term effect is press misalignment under pressure that is much the same in its effect as that of a slug under the die shoe. The quality of work and number of pieces produced between die sharpening will be less than would be the case if the load were centered. The long-term effect is uneven press wear. Bearings will not wear evenly, resulting in an out-of-parallel condition in the future. The gibbing will be subjected to high localized pressures resulting in rapid uneven wear and scoring.

Mishit Damage

Many dies are designed with balanced cutting action. This is done to avoid side thrust, which can change die clearances. If the part design or stock layout will not permit balanced cutting action, very large guide pins or heel blocks are required to limit the side movement to an amount that is acceptable.

Badly positioned stock can damage the die. The unbalanced loading can be so severe that the stock is flanged rather than cut. Out-of-location stock is flanged into a large cutting-die opening due to the unbalanced cutting action resulting from mislocated stock (see Figure 20-18). Because the cutting clearance is only 10% of the stock thickness, the side displacement is so great that the punch and die-cutting edges hit upon die closure and are damaged.

It is very important that everyone be made aware of how mishit damage occurs and how it may be avoided. Honest reporting of any damage of this type is important too. The damage shown in Figure 20-18c could be reported as a broken punch. Replacing the small punch in the press would not fix the damage to the other stations. Press time should not be wasted. A die damaged this badly should be taken to the die room for extensive repair.

REFERENCE

Smith, D. 1994. *Fundamentals of Pressworking*. Dearborn, MI: Society of Manufacturing Engineers.

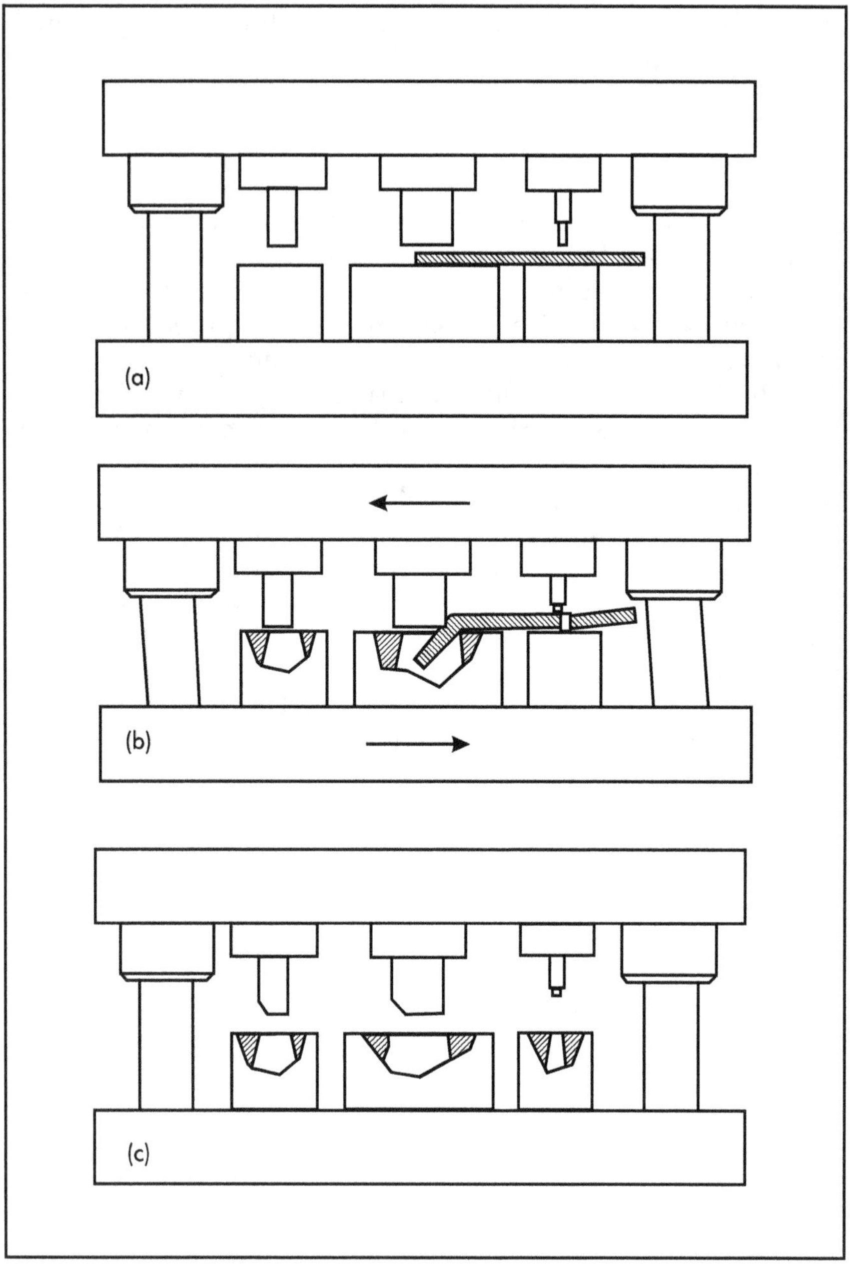

Figure 20-18. (a) Out-of-location stock can cause die damage; (b) a partial cut results in the stock being flanged into a die opening, causing the die shoes to be displaced sideways; (c) the damage caused by the displacement.

21

Troubleshooting Large Panel Stamping Dies

DIE BOLTING

Each manufacturer's engineering and safety department should determine the requirements for the type and number of fasteners used to bolt the die in the press. In the case of large dies, this is determined in the die-design stage and is based on die engineering standards.

Tie-down and handling slots that are cast in place facilitate the safe handling of cast die shoes. Dies made of fabricated steel plate have slots flame cut on locations specified on the die blueprints.

Most large presses made for use in North America have standard 1-in. (25.4-mm) Joint Industry Council (JIC) T-slots spaced with 6 in. (152.4 mm) centers. The standard used determines the T-slot locations when building the die.

The die should have a positive locating method, such as bumper pins, to assure that it will be in a position to line up with the bolster and ram T-slots. The die design and building process should result in a die that can be located easily and bolted securely. This is especially important if powered die clamps, rather than conventional T-bolts, secure the upper die. Figure 21-1 shows an air-moving bolster prestaged and ready for changeover on a two-ram transfer press. Hydraulically powered swing clamps secure the upper die.

Figure 21-1. Shown is an air-moving bolster prestaged for changeover on a two-ram transfer press. It is very important to make sure that the dies are properly located and line up with the clamp slots in the adapter plates. (Courtesy Ford Motor Company)

Correcting Misalignment

In the event of T-bolt tie down or die-clamp slots not lining up, an action plan should be agreed upon to effect correction. Possible causes of misalignment include:

- The positive die-locating method used is not being machined on the correct location.
- The dowels or keys required to line up die-buildup components, such as adapter plates or subplates, may be missing.
- If several of the U-shaped T-bolt cutouts do not line up, it is probably because the cast, flame-cut, or machined slots are incorrectly located.

Dies that have buildup components, such as cast or fabricated risers, subplates, and adapter plates, rely on keys and/or dowels to maintain proper alignment. If several parallels are used, it is usually sufficient to dowel each end parallel to the die shoe and the subplate or adapter plate at two locations each.

Action Plan

Figure 21-1 shows a large air-moving bolster double-ram transfer press. To achieve fast die changes, all clamps must line up and work properly. If there are alignment problems, an action plan is required to identify and fix them. If only one or two clamps fail to engage, the problem may be the alignment of the clamps or the slot the clamp engages. If most clamps fail to engage, there are several possible causes. These include a pressure-source problem or a bolster shot-pin misalignment (Smith 1990).

A team effort is needed to address any delay factors that get in the way of setting the die on location and securing it in the press rapidly. Finger pointing between departments that may have some responsibility for the problem is unacceptable. Here are some common-sense items to check:

- Use a square and/or an angle plate with a straight edge to be sure the bolster and ram T-slots are in line with each other. If not, the bolster is probably out of location.
- It is also possible that the guiding system, known as the gibs or gibbing, is not set to center the ram.
- Carefully measure the positive die locators to make sure they are in the correct position.
- The die tie-down and clamping locations require correct machining. Noting the amount and direction of error carefully will insure that reworking the locations will solve the problem.
- If only a few die tie-down locations are in an incorrect location by a slight amount, scribe and center-punch a layout to permit milling the bolt or clamp interference away. Note that it is unacceptable to grind a bolt or clamp to allow a die to be set.

NITROGEN PRESSURE SYSTEMS

Spacer blocks attached to the draw ring should be machined so that the draw ring is even with or slightly below the height of the draw punch to determine the exact dimension in tryout. The spacer blocks may be stepped in height to permit gradual engagement of the nitrogen cylinders and reduce the press load high in the stroke.

Usually only four to eight of the units in a typical 24-cylinder system should supply pressure at the initial contact of the upper blank holder. The working surfaces of the punch and blank holders are either hard chromium plated, or ion nitrided for wear resistance. This lessens dirt pickup of the electrogalvanized zinc coating used on the blanks.

Most nitrogen fittings are actually high-pressure hydraulic fittings that have been adapted to nitrogen service.

It is very important to follow the manufacturers' recommendations for safe operation and maintenance of high-pressure nitrogen systems (see Chapter 19).

Marking Nitrogen Pressure

The exact amount of nitrogen pressure required should be stamped on a metal tag next to the charging console. Such metal tags are obtained from sources such as the following:

- the nitrogen manifold or system manufacturer;
- the die builder; or
- stampers with their own standardized-data tag, which has a space for nitrogen pressure and other information.

A SYSTEMATIC APPROACH TO DIE TRYOUT

New dies that have never been operated under production conditions usually need some minor rework of the blank holder. Often such work allows the die to operate at lower blank-holder pressure. The goal is to draw or form the metal with minimum force and material thinning. Skilled workers typically need 12–24 hours to fine-tune the blank-holder surfaces.

Figure 21-2. A cartoon from the Australian Sheet Metal Forming Group shows a rather unscientific approach to die tryout. (Courtesy Stuart Keeler)

It is extremely important to follow good procedures when trying out new or reworked dies. Figure 21-2 illustrates the alternative, which shows an unscientific approach to new die tryout. Today, good predictive and diagnostic die tryout tools can avoid the costly delays associated with the old "guess and try" approach.

Figure 21-3 illustrates the advantages of applying a systematic approach to die tryout. A systematic approach to die tryout and die troubleshooting includes the use of circle grid analysis (CGA), the forming limit diagram (FLD), and taking ultrasonic thickness measurements. Use of such tools has taken die tryout procedures from a black art to an engineering science. FLD is used to plot the deformed circles in the gridwork pattern.

The following types of systematic work are normally performed during the die tryout process:

- Inspect the press to be sure it is capable of maintaining parallelism under the required tonnage at the bottom of the stroke. Otherwise, the work done to the die will reflect the inaccuracies of the press. It is preferable to use the actual home production press if possible.

Figure 21-3. A cartoon illustrating the advantages of using a systematic approach to die tryout. (Courtesy Stuart Keeler)

- Verify that the steel used meets all specifications for the production material required for the type and severity of deformation involved.
- The entire blank, or at least critical areas, can have a grid of circles applied by electrolytic etching or CGA. The CGA process is intuitively simple to learn. Attending a class in sheet-metal formability suitable for tradespersons, engineers, and managers is wise.
- The shut height is raised so the inverted draw punch in the lower die makes minimal or no contact with the part. Full draw-ring pressure is maintained.
- A blank is coated with Prussian blue (also known as spotting blue) around the areas that will contact the blank holder. A binder hit is then made.
- The blued-blank panel is examined on both sides for hard marks and a full impression of any draw or lock beads if they are used. This panel is marked as the first hit. The time and date are also noted. The panel is then set aside for reference.

- The shut height is gradually lowered a little at a time with a blank in the die and the press cycled. Any tendency of the metal to thin, neck, or fracture is noted.
- Constant caution is required to avoid over-bottoming the press, which can result in the press becoming stuck on the bottom, as well as press and die damage. The tonnage-monitor readings should be carefully watched. A sudden increase indicates that the die is on the bottom.
- The minimum nitrogen pressure required to prevent metal movement while maintaining the locked condition should be used.
- It is important not to exceed the press's force capacity above center. Press-tonnage curves that provide available force at a given distance from bottom dead center (BDC) should be consulted. Exceeding the press-tonnage-curve-force capacity with nitrogen blank holders or draw rings is frequently a cause of press damage.
- To minimize draw-ring-pressure requirements and to control metal movement, draw beads are frequently used. Draw beads retard metal movement on the blank holder by increasing the amount of pulling force required to draw the metal. This is done by forcing the metal to bend and unbend as it is drawn over the draw beads. Sometimes more than one bead is used.

For a part like a roof panel, which is a semi-rectangular shape, the metal will thicken at the corners. This is caused by the metal being compressed in the corner areas as it is drawn into the draw cavity. This is called *circumferential compression*. The amount of thickening can be as high as 25% of the metal thickness or more.

Thinning Analysis

When using CGA, the deformed circles may be measured to determine both the major and minor strain. In addition, the amount of thinning may be accurately determined by calculating the change in area of the deformed circle, which is usually elliptical. The change in the deformed circle area relates directly to the change in metal thickness. This is easily calculated using the mathematical formula for the area of a circle or ellipse. This method works because there is no volumetric change in the area of the circle before or after deformation.

Measuring with an Ultrasonic Thickness Gage

An ultrasonic thickness gage is capable of accurately measuring small areas of formed parts. Measuring with an ultrasonic thickness gage is a nondestructive procedure that is very fast and avoids cutting parts for measurement with a micrometer.

Ultrasonic thickness gages work on the same principal as sonar water-depth measurement. A high-frequency acoustic pulse is sent from a small transducer on the end of a convenient hand-held tool. The hand-held transducer connects to a small battery-powered package containing the electronic circuitry and a digital readout device.

The transducer is held against the part location to be measured. The instrument generates precise ultrasonic pulses, which are transmitted into the part. The ultrasonic energy pulse is reflected back toward the transducer when it reaches the opposite side of the stamping. The instrument starts an electronic clock when the pulse is transmitted. The clock is stopped when the transducer detects the reflected pulse. The amount of time recorded by the electronic clock is compared to the speed of sound in the metal, and the result is provided as a direct readout in thousandths of an inch or hundredths of a millimeter. Most instruments have a user-selectable feature to provide a reading in either inch or metric units.

Field calibration of the instrument is easily accomplished by measuring a material sample and adjusting the instrument for the correct reading. The transducer is often fitted with plastic facing. This protects the transducer from damage. In addition, the plastic delays the transmitted and received signals by a known amount of time. This delay is helpful to the operation of the instrument since the speed of sound is much faster in metals than air or water. The fixed delay provided by the plastic facing provides time for the transducer to switch to the receive mode. Time is required for the transducer to cease vibrating from the transmitted sonic pulse. This permits accurate detection of the return echo signal.

Ultrasonic thickness measurements provide fast, accurate measurements of the metal thinning and thickening that must be optimized in draw die tryout. Ultrasonic thickness determinations should be recorded for future reference along with CGA and material FLD data. This information, when compared to current data, can help pinpoint what may have changed in case difficulty occurs during normal production.

Ultrasonic thickness measurement also should be used in the production shop for process control and detection of possible thinning or necking failure. Many stampings are formed by drawing, stretching, and a combination of these and other processes. Stampings prone to necking and fracture failures have only one or very few localized areas that thin and fail. By measuring only the problem areas, data acquisition labor is reduced and the results are easily tracked.

DRAW BEADS

The function of draw beads in the blank holder of a draw or redraw die is to prevent or provide a controlled resistance to metal flow, thus controlling the movement of the metal into the die cavity. This reduces the amount of blank-holding pressure required when compared to a plain blank holder. It also provides accurate control of metal movement into the draw cavity and draws a tight panel while saving metal.

Blank Holders without Draw Beads

Pressure applied with a flat blank holder is often used to retard and control metal movement into the draw cavity. This simple method usually works well, especially when simple cup drawing is needed. There are two main success factors. First, the pressure is sufficient to prevent excessive wrinkling of the metal on the blank holder. Second, the pressure required for metal control does not result in metal pickup or galling of the blank holder.

Bending and Unbending

With the exception of lock beads that "lock" the metal into place with the blank holder, draw beads act to control or retard metal movement into the draw cavity. As the punch enters the draw cavity, the energy required to bend and unbend the metal passing over the draw bead retards the blank movement. This is often required to stretch the metal around the punch. The height and placement of the draw beads is critical to success. Control should be provided by the work of bending and unbending the metal as it is drawn over the beads. Providing the correct retardation by the beads is essential for a stable repeatable process.

Commercial Steel Draw Beads

Cast-iron dies for large-part production may be constructed with commercial rolled-steel sections for inserting draw beads into the blank holder. This practice is essentially obsolete. This type of draw bead is retained by special alloy tapered rivets driven through the bead into the iron blank holder. A problem with this type of bead is that it is difficult to install correctly. Since it is soft, it tends to wear rapidly. Galling or metal pickup is another problem with this design.

High-volume forming and drawing dies use inserted tool-steel beads. Except for simple small dies, these inserts are made in several sections. Tool-steel inserts are held in place with screws or keys to permit easy removal for repair or adjustment. This type of bead permits metal movement in the draw cavity. The amount of draw-bead retardation of metal movement is determined by the height and shape of the draw bead. For example, a higher bead will increase metal retardation. Decreasing the corner radii also increases the retardation effect. Additional blank-holder pressure will increase friction if the flat blank-holder surfaces are in tight contact with the metal.

Lock Beads

A *lock bead* is a special type of draw bead that locks the metal so no movement on the blank holder can occur. Dies with lock beads stretch metal, rather than draw it. Such dies are correctly termed *biaxial-stretch-forming dies*. In some instances they are called "stretch draw," although no true cup drawing may be occurring in the process. Dies used to form large complicated irregular parts, such as automotive quarter panels, typically will use lock beads and conventional draw beads—a combination of several forming processes occurring simultaneously.

Blank-holder lock beads are used to lock or prevent metal movement on the blank-holder surface. Draw dies may use lock beads in some blank-holder locations to lock or prevent metal movement in selected areas of the draw cavity. Dies with blank holders that do not permit any metal movement into the die cavity are called *stretch-forming dies*. Large automotive parts, such as roof and hood panels, are typically formed in biaxial-stretch-forming dies that use lock beads around the entire blank-holder periphery.

Cast Draw Beads

To reduce cost and die construction time, the blank holder is often made of cast iron or steel. Including the draw beads in the blank holder of the casting further reduces cost. The castings contain alloying constituents to increase wear life and reduce metal pickup. Additives such as chromium and nickel enhance the ability to flame harden the blank holder and increase wear resistance.

One-piece cast blank-holder designs are often hard chrome plated or ion nitrided to protect against wear. These hard surface treatments greatly reduce pickup of the zinc coatings used today to provide corrosion protection. The plating and nitriding processes would require the removal and separate treatment of any inserted-type beads.

Hard chromium plating may cost less to apply than ion nitride. It is a cold process and may be applied selectively to the areas requiring plating. Ion nitride offers some important practical advantages over hard chromium plating. One advantage is its ability to use oxy-fuel gas torches to heat a slug mark. Slug marks occur by hitting double metal in the draw die. Localized heating of the resulting low spot can relieve the compressive strain and remove the slug mark without destroying the coating.

Multiple Draw Beads

Two or more beads may be placed in areas requiring greater control of the metal as it goes into the die cavity. While the size and location of the beads can be determined in the die tryout, dies for producing similar parts may be used as a guide. Often, a single bead is placed around the die cavity and additional beads are placed in local areas only as required. In tryout, it may be found that the single bead must be reduced in size or eliminated in some areas.

Metal-forming simulation software can aid in determining the amount and location of restriction required on the blank holder. In tryout, if additional beads are required, the beads are built up by welding and grinding to shape. Successful drawing of complicated shapes usually depends on impressing the shape of the punch into the blank before metal movement starts on the blank holder.

The draw beads restrict metal flow by means of the work required to bend and unbend the metal. If a pressure pad or blank holder without draw beads does not have enough pressure, adding draw beads may not work because:

1. Sufficient blank-holder force is not available to set the beads.
2. Once the beads are impressed into the metal on the blank holder, there must be enough force to keep the blank holder from opening up as the metal is drawn into the draw cavity at an angle. This angular force vector will tend to pull the blank holder open because it often takes more force to keep the blank holder closed than that used to form the beads.
3. Double-action presses are typically designed with six units of force available with the blank holder for each ten units of force on the draw slide or ram. If such high force is needed, it is hard to achieve with springs or nitrogen cylinders in single-action press applications.
4. Hydraulic presses with hydraulic cushions can provide precise force control throughout the working stroke. They are superior to mechanical presses for difficult forming applications.

5. Hydraulic presses do not have a force-curve limitation—full force is available at any distance from the bottom of the stroke.
6. With the exception of lock beads, draw beads are semi-circular in cross section and they must mate with a correctly located clearance in the blank holder. The draw beads should not pinch the metal.

DOUBLE-ACTION PRESS OPERATIONS

Double-action presses are ideal for most large sheet-metal-drawing operations in the automotive and appliance industries. The blank holder or the outer slide dwells on the bottom during the time that the drawing action of the inner slide occurs. Double-action dies often have an equalizer block on each corner of the die that is shimmed to grip the blank differently in specific areas to control metal flow. However, relying on frequent fine adjustments to deal with normal material variations wastes production time. The dies can be designed to use die cushions or nitrogen cylinders to exert blank-holder pressure.

DIRT PROBLEMS

Getting a good drawn or formed panel out of the first operation is essential. If the part is not properly formed or has any defects, there is no good way to correct them in subsequent operations other than by metal finishing. For occasional salvage of a panel or two with a slight burr or dirt pimple, metal finishing is acceptable. However, in world-class pressworking operations, extensive metal finishing is not permissible.

Sources of Contamination

Dirt problems with class one or class "A" body panels are a serious quality and line delay factor. Dirt is generated in a number of definable ways, which can be addressed through good stamping process control. Sources of dirt include the following:

- grinding and stoning dust from die repair operations;
- dirty steel from the steel supplier, which is especially a problem with electrogalvanized and galvannealed sheet products;
- loose coating from electrogalvanized products;
- blanking die or blank cut-to-length line causing excessive burr-height problems;
- incidents of contamination with oil absorbents, such as floor-dry-type products—material that should never be allowed in a stamping plant (custodians should stop buying it and the money should be spent on correcting oil leaks and/or buying mops);
- improper blank-stacking techniques that cause the cut edge of a blank to scrape coating material off the blank under it when stacked;
- worn and frayed blank stacking and destacking belts;
- failure to properly wash, oil, and cover draw, form, and redraw dies with plastic sheeting while in storage (when washing any punch, vent-hole tubes should be cleaned and checked);
- when a floating draw punch is used, dirt can be generated by fretting corrosion caused by lateral movement of the inner ram against the punch plate (this is caused by a press problem such as worn or damaged parts or incorrect alignment and adjustment of the press);
- dirt generated in the press, often from wear particles such as bronze;
- steel, iron, or bronze particles from wear plates or wear surfaces entering the die;
- problems with blank washers such as inadequate washing, filtering of the solution, and dirt generated by squeegee rollers (and other parts subject to wear) that contact the blank;
- airborne dirty particles from anywhere inside or outside of the plant;
- particles of galvanized, galvannealed, electrogalvanized, or Zincrometal™ coatings adhering to the working surfaces of draw and stretch-form dies (one way to reduce this problem is by using die materials, lubricants, and coatings that reduce friction; for nodular iron-alloy dies, ion nitriding and chromium plating are the two most popular surface treatments) (Smith 1990, Smith 1994).

Blanking plants and steel suppliers that specialize in producing class one or "A" blanks often have no windows. Locating the plants in areas that are relatively free of airborne dirt is helpful. The air supply into the building is often filtered, and a positive pressure is maintained inside the building.

Finding the cause of dirt involves detective work. Often multiple sources are involved. In some cases, the cause is obvious and easy to identify. For example, if rubber particles are found to cause dirt pimples, suspect the rollers on the blank washer or loader magnetic belts. Here the evidence will be worn rubber parts and particles of rubber.

In extreme cases, spectrographic analysis should be used. Very small dirt particles on a draw or stretch form punch for automotive outer skin panels can cause unacceptable visual defects. Spectrographic analysis identifies the chemical elements that make up dirt particles. This greatly aids the source identification process. For example, copper and tin alone or with lead indicate the dirt is caused by wear particles from bronze bushings and gib liners. If silicon and aluminum are identified in addition to copper, then the wear particles are probably from a bearing made of a silicon-aluminum-bronze alloy, such as AMPCO™, a proprietary bearing material. Silicon and carbon as the main constituent elements tend to indicate silicon carbide abrasive dust. Aluminum and oxygen indicate contamination with aluminum-oxide abrasive dust.

FORMING RIBS AND EMBOSSMENTS

The forming of stiffening ribs, character lines, embossments, and similar features should be accomplished by bending and/or stretching metal. The bending and stretching processes normally require over-forming the metal to compensate for elastic recovery, also known as springback.

If abnormally high tonnage is required to form stiffening ribs and embossments, this is probably because the die's forming surfaces have worn excessively (see Figure 21-4). To compensate for the wear, coining pressures are required to bring the thickness dimension of the rib or embossment up to the yield point of the material.

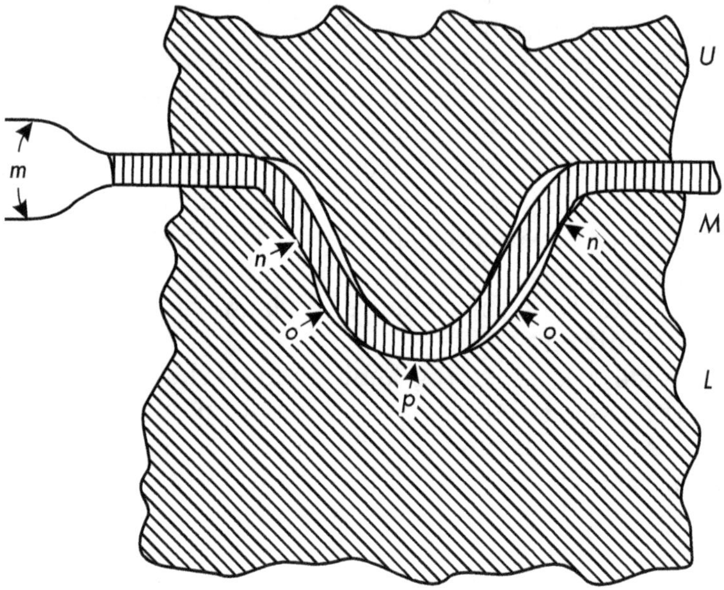

Figure 21-4. Failure to provide and maintain the needed clearance can result in improperly formed beads and excessive tonnage (Smith 1896).

Whenever a quality problem involving gage fit of the angle or height of an embossment or stiffening rib is found, die wear should be suspected as the root cause. If additional tonnage seems to correct the problem, the rib or embossment surface should be examined for shiny coined marks. The quality control department should never specify the tonnage required to get gage fit. The correct action to take is to fix the die. Solutions may include:

- welding up and refinishing the worn surfaces;
- using wear-resistant welding alloys, which may provide increased die life; and
- applying a wear-resistant coating or treatment to the die surfaces if hot-rolled picked steel with only a mill oil lubricant is to be used.

Good nodular iron-alloy dies respond well to flame hardening and ion nitriding to extend wear life. Hard chromium plating also will reduce wear and dies can be easily replated if they wear through. In extreme wear conditions, wear-resistant tool-steel inserts with coatings, such as titanium nitride or carbide, may be required. The cost is much less than the damage that can occur to the press due to an overload condition.

KNOCKOUT BARS

The purpose of knockout bars in the ram is to strip or knockout the part at the top of the press stroke. The part is then removed from the press by means such as an air blast or shuttle unloader.

Usually the knockout bar has captive pins that extend through the upper platen of the press. The bars are usually supported by springs so the dead weight of the bars and attached pins do not cause the part to be ejected prematurely. The knockout pins in the press engage a plate recessed into the upper die shoe. In the case of die sets having punch stems or shanks, a single knockout pin is fitted in the shank.

Adjustment

The usual method of actuating the knockout bars is to attach adjustable-length rods to the press frame or crown that contact the knockout bars at the top of the press stroke. The stationary rods are adjustable in length by means of a threaded extension, which is locked in place with a jam nut. Correctly adjusted, the system provides for positive knockout action.

Avoiding Errors

If the adjustable-length rods do not properly engage the knockout bars, the parts may not be dependably ejected from the upper die. This can result in multiple parts being retained in the upper die. The result is often serious die damage.

If the adjustable length rods are set too long, the rods, knockout bars, and die may be damaged. It is very important that the die setter make sure the jam nuts on the fixed rods are tightened properly. Otherwise, the adjusting screws may work downward, resulting in excessive knockout forces. Should the ram adjustment be raised for any reason, it is necessary to first shorten the adjustable-length rods to avoid damage.

REFERENCES

Smith, D. 1994. *Fundamentals of Pressworking.* "Die Materials and Lubricants." Dearborn, MI: Society of Manufacturing Engineers.

——. 1990. "Adjusting Dies to a Common Shut Height." Stamping Quarterly (spring edition). Rockford, IL: Fabricators and Manufacturers Association, International.

——. 1990. *Die Design Handbook.* Dearborn, MI: Society of Manufacturing Engineers.

Smith, O. 1896. *Press-Working of Metals.* New York: John Wiley and Sons, Inc.

22

Waveform Signature Analysis

In pressworking, a *waveform signature* is a pictorial or graphical representation of relative movements or amplitude displayed as a rectangular coordinate line chart. The *X* or horizontal axis represents units of time, distance, displacement, or degrees of crankshaft rotation. The *Y* or vertical axis usually represents force or amplitude.

STRESS-STRAIN CURVES

Stress versus strain curves or signatures are produced by a tensile-testing machine. Figure 22-1 is a simplified curve of a typical stress-strain signature. Test samples or *coupons* are increasingly loaded until the sample pulls apart.

Tensile-test curve signatures provide important information about the formability and strength of the test coupon. The point on the vertical axis where pronounced elongation or movement to the left occurs is the *yield strength* of the material. The peak-unit stress or applied load is the *ultimate tensile strength.*

The amount of elongation that occurs before fracture is an important measure of material formability. The material yield and ultimate tensile strengths are useful to predict the force required to form the material. The stress-strain curves or signatures for various materials are available in reference publications (Boyer 1986).

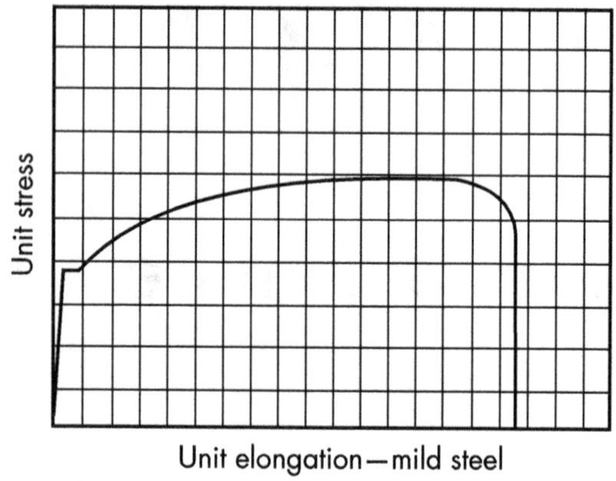

Figure 22-1. A chart shows the stress-strain signature of a test coupon stretched to failure.

Charting Tensile-testing Machine Waveforms

Tensile-testing machines have a load cell directly in line with the test coupon. The load-cell output signal is proportional to the stress or force applied to the sample at any moment.

A sensor provides a signal proportional to the distance the coupon is stretched and measures the strain or displacement of the test coupon. Zero force (stress) and distance (strain) correspond to the lower left-hand corner of the chart.

Tensile-testing machines are of robust construction and operate smoothly. There is very little shock or machine vibration superimposed on the stress-strain waveform chart.

Except for superimposed noise signals, similar stress-strain relationships to those observed with a tensile-testing machine occur when punching or shearing metal. In each case, the process results in a fracture or separation of the material.

CASE STUDY: WEBSTER INDUSTRIES

A customized training program has helped assure active employee involvement at Webster Industries in Tiffin, Ohio. Due to problems with tooling and new technology, a great deal of additional shop floor training was required to improve employee confidence in the procedures. The die setters at Webster who build tooling from standard components now have an excellent insight into the physical principles that underlie successful pressworking. Tonnage-meter readings and chart-recorder data are both trusted to give the needed process-control information, making it easier to consistently produce high-quality parts, while avoiding equipment damage.

Early on, management gave its unequivocal support to the recommended changes in maintenance and manufacturing methods required at Webster. It prioritized training as an immediate requirement. Factors insuring success included all of the following:

- the absolute support of top management;
- analysis of the results of confidential pressroom employee interviews and adoption of the recommended action plan;
- management attended the training sessions;
- there was a skilled, literate work force to train;
- attendance at training sessions was mandatory;
- a project-oriented individual was assigned to act as an in-house facilitator for training and project implementation;
- every employee was required to follow written procedures detailing the correct and safe accomplishment of every production and maintenance task;
- support for management's training objectives existed from the labor union; and
- use of the chart recorder was required to insure that previous setups were duplicated and that there was no excessive snap through during setup.

A portable chart recorder was selected to check every setup to insure that optimal die timing was used. The tonnage-monitor manufacturer offered a standard modification that provided analog outputs. Some manufacturers offered equipment with built-in waveform displays.

To simplify chart interpretation, colored plastic overlays on overhead projector film stock were used to ease the evaluation process. These overlays (see Figure 22-2) allowed workers to quickly evaluate peak positive and negative loads. While the machines are

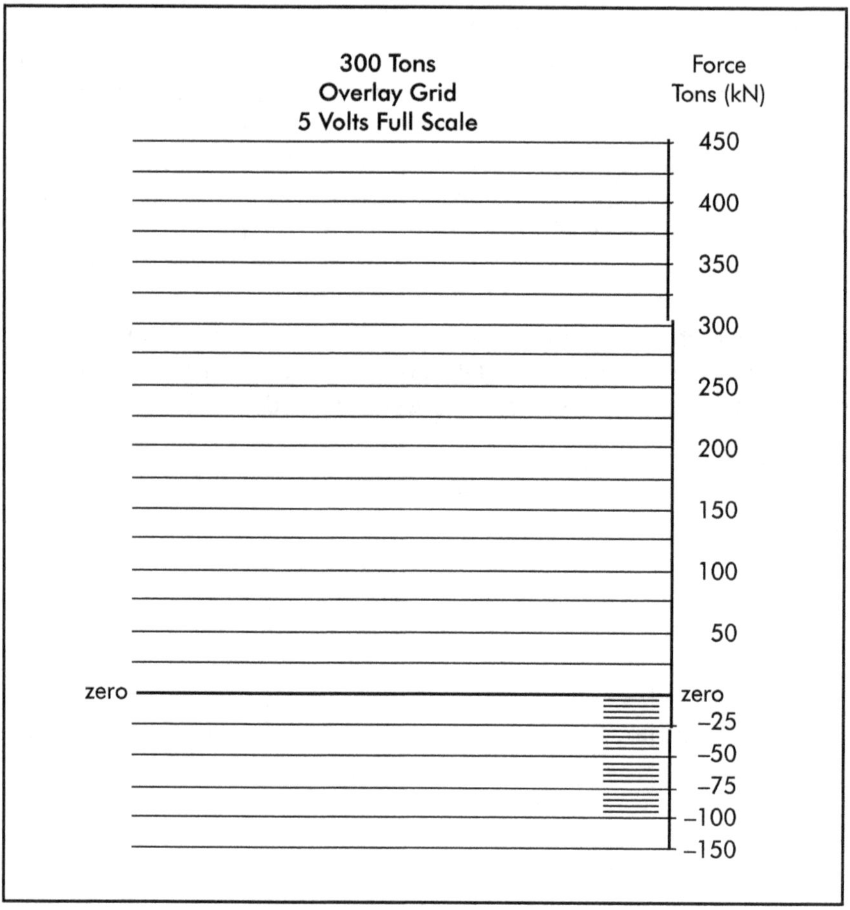

Figure 22-2. Shown is a plastic overlay printed on overhead projector film stock for evaluating peak positive and negative loads.

quite robust, to avoid maintenance problems, the reverse or snap-through load is maintained to within 10% of press capacity.

Chart Recorder Selection

Required bandwidth or frequency response and the ease of use are the two most important considerations when selecting a chart recorder for signature analysis. The nominal frequency response of a tonnage meter is at least 1 kHz.

Electromechanical pen-and-ink recorders typically have a frequency response that extends to approximately 125 Hz. This is sufficient for observing slow phenomena such as conventional drawing, embossing, and coining at low press speeds. For analysis of fast rise-time events, such as those generated by snap-through release, as well as cam and draw-ring impact, a chart recorder with a frequency response of at least 1 kHz is desirable. Lower-frequency responses will not display the magnitude of shock and impact problems.

A two-channel recorder is a minimal requirement for comparison of the coincidence of events on two different channels. An important application in press troubleshooting is making sure that all corners of a slide reach bottom dead center at exactly the same time. Timing problems caused by a twisted crankshaft or a partly sheared key are identified by an A-to-B comparison of when peak tonnage was developed on two or more channels. A four-channel recorder permits observation of four channels simultaneously, which often speeds signature analysis work.

At Webster, the requirements decided on were:

- thermal paper to avoid the mess and clogging problems associated with pen-and-ink recorders;
- at least two channels;
- automatic identification of voltage range, paper speed, date, and time;
- at least 1 kHz real-time frequency response;
- a wide range of speeds and voltage-input levels to permit other uses in the plant such as power-line monitoring;
- portability; and
- battery-powered operation.

An existing oscilloscope with camera was considered, but it was decided that the cost of film to record the amount of data to be gathered would be prohibitive. Further, a chart recorder would permit real-time recording of a number of sequential events.

Digital signal acquisition and storage on a personal computer were also considered. These were ruled out because they would require the operators to be trained in basic computer skills, in addition to signature analysis. Further, the amount of equipment requiring portable operation would be greater.

Upon review of available equipment, a four-channel 25-kHz real-time chart recorder, having all of the required features, was selected. Upon installing recorder interfaces on each of the monitors, and placing the recorder on order, training classes started. An instrument cart with pneumatic tires was also purchased. A protective cover of 0.250-in. (6.4-mm) clear acrylic plastic was fabricated in the tool room. To avoid the need to plug the unit into an AC power source, which could cause ground-loop noise pick up, a 12-volt automotive storage battery was placed on the bottom shelf and connected to the recorder by a cable with a quick-disconnect plug.

The battery-powered recorder is normally mounted on a cart having all the needed supplies. It plugs into the press's electronic-force meter to obtain readings. To speed the analysis, laser-printed, plastic overlays are used to interpret the readings.

Tonnage Monitors

Tonnage-monitor manufacturers today provide instructions for modification of tonnage meters. This allows a worker to obtain analog signals from each channel, as well as the sum channel. The analog direct current (DC) voltage level is 0–4 volts DC for 0–100% tonnage. An Amphenol®-type connector is recommended for the stamping shop environment.

To avoid the need to re-zero the chart recorder each time it moves to a different press, a 4.7-μF DC-blocking capacitor is placed in the signal line from each channel. A larger value of capacitance may be necessary for very slow operations. This value can be calculated by using the time-constant formula, based on the number of seconds a press cycle takes, and the input impedance of the chart recorder.

Long-term Results

Every job at Webster is checked with the chart recorder, and records of all results are maintained. This information is invaluable when responding to rush orders for the hundreds of types of engineering-class chains produced. The following positive results have been achieved:

- The annual cost of perishable-tooling breakage dropped from over $18,000 to $0.
- By optimizing die timing, all jobs could be run at full press speed, resulting in at least a 25% productivity improvement.
- Freedom was gained from the occurrence of double breakage in pierced holes because of faster press speeds.
- Improved setup repeatability resulted from verification of the exact duplication of a previous setup that produced high-quality chain components.
- Optimizations of punch-to-die clearance on all pierce-and-shave operations were enabled by means of waveform signature analysis.
- Enhanced record keeping enabled analysis of the factors needed to duplicate quality part runs.
- The press shop became a quieter place to work due to the snap-through energy being under control.
- There was an elimination of press damage beyond normal wear, resulting in a savings of over $100,000 per year.
- Improved scheduling flexibility occurred, due to freedom from press breakdowns.
- The pressroom was no longer a production bottleneck.
- Throughput of stamping increases every year without the requirement of additional labor or machinery.

DIAGNOSING HYDRAULIC OVERLOAD PROBLEMS

Process variation problems often occur in presses equipped with hydraulic overload systems. Waveform signature analysis is an excellent tool to help pinpoint the causes. Waveform signatures are taken directly from load cells placed on strong supports in the die space. Charting successive hits from the analog-voltage output of a dedicated press-force meter can also show variation from stroke to stroke.

Triple-action Toggle Press Case Studies

In the first case study of a triple-action toggle press with hydraulic overload, the complaint was a serious variation of the depth of the embossed features in automotive inner doors. The machine studied was a 1,000-ton (8,896-kN), 120-in. (305-cm) wide, triple-action toggle press.

The outer slide, or blank holder, and the inner slide have toggle mechanisms to provide dwell at the bottom of the stroke. Both slides have hydraulic overload protection. In the instance of an overload, hydraulic cylinders in the slide of this under-driven machine are designed to dump, providing approximately 0.750 in. (19.05 mm) of additional shut height. Figure 22-3 illustrates a waveform signature taken from one of four 250-ton (2,224-kN) load cells used to test the inner slide.

As the toggle mechanism cycles, the slide dwells on bottom by means of idle points in the driving linkage. A double-hump variation with pressure of equal amplitude during the stroke is normal for this type of slide actuation.

The waveform in Figure 22-3 is abnormal. Reading the illustration from left to right: during the first waveform (a), the cylinder is leaking down under the load, resulting in a loss of force during the dwell portion of the cycle. Near the end of the second cycle (b), the relief valve dumps suddenly. During the third cycle (c), the cylinder is recharging. The pressure is seen increasing during the dwell on bottom.

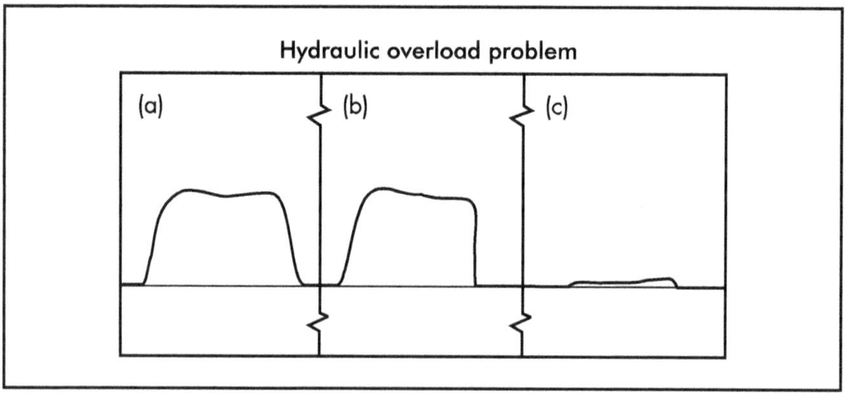

Figure 22-3. A waveform signature illustrates a partial dumping of a hydraulic-overload system on a toggle press during three successive cycles. The result is serious product variation.

Clearly, the force variation is caused by the hydraulic overload problem. This is responsible for the inconsistent depth of part embossment. A number of factors contribute to this. First, the press is several decades old, and any maintenance performed is generally in response to a breakdown. In addition, the lube oil, which is also used to charge the hydraulic-overload cylinders, is often not changed unless the press is torn down to repair a broken part. The recommended anti-foaming additives are not used. Contamination with water-based drawing lubricants is another contributing factor to inconsistent depth of part embossment. Another may be that the cylinder packing used is the least costly obtainable, rather than what the manufacturer recommends. There are many other related causal factors responsible for the pressure-variation problem and the overload system tripping erratically.

The most serious maintenance error found is a jumper that defeats the limit switches signaling the press to stop if an overload dumps. This can result in unsafe press overloading, as well as machine and die damage. In this case, a direct result is the production of thousands of dollars in scrap parts, some that may eventually be assembled into automobiles.

In this case study, portable instrumentation was used to highlight the problem. The overall root cause, however, was the poor maintenance management practices and the lack of systematic employee training in machine maintenance skills.

In the second case study, the same model and size of press showed ample evidence that it had been abused by overloading and exhibited lubrication failures. Figure 22-4 indicates these problems. It was made with a pen-and-ink, two-channel recorder.

In this particular case, management tolerated the practice of jumpering important safety systems, such as slide-position limit switches, lube-pressure switches, and hydraulic-overload dump switches.

Figure 22-5 shows an example of a severe hydraulic overload leakage or bleed-down problem. Note the pronounced tilt shown as the inner slide is on dwell. The probable cause was a cylinder-packing failure or dump-valve leakage. This machine also shows a severe misalignment problem (see Figure 22-6). The punch plate was shimmed 0.105 in. (2.667 mm) on the left side to uniformly set embossments in automotive package-tray and inner-door panels.

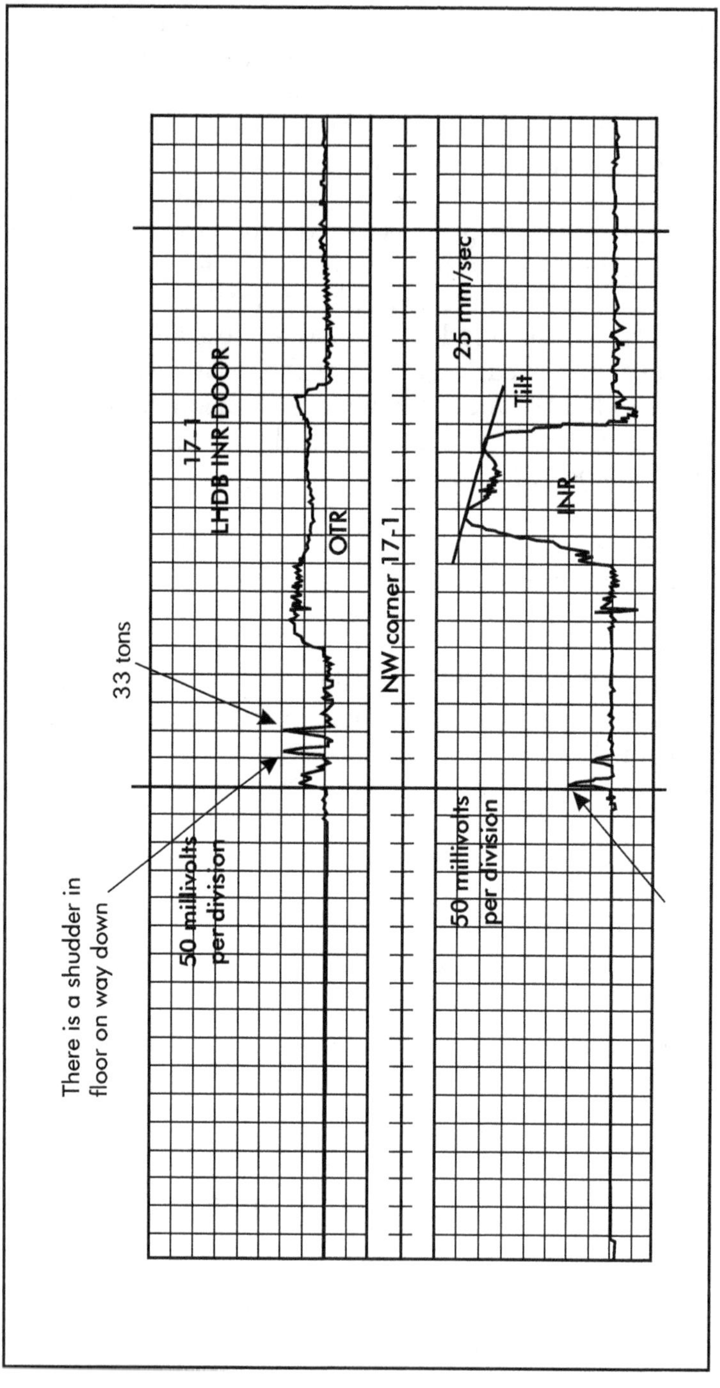

Figure 22-4. Waveform shows a number of mechanical and hydraulic overload problems.

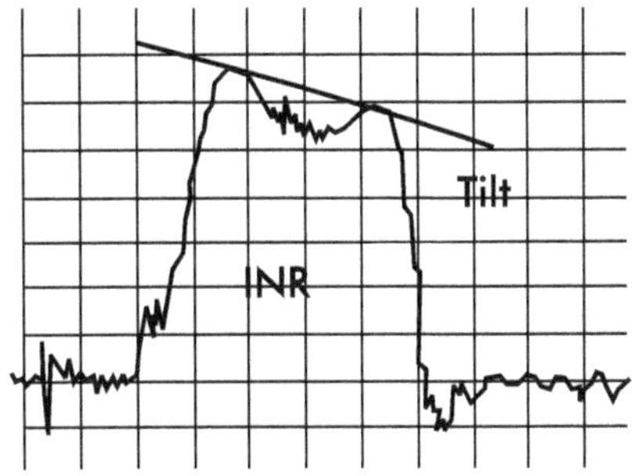

Figure 22-5. A severe hydraulic overload leakage or bleed-down problem is evidenced by the tilt shown as the inner slide is on dwell.

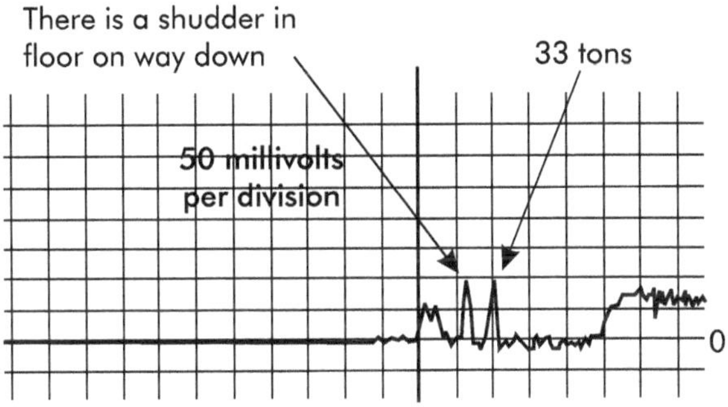

Figure 22-6. A severe mechanical bind in the press gibbing occurs due to press misalignment.

PRESS-TONNAGE CURVES

Large, conventional, double-action drawing presses, used to stamp automotive body panels, are designed to develop full force at a short distance from the bottom of the stroke. The blank holder is typically designed to exert full tonnage at only 0.250–0.500 in. (6.35–12.7 mm) above bottom dead center (BDC).

Force-monitoring Equipment

Figure 22-7 illustrates the waveform of an inverted stretch-form die running in a conventional double-action press. Both the inner- and outer-slide signatures are shown, with the press-tonnage curve limits superimposed (Mettert 1990).

The press-tonnage curve limits of both the inner and outer slides are compared to the force of each slide at many points during the press cycle. This is accomplished as follows:

- The tonnage curve data supplied by the press manufacturer, in terms of maximum force versus distance from BDC, is programmed into non-volatile digital memory in the press tonnage or force monitor.
- A rotary resolver connected to the press crankshaft sends angular position data to a specially adapted press tonnage or force monitor.

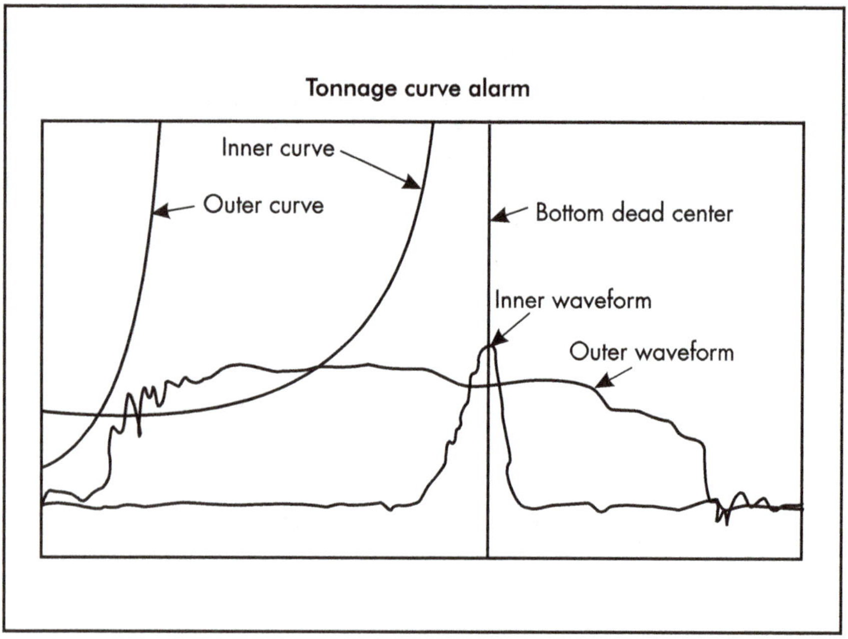

Figure 22-7. Shown is a double-action press waveform of an inverted stretch-form die. Both the inner- and outer-slide signatures are illustrated with the press-tonnage curve limits superimposed. (Courtesy Toledo Transducers, Inc.)

- Force data taken from strain gages or sensors on the press-pull rods, pitmans, or eccentric straps is converted to an accurate force value for both the inner and outer slides.
- When the force throughout the critical die-closure portion of the cycle violates the maximum value specified by the press manufacturer, an alarm signal is sent to stop the press.

In the event that a force or tonnage-curve overload condition is detected, waveform analysis is used to pinpoint the problem. A display such as the one illustrated in Figure 22-7 is acquired and shown on the screen of a portable personal computer. This data can be stored on a magnetic disk for detailed analysis or it can be hard-copy generated at the press on a portable printer with graphics capability.

Press Controls with Waveform Signature Readout

Figure 22-8 illustrates the operator terminal of a press control system with an integrated force meter. In addition to press and auxiliary equipment-control functions, the force-meter portion of this system has useful capabilities that include:

- a built-in screen with waveform-signature capabilities;
- selectable high limits at a number of crankshaft angular positions in the press cycle, and
- a reverse-load alarm feature that is capable of being preset.

A system of this type is especially useful for progressive die and other combined operations where cutting, snap through, and forming forces are present. The alarm-window set points can be stored by job number to speed up die-setting operations.

Plant-wide Monitoring Capability

The Toledo Transducers waveform signature analysis system performs critical tonnage-curve monitoring. In some plants where this system is installed, the waveform signal is collected from presses plant-wide by daisy chaining all monitors together with data-transfer cable by means of the industry standard RS422 data protocol.

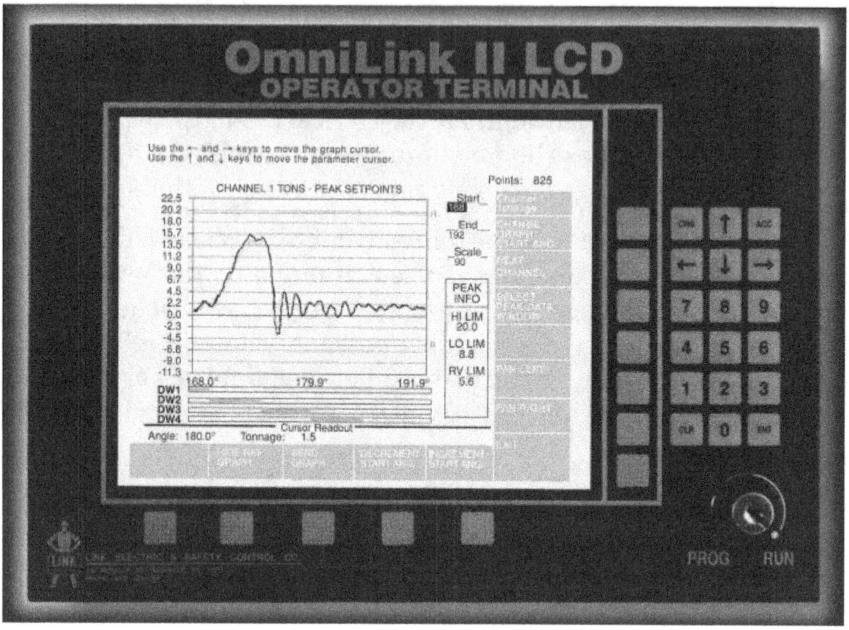

Figure 22-8. The operator terminal of a press-control system with an integrated force meter has a built-in, flat-panel screen that provides waveform signature display in color and multiple-set-point alarm capability. (Courtesy Link Systems, Inc.)

The RS422 cable is terminated in a computer, which can perform plant-wide monitoring. If the computer is configured to act as a file server, the waveform signal can be observed at any computer as long as it has the copyrighted Toledo Transducers software and is connected to the plant's digital communications network.

Press force monitors with remote monitoring capabilities are available from several manufacturers. Force data and waveform signatures can be exported to remote monitoring locations. While the technology is quite feasible, the desirability and cost effectiveness of gathering such data in a central-command center is not universally accepted as a useful management tool (Smith 1988; Smith 1991; Smith 1993).

Some managers believe force-monitoring equipment is used solely to prevent press-overloading damage. Experience has shown that efforts to use the meter for this purpose alone usually fail.

The typical result is that the equipment is disabled and then unable to be used. Properly used, the tonnage or force monitor is a tool for the pressroom work force to set up, troubleshoot, and control the stamping process.

Correction Measures

If a tonnage-curve violation is found, one or more of the following corrective measures should be taken:

- The engagement of nitrogen cylinders can be stepped to lessen the initial force.
- If the press is equipped with a constant-energy drive, the press may be slowed down on initial draw-ring contact. This will lessen problems with bounce or rebound.
- Die modification to lower the draw ring should be carried out if the ring has more travel than necessary.
- A press with a better force, versus a better distance from bottom, may be used if available.
- The press may be modified, in some cases, by installing stronger parts to increase the tonnage-curve capability.
- A hydraulic press that has no tonnage-curve limitation can be used.

DETECTING WAVEFORM CHANGES

There is a patented method available that measures small deviations occurring within a repeating series of waveform signatures. The simplest forms of tonnage monitors provide only a peak-force reading and an electrical output that can be used to stop the press if a preset force is exceeded and/or a press overcapacity condition occurs.

The method establishes process-control boundaries, which function as high and low, and alarm set points around the entire waveform signature. Figure 22-9 illustrates the press-force signature running properly within a mask of control limits.

Figure 22-10 illustrates discrete signature points and corresponding high and low limit points. The data-sampling rate depends on

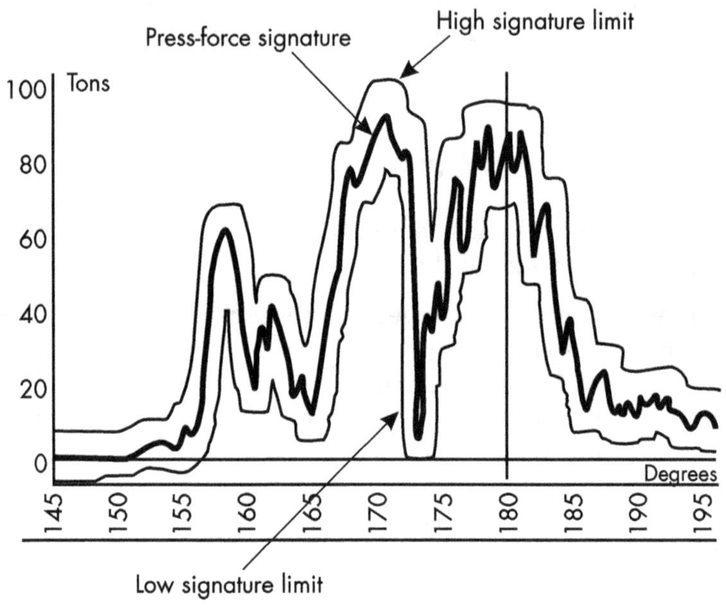

Figure 22-9. A patented signature analysis system plots within mask or control limits with both high and low signature limits. (Courtesy Signature Technologies, Inc.)

the machine speed and amount of detail in the operation being performed. As shown, acquisition resolution of individual data points within 0.022° is possible, depending on the signature window size and press speed. In addition to trip or machine shutdown functions, additional high- and low-limit points can be provided to warn of changes in process operating parameters that indicate a drift of any of the signature points.

The system features a color computer-monitor display. The signature, together with the warning and limit points, are shown with machine counts and many other types of useful data.

Figure 22-11 indicates an alarm condition resulting from a signature fault on the left (downstroke) side of the signature. Note the record of the conditions of the fault displayed on the left side of the screen.

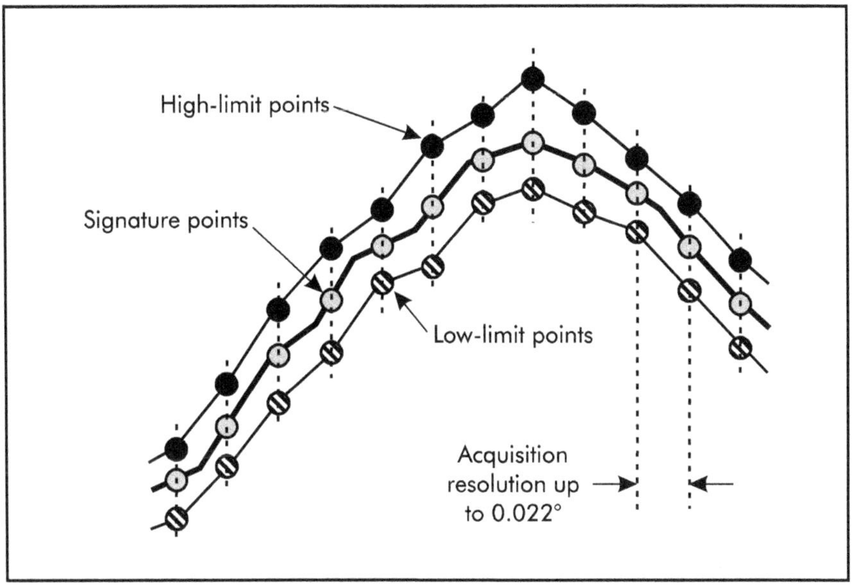

Figure 22-10. Shown are discrete signature points and corresponding high- and low-limit points. The data-sampling rate required depends on the machine speed and complexity of the signature. (Courtesy Signature Technologies, Inc.)

DIE- AND PRESS-PROTECTION SYSTEMS

Some force-monitoring systems provide alarm outputs for additional conditions, which include in part:

- excessive amounts of reverse or snap-through load;
- violation of available press tonnage as a function of crankshaft degrees of rotation or of distance from the bottom of the stroke;
- failure to achieve a preset minimum force, and
- exceeding a preset load within a user-definable portion of the pressworking stroke.

Several types of die and press protection systems have been in use for many years. In addition to force monitors, the use of hydraulic-overload systems, stretch-links, and shear collars are employed.

Misfeeds and mislocated stock in progressive die and transfer press applications can cause tool damage when the overload set

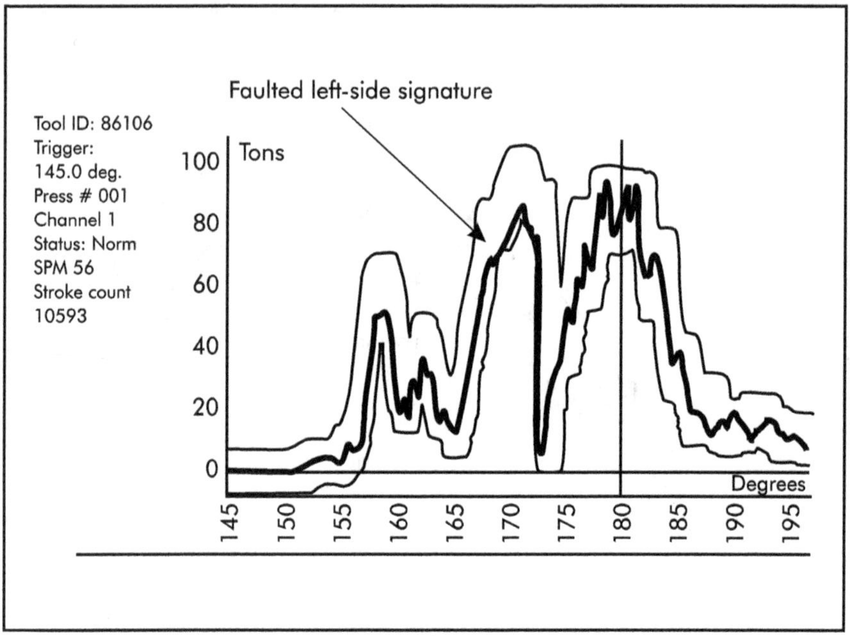

Figure 22-11. Shown is an alarm condition resulting from a signature fault. Note the record of the conditions displayed on the left side of the screen. (Courtesy Signature Technologies, Inc.)

point on a force meter is tripped. In many operations, just one bad hit can damage the die. The result is an increase in product variation, and a need for prompt bench repair. Double hits involving very heavy stock can result in catastrophic press and die damage.

Rapid energy release, typical of snap through at the completion of heavy cutting, does not necessarily cause a negative load on the press. The press's structure stores energy as a form of deflection. This avoids the sudden release, even though a forming load is in process through optimal die timing of the cutting load. Figure 22-12 illustrates a rapid release sufficient to damage press parts.

Hydraulic Overload Protection

Hydraulic overload systems are designed mainly to protect the press. The trip points are often adjusted to loads of 50% or more of press capacity.

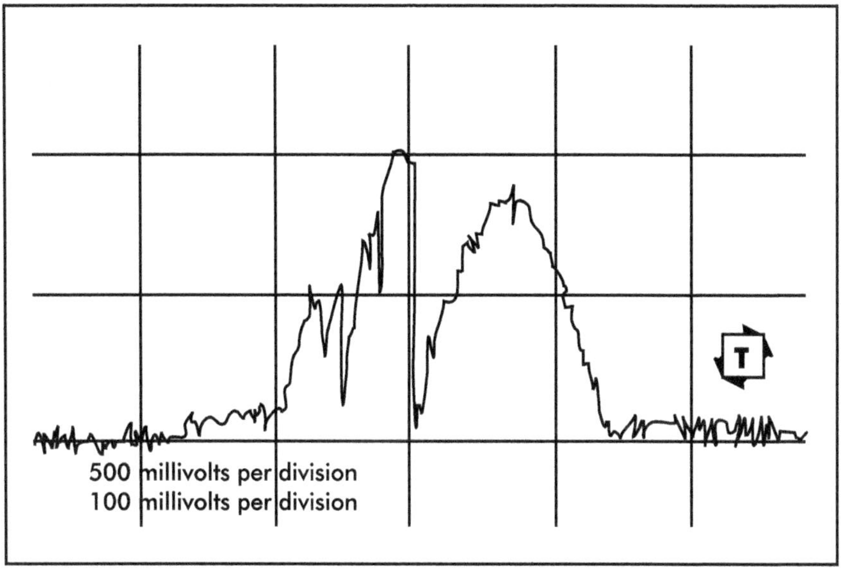

Figure 22-12. Shown is an example of sudden snap-through energy release occurring at the start of a large forming load. Although the sudden downward excursion does not become negative, the energy release damages the press. (Courtesy Toledo Transducers, Inc.)

The value of hydraulic overload systems is undisputed. Indeed, such systems should be specified wherever possible for gap-frame and solid-frame straight-side presses that are liable to become stuck on BDC. However, high forces may cause expensive damage to delicate tooling even if press damage is avoided.

Shear Collars and Stretch-links

Both straight-side and gap-frame presses may incorporate a replaceable steel shear ring. The ring has a stepped diameter and is placed under each connection. One side is machined out to provide an area that will shear under a press-overload condition, causing the ring to collapse.

Under-driven presses may incorporate a stretch-link in line with each pull rod. This device is shaped like a dumbbell used for calisthenics. It is retained by split collars, which are designed to facilitate replacement.

Shear collars and stretch-links provide a degree of protection for the press. However, because they are subjected to large cyclical loads, the force required to deform the device changes with time. There is a serious problem with this protection on presses with multiple connections. If only one device yields, the slide will be severely tipped out of alignment. This may not be noticed until severe scoring of the gibbing and/or die damage has occurred.

CONCLUSION

In the author's experience, training *all* pressroom personnel on the fundamentals of pressworking operations is a part of the process of improving efficiency. This training must impart knowledge of how presses, dies, and auxiliary equipment work. Employee training that leads to increased responsibility is the goal. In many cases, it is far better economy to train employees to control the process at the press, than to have the ability to transmit and analyze the data at a remote location. The latter approach often is wasteful because it does not add value to the product.

REFERENCES

Boyer, H. 1986. "Atlas of Stress-strain Curves." Materials Park, OH: ASM International.

Mettert, B. 1990. "Load Sensor Placement and Tonnage Data for Underdrive Presses." Technical Paper MS90-384. Dearborn, MI: Society of Manufacturing Engineers.

Smith, David. 1988. "How to Improve Hit-to-hit Time with a Tonnage Monitor." Technical Paper TE88-780. Dearborn, MI: Society of Manufacturing Engineers.

——1990. *Die Design Handbook.* Section 4, Shear Action in Metal Cutting. Dearborn, MI: Society of Manufacturing Engineers.

——1990. "Using Waveform Signature Analysis to Reduce Snap-through Energy." Technical Report MF90-11. Dearborn, MI: Society of Manufacturing Engineers.

——1991. *Quick Die Change*. Chapter 28, Instituting a Tonnage Meter Program. Dearborn, MI: Society of Manufacturing Engineers.

—— 1993. "Visual Indicators in the Workplace." *The Fabricator*, January-February. Rockford, IL: Fabricator's and Manufacturer's Association International.

23

Repairing Damaged Die Steels and Castings

Gray iron, as well as iron alloy and steel die castings, occasionally become cracked or broken. Typical causes of failures are careless handling, die-setting shut-height errors, and multiple parts in the die due to part-ejection failures. Cracked or broken castings must be repaired quickly. Most dies subject to this type of damage are of the type used to produce large, irregularly shaped parts for the automotive and appliance industry.

Dies with large special castings are usually unique with patterns that are almost universally of the lost Styrofoam™ type, since only one casting is required. This type of pattern is destroyed during the casting process. If a replacement casting is needed, a new pattern must be made. Typical casting weights are from a few hundred pounds to 20 or more tons.

MAKING THE REPAIR DECISION

When a mishap results in a broken die casting, a decision must be made quickly and a plan must be acted on immediately to start repairs. Factors to consider when making a repair decision include:

- the base metal of the casting and its suitability for electrical-arc welding or oxygen fuel-gas-brazed repairs;
- welding and brazing processes' requirement of careful preheating, post heating, and often some remachining and refitting to

correct distortion—this takes time to do properly and insure a lasting repair;
- the nature of the damage, and if mechanical repairs, such as bolting the die shoe to a backing plate and the use of a tie-rod and/or repair plate fitted into a machined pocket, known as a Dutchman, may be used;
- if a new casting is required to replace a damaged one, even if emergency repairs are possible, and
- the thickness of the casting walls and webbing, which may dictate the repair procedure, especially in lightweight designs.

Lasting Repairs

The wall thickness and size of reinforcing ribs of the casting are important factors to consider when making repairs. Heavy sections permit long-lasting mechanical repairs that would be difficult on a lightweight cost- and weight-saving casting. However, the lighter casting is easier to preheat and requires less filler metal than a casting with thick sections.

The lead time for making a large new casting in an emergency situation is a minimum of approximately two weeks, if the necessary drawings for making the pattern are available. Usually a considerable amount of machining is required on a large casting. In addition, parts such as wear plates and other details must be attached. Fitting to existing parts must also be accomplished. Finally, the assembled die must be tried out to insure it functions properly and that the parts produced meet all specifications.

If a successful mechanical or welded repair can be made, there is little reason to order a new casting. Mechanical repairs in dies that are subject to die-setting errors or multiple-part damage may fail if an overload occurs. The bolts, tie-rods, and Dutchman used to make the original repair will tend to stretch or break. This will limit further damage to the casting. Thus, if the cause of the failure recurs, as is often the case, it is much easier to repair the second time. This is typically accomplished easier than fitting a new shoe casting, even if a casting was ordered as a spare when the previous failure occurred.

MECHANICAL REPAIR TECHNIQUES

Figure 23-1 illustrates three techniques useful for the repair of broken cast die shoes:

1. Tie-rods tensioned with SUPERBOLTS™ (see Figures 23-2, 23-3) are used on both side-through openings in the casting ribs.
2. A securely attached steel plate covers the entire area of the upper-die shoe.
3. A Dutchman is applied to the casting wall to assist in holding the break securely together under load.

As shown in Figure 23-1, the die is inverted or in an upside-down position on the floor. The die is a trimming die. Dies of this type are used for trimming formed, irregularly shaped, sheet metal parts. Any number of reasons could cause this crack. Typical causes of the failure are careless handling, die-setting shut-height errors, and multiple parts in the die due to part-ejection failures.

Figure 23-1. A large die for trimming formed, irregularly shaped, sheet metal parts is shown. The die is in the inverted position. The die shoe resting on wood timbers is the top shoe. Note the steel plate added to reinforce the top of the shoe. In addition, there is a tie-rod running the length of the die to hold the two broken halves in compression. (Courtesy SUPERBOLT, Inc.)

Figure 23-2. End view is shown of tie-rod tensioned with a SUPERBOLT. Note the use of a thick steel plate to spread the load evenly. (Courtesy SUPERBOLT, Inc.)

Figure 23-3. Close-up view is shown of the SUPERBOLT tie-rod tensioner in Figure 23-2. This is used to pretension a tie-rod on each side of the die shown in Figure 23-1. Note the tightening sequence marked on the tie-rod end. (Courtesy SUPERBOLT, Inc.)

Preparation of a Cracked or Broken Casting

Gray cast irons vary greatly in tensile strength. They are produced in ASTM classes 20, 25, 30, 35, 40, 50, and 60, with each class corresponding to the minimum tensile strength in thousands of pounds-per-square inch (ksi). This range spans the metric-equivalent tensile-strength range of approximately 138–414 MPa.

Other cast-die materials include the ductile and nodular irons and steel. These materials, as well as the higher-class gray irons are subject to elongation prior to fracture. Elongation, especially in the irons, tends to be localized in the area of the fracture. Elongation can cause substantial distortion of the casting in some cases.

However, in most cases, the stresses leading to failure are concentrated in one or more areas of the casting. There are two casting failure modes; each requires a different repair preparation. These modes are:

- Repeated cyclical strains initiate and result in failure through gradual crack propagation. In these cases, there is negligible or very slight elongation or distortion in an area other than the area of failure.
- Suddenly exceeding the yield and ultimate tensile strength of the casting material can be evidenced by a rough tearing of the metal that is localized to one or more areas. These rapid failures are likely to result in some deformation of the casting, especially in areas near the fracture.

Casting Distortion

If elongation or distortion of the cracked or broken casting is suspected, it should be placed on a surface plate to permit measurements of shoe flatness and correct dimensions. Large heavy castings may be tested for fit, as well as dimensional accuracy, by placing parts on the bed of a large milling-machine table to determine how accurately they nest or fit together.

If the crack is a gradual one extending partway through the casting, one of several mechanical repair methods may prove suitable to prevent further crack propagation and movement in the cracked area known as *breathing*.

In sudden-overload fractures, there are usually projections of torn metal that can prevent the casting from being fitted back together for mechanical repairs unless the projections are carefully ground away. In lower-grade gray irons, there is usually a clean fracture with little elongation. Carefully grinding the high points may provide proper preparation for a mechanical repair. However, if there is elongation or distortion of the casting, a hand-grinding process is often the best way to fit the broken pieces together in correct alignment. Skillful work is required to fit the parts together correctly. This work is best done by hand, using Prussian or spotting bluing to achieve a close fit. Because of the irregular fracture paths, machining is not a good way to fit the pieces together.

Using Tie-rods to Hold Broken Castings in Assembly

Tie-rod Materials

The most commonly used steel shafting for making die-repair tie-rods is AISI 1018 cold-drawn steel. This material has a yield strength of approximately 45 ksi (310 MPa), depending on the extent of cold working. Generally, fasteners including tie-rods are pretensioned to no more than 70% of the yield strength.

Other materials are available with higher tensile strengths, such as cold drawn AISI 1040, and 1140 material, with tensile strengths in the 75–90 ksi (517–620 MPa) range. Small diameters tend to be cold worked more severely and, therefore, attain higher tensile strengths. Another readily available material is AISI 4140, which can be heat treated to achieve yield strengths in excess of 140 ksi (965 MPa).

The maximum straight-line compressive loading is approximately three times the tensile strength for class-20 material. It is twice the tensile strength of the class-60 material. Therefore, tie-rods can be preloaded to high tensile forces to hold a broken casting in compression. However, a substantial reduction in preload must be allowed if there are ribs and voids, which can deflect or fail under compressive loads.

Pretensioning Tie-rods

Heating can be used to pretension the tie-rods shown in Figure 23-1. This is done by using large, oxygen-fuel gas torches applied according to engineering formulas similar to those used for pretensioning press tie-rods (Smith 1994). It is workable in this case because the tie-rod is exposed for heating. However, an error resulting in over-tensioning could break the end ribs.

The following data is based on a linear coefficient of expansion for carbon steel of 12 times 10^{-6}, per °C. Carbon steel expands approximately 0.00000667 in. per °F, which is 0.00008 in. per foot, per °F. To attain the preload of 20 ksi (138 MPa) typically used to pretension press tie-rods, the tie-rod would need to be uniformly heated to 105° F (41° C). This would expand the tie-rod by 0.0007 in. per inch, which is 0.0084 in. per foot. Uniform heating of long die-repair tie-rods is difficult at best.

Other heating approaches can be used if the amount of pretension preloading is accurately known in advance. The preload must be enough to securely hold the broken casting together without causing excessive deflection or any risk of breaking ribs or thin sections. The rod can be heated to a higher temperature than necessary and a conventional nut can be used to hold the casting in the correct compression. This is done by turning the nut a calculated number of degrees or distance. This procedure is very unforgiving, especially when a heated tie-rod must be placed in a long drilled hole through solid metal.

To accurately control the amount of preload and to add stiffness to the assembled die, a tubular spacer may be placed around the tie-rod. The spacer should be 1.5 times the size of the area of the tie-rod. In addition to adding stiffness, the spacers reduce the alternating load in the tie-rod thread. The tubular spacer must be made slightly longer than its final compressed length. Providing a short section of tubing on one end of the spacer is a good idea. It can be easily removed for shorting during fitting to save time. In the event that a longer spacer is needed, it is easy to supply and fit one.

As shown in Figures 23-2 and 23-3, the SUPERBOLT is used much like a hydraulic tie-rod nut to pretension the tie-rods used to hold the broken die shoe together. However, unlike a conventional tie-rod nut, no heat or hydraulic pressure is required to pretension the device. Figure 23-4 illustrates an isometric cut-away view of the SUPERBOLT. The screws known as *jackbolts* are tightened in a specified sequence with an accurate torque wrench.

The amount of actual tie-rod tension can be applied to a close tolerance by knowing the amount of torque applied to the jackbolts. It is important to use the lubricant specified by the manufacturer to achieve a known value of jackbolt-torquing tension.

An important feature is being able to adjust and easily remove the die-casting repair tie-rod. Tension can be easily checked and adjusted in the production press. The SUPERBOLTS are acces-

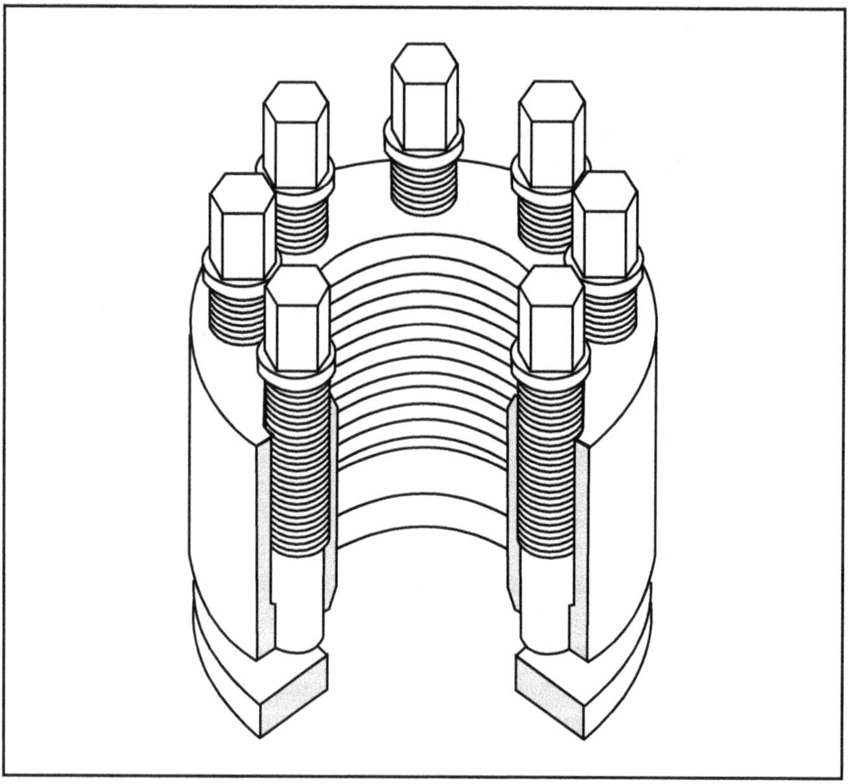

Figure 23-4. Cut-away view of the SUPERBOLT tie-rod tensioner shown in Figures 23-2 and 23-3. Note the placement of a hardened-steel shim plate under the tensioning device supplied by the manufacturer. (Courtesy SUPERBOLT, Inc.)

sible. Be certain that the proper lockout procedure is followed when working in and around power presses.

The repair illustrated in Figure 23-3 did not require the full torque capacity of the SUPERBOLTS and the tie-rods. The final torque value can be applied in steps until full crack closure is assured under normal operating forces as the press is cycled.

SUPERBOLTS have been used a number of times with successful results for tensioning die-repair tie-rods. The cold-tensioning method has many advantages when compared to the uncertainties of heating tie-rods to obtain tension.

Adding a Steel Plate to Reinforce Broken Die Shoes

Figure 23-1 illustrates a steel plate that is approximately 2-in. (50.8-mm) thick and bolted to the cracked upper-die shoe (shown in an inverted position). A safety feature of this portion of the repair is that the die shoe has bolt tie-down slots that line up with those of the broken shoe. This permits the bolts that attach the upper-die shoe to the press ram to hold the sandwiched plate-and-shoe assembly in compression.

The plate should be doweled to the pieces of the broken die shoe for locational fit only. Avoiding lateral movement of the plate relative to the shoe pieces cannot be assured by the use of dowels alone. Dowels are for locational reference only.

If the clamping force of the screws to prevent movement by friction between the shoe pieces and the plate is insufficient, then rectangular keys should be used in tightly fitted machined pockets.

It is important that the press ram impart no lateral movement to the upper die. This is especially true of one that has been repaired. Lateral movement can occur if the press is not aligned properly. This is especially a problem on twin-end-drive geared presses that are out of synchronization.

Applying Clamping Force to a Crack with a Dutchman

Figure 23-5 illustrates a clamping device for repairing die castings known as a *Dutchman*. As shown, it is shrink-fitted with heat.

Figure 23-5. Shown is a steel plate with projections on each end that mate with milled pockets on either side of a crack in a die casting. The device retained with screws is a Dutchman. It is shrink-fitted in place by heating and tensile preload develops as it cools to ambient temperature. (Courtesy SUPERBOLT, Inc.)

The same thermal-expansion formulas discussed earlier are used to determine the correct preload.

A seldom-used alternative is to provide a calculated draft angle in the milled pocket and mating projections to obtain the required preload without the application of heat. This procedure also permits easier removal.

It is suggested that additional screws be used to fasten the Dutchman in place, especially if the cold process is used. A staggered screw pattern should be used to spread the load and avoid further weakening of the casting.

Other Cold Crack Repair Techniques

There are several methods for inserting repair blocks into holes machined into pockets at the location of a crack or fracture. Hand drills and chipping hammers are often used to cut the pocket. Steel blocks are then fitted into the pockets. The repair blocks are firmly held in place by threaded studs that are retained by drilling, tapping, and screwing in place, and by peening for retention at the juncture of the repair insert and the pocket.

Another method makes use of many screws placed directly in the break in the casting at an angle to pull the crack together. Since this requires considerable skill and experience to carry out successfully, extensive specialized repairs of this type are best left to contracted outside experts. Unless die castings are broken very frequently, the necessary skills are not likely to be learned by many repair persons. The repair providers who specialize in casting repairs of this type are usually willing to do the work at the client's plant on short notice in an emergency.

As a common-sense rule, careless practices that damage die castings and the care, patience, and skill needed to carry out successful repairs are unlikely to be found in the same shop. A plant culture that tolerates carelessness, thus fostering the need for highly specialized repair skills, is a contradiction in expected management practice.

ELECTRICAL WELDING REPAIRS

In die welding, wear-resistant alloys applied to the surface of dies increase service life and avoid downtime to rebuild or repair dies damaged in service. Welding is useful to correct machining errors and to increase the wear resistance of die surfaces. Varying degrees of hardness, toughness, and wear resistance are available in welding alloys, depending upon the application.

Three electrical welding methods are used for die repair. The American Welding Society classifies them using the following lettering system:

- SMAW—shielded-metal-arc welding or stick welding;
- GMAW—gas-metal-arc welding, also known as MIG; and
- GTAW—gas-tungsten-arc welding, also called TIG or heliarc welding.

SMAW or stick welding is the most popular method since it is extremely versatile. This method also offers a wide selection of filler metals. The GMAW or MIG process is usually superior in terms of speed and quality, however. The filler metals are available as either solid or tubular wire. The tubular-wire form is available in the widest range of alloy selection.

The GTAW or TIG process offers the operator optimum control and inclusion-free welds. The selection of bare-metal filler rods is somewhat limited.

Steel Dies

When the base metal for an intended application is carbon steel, there are no particular restrictions in the choice of an alloy tool-steel welding material (Smith 1990; Smith 1994). There are, however, some restrictions with other base metals, including stainless steels. Tool-steel welding requires precautions similar to those necessary with any other type of welding process. This will prevent cracking of the base metal during heating and cooling.

Water-hardening, oil-hardening, and air-hardening steel are the tool steel groups most commonly repaired by welding. The degree of welding difficulty is dictated by the alloy content of the base metal. Usually steels with a high-carbon content require higher preheats before welding, more care during welding, and consideration of tempering temperatures after welding.

All base metals with carbon content higher than 0.35% should be preheated and postheated to decrease brittleness in the base metal near the heat-affected zone. High-carbon alloy steels, such as air-hardening tool steels, are more difficult to weld because of the likelihood of cracking. Tool-steel welding is used successfully in most instances for the repair of tools made of air-hardened steel. This type of repair is successful for hot-extrusion tools.

Welding Materials

Tool-steel welding materials are normally heat-treatable alloys. Alloys for SMAW (stick) welding have a wire core covered with a flux coating, which may also contain alloy constituents, including rare earth elements. Alloying elements in the flux coating pass across the arc and form a homogeneous weld deposit.

GMAW arc-welding alloys for semi- and fully automatic MIG welding are available as tubular and solid wire products. This process requires an external shielding gas. Additional alloying elements can be incorporated in the tubular flux. Solid MIG welding wire does not provide additional alloying.

GTAW (TIG) alloys are made of bare wire cut to 36-in. (914-mm) long. Small-diameter sizes are available that permit the operator the utmost control for precise buildups and contours. All basic tool-steel alloys are available.

Cast-iron Dies

Cast-iron dies can be welded, but the method of application of the weld differs from that for steel because cast iron melts at a lower temperature. Nickel welding alloys with lower melting temperatures are recommended for underlay on all cast irons.

Generally, after careful preparation involving grinding out the area to be repaired and preheating the work, a deposit of nickel-alloy weld is made, followed by the welding material required to obtain the needed surface properties. Where a very high-strength bond with the cast iron is needed, the iron can be studded with mild-steel threaded inserts. An easy way to do this is to drill and tap the iron casting for mild-steel cap screws. The screws are tightened and then cut off flush with a cold chisel. It is very important that alloy-steel screws are not used, because a brittle weld may result. This method is excellent when repairing old, oil-soaked cast iron.

Welded-trim Edges

The dies used to trim large, irregularly shaped drawn panels are made of gray cast iron. The cutting steels on the punch shoe are either of cast-alloy steel or wrought composite tool-steel construction. A gray-iron casting called a post is mounted to the lower-die shoe. The part to be trimmed is placed on the post. The post has a milled ledge or pockets where the tool-steel lower-trim inserts are fastened as shown in Figure 23-6.

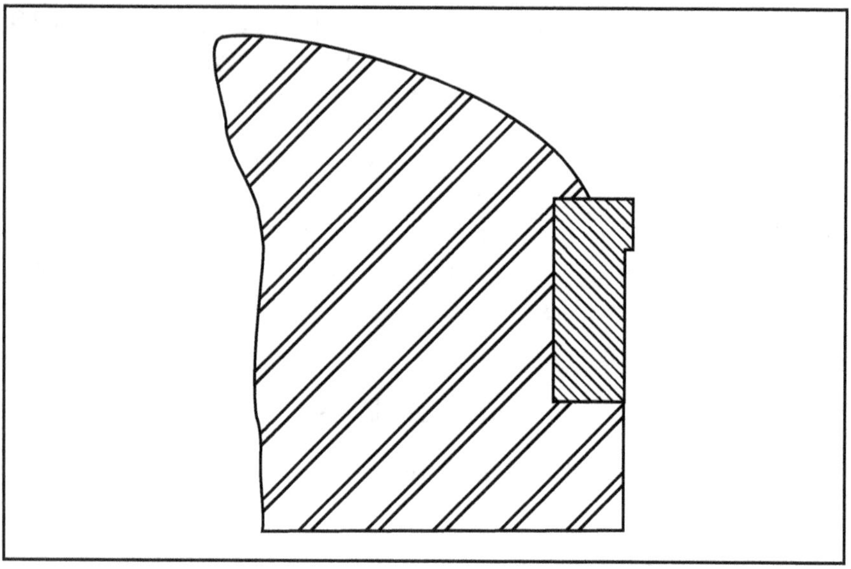

Figure 23-6. A sectional view of a conventional cast-iron trim-die post with a tool-steel insert fitted into a milled pocket. (Courtesy MPD Welding, Inc.)

An alternative construction method is to eliminate the post inserts and provide a tool-steel cutting edge on the gray-iron post by welding as shown in Figure 23-7. Success in producing a cutting edge by the welding method depends on adherence to strict procedures in preheating, the selection of welding materials, welding technique, peening, and cooling.

A U-groove is first ground or machined around the post. This is readily done on the CNC-profile mill used to machine the finished contour of the post. A vertical land is provided as a reference of the trim-line location for finish grinding after welding.

To avoid producing a layer of brittle, white, chilled iron under the weld, the casting is preheated to 750° F (399° C). A good method for heating is to place the casting on a special table fitted with a number of small gas burners. To conserve heat, the casting is covered with mineral-fiber blankets except for the area being welded.

Temperature control can be provided by a thermocouple probe, which can also automatically regulate the burners. Many skilled operators control the burners manually and determine the heat with temperature-indicating marking sticks.

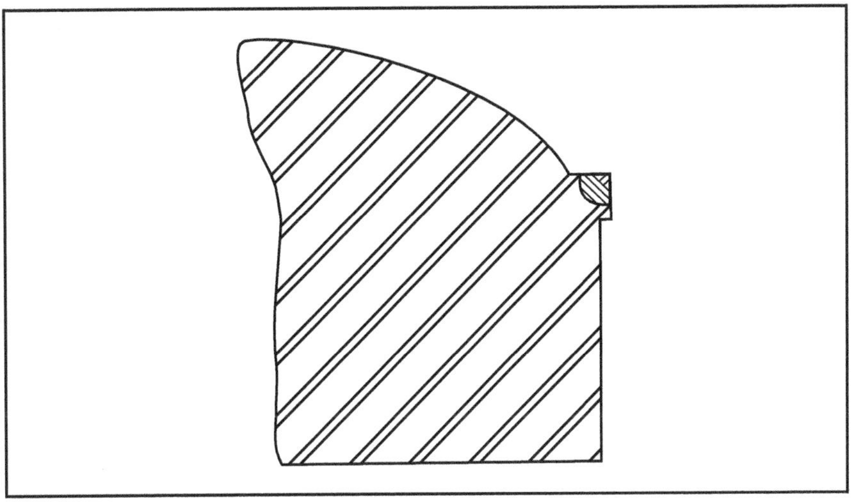

Figure 23-7. A welded trim edge is built up on a cast-iron trim-die post. (Courtesy MPD Welding, Inc.)

Two types of weld are deposited. A layer of unique nickel-alloy weld is deposited to act as a buffer for the hard-weld material that forms the cutting edge. Materials commonly used include AISI-S7, H12, and H19. The buffer layer serves to prevent excessive carbon and other elements in the gray iron from mixing with the air-hard weld and changing its properties. The nickel-alloy layer also provides a good bond with the casting.

The buffer layer is carefully deposited on one small area at a time. To avoid the production of white chilled iron, several passes are deposited in sequence. Each successive pass serves to anneal the iron under the weld.

To relieve the tensile stress that would occur upon cooling, the weld metal is upset by careful peening with a pneumatic hand tool between each pass. The finished weld must be as stress free as possible, although slight compressive stresses are not considered harmful.

The final cutting edge is built up with an air-hard weld. After cooling to room temperature, it is carefully tempered.

If trim changes or repairs are needed, the casting should be ground out to expose sound metal and the initial welding procedure, including preheating and slow cooling, should follow. Small chipped spots can be repaired by TIG welding and peening as emergency measures, but the quality of the repair is compromised.

Brazing Repairs

Large, broken, iron die castings can be repaired by brazing. Preparation by machining or grinding is required to provide a large area for the adhesion of the braze material. Careful preheating and postheating is an absolute necessity. Charcoal is an excellent fuel source for occasional repairs on large castings. Brazed repairs are especially useful on lightweight cast die shoes that cannot be repaired easily or permanently by mechanical means.

Worker concerns. The repair of a large casting by brazing may take several days of around-the-clock work. To avoid both human and mechanical stresses, it is often required to have two welders working at one time on opposite sides of the casting. A second pair of welders should be resting and ready to alternate places periodically so that the work is carried on continuously.

Proper protective clothing and respirators are essential for this work. Metal-fume poisoning is a problem. Zinc is a necessary constituent of the brazing-filler metal and its fumes are usually the main health concern. However, minute amounts of cadmium in the brazing-filler metal are also an especially worrisome impurity from a worker-health standpoint.

It is common to deposit several hundred pounds or more of filler metal to repair a large casting. To add strength, the braze may be built up above the surface of the casting and extended to each side as long as mechanical interference in the functioning of the finished work is not a problem. Peening of the filler metal may be required to relieve any residual tensile stresses that may develop upon slow cooling of the work.

REFERENCES

Smith, David. 1994. *Fundamentals of Pressworking*. Dearborn, MI: Society of Manufacturing Engineers.

——. 1990. *Die Design Handbook*. Dearborn, MI: Society of Manufacturing Engineers.

24

Die Materials and Treatments

Tool steels are used to construct the die components subject to wear. They are used in a variety of pressworking operations. These steels are especially designed to develop high hardness levels and abrasion resistance when heat treated (Jarvis 2000).

The plain-carbon and low-alloy steels are readily machinable and weldable. These low-cost steels are used for machine parts, keys, bolts, retainers, and support tooling. Cast-steel dies are used for large drawing and forming dies where maximum impact toughness is required. At carbon levels of 0.35% and higher, cast-alloy-steel dies can be effectively flame-hardened at points of wear.

Cast irons are available in many different compositions with special properties. These compositions are used for items such as shoes, plates, dies, adapters, and other large components. Die irons are often alloyed to permit flame hardening when used for the wear surfaces of large sheet-metal drawing and forming dies. The ductile (nodular) irons retain the casting advantages of cast iron, while having toughness, stiffness, and strength levels approaching those of steel.

In addition to ferrous die materials, varieties of die components are made of nonferrous metals such as zinc and copper alloys. Elastomer products find widespread application as die pads, rubber springs, and automation components. Even wood and wood-fiber products are used for low-cost dies.

CHARACTERISTICS OF TOOL AND DIE STEELS

The steels listed in Table 24-1 are used in the great majority of metal-stamping operations. The list contains 27 steels, which are available from many tool-steel sources. Some of these steels have slight variations for improved performance under certain conditions (Smith 1990).

AISI-SAE letter and number designations identify the steels listed in Table 24-1. The letter represents the group of the steel. The number indicates a separation of one grade or type from another. The use of standard industry identification is encouraged, rather than shop jargon and trade names used to describe the steels.

Some high-alloy steel, made by the powdered-metallurgy process rather than cast ingots, produces a product of uniform composition and grain structure. This process permits the high-volume production of tool steels that would be very difficult to make in any other way.

CHOOSING TOOL STEELS

Making a wise choice of tool steels is important when dies must insure good wear performance. Specifying more costly tool steel than is justified by die-wear requirements is wasteful. Tool steels cannot be selected based on dimension alone. The designer must consider the type of steel used, with particular regard as to whether the steel is quenched in water, oil, or air. Generally, liquid-hardened tools must be conservatively designed, while air-hardened tools can incorporate features that would cause cracking during liquid quenching. The design of a tool must also take into account the equipment available for heat treatment. Table 24-2 lists hardening and tempering treatments for common tool and die steels.

Expensive die details should be designed for long wear. Likewise, die parts that wear rapidly—requiring downtime for replacement and high repair costs—should be designed for good wear resistance. Table 24-3 gives a comparison of the basic characteristics of common tool and die steels. Table 24-4 gives examples of applications of tool steels for cold-work operations.

Table 24-1. Tool and die steels for metal stamping operations (*Courtesy Sun Steel Treating*)

AISI Steel	Nominal Steel Composition Percentage							
Type	C	Mn	Si	W	Cr	Mo	V	Other
W1	1.05	0.25	0.20	—	0.20	—	0.05	—
W2	1.00	0.25	0.20	—	—	—	0.20	—
O1	0.90	1.25	0.30	0.50	0.50	—	—	—
O2	0.90	1.60	—	—	—	—	—	—
O6	1.45	0.80	1.10	—	—	0.25	—	—
A2	1.00	0.60	0.30	—	5.25	1.10	0.25	—
A8	0.55	—	1.00	1.25	5.00	1.25	—	—
A9	0.50	—	1.00	—	5.20	1.40	1.00	1.40 Ni
D2	1.55	0.30	0.40	—	11.50	0.80	0.90	—
D3	2.10	0.40	0.90	0.80	11.70	—	—	—
D4	2.25	0.35	0.50	—	11.50	0.80	0.20	—
D7	2.30	0.40	0.40	—	12.50	1.10	4.00	—
S1	0.50	—	0.50	2.25	1.30	—	0.25	—
S5	0.60	0.85	1.90	—	0.20	0.50	0.20	—
S7	0.50	0.75	0.30	—	3.25	1.40	—	—
T1	0.75	—	—	18.00	4.00	—	1.10	—
T15	1.55	—	—	12.25	4.00	—	5.00	5.00 Co
M1	0.80	—	—	1.50	3.75	8.50	1.05	—
M2	0.85	—	—	6.25	4.15	5.00	1.90	—
M4	1.30	—	—	5.50	4.00	4.50	4.00	—

Table 24-1. *(continued)*

AISI Steel	Nominal Steel Composition Percentage							
Type	C	Mn	Si	W	Cr	Mo	V	Other
M7	1.00	—	—	1.70	3.75	8.75	2.00	—
M42	1.08	—	—	1.50	3.75	9.50	1.10	8.00 Co
H12	0.35	—	1.00	1.25	5.00	1.35	0.30	—
H13	0.38	—	1.00	—	5.20	1.25	1.00	—
H21	0.30	0.25	0.30	9.00	3.25	—	0.25	—
H26	0.53	—	—	18.00	4.00	—	1.00	—
L6	0.75	0.70	—	—	0.80	0.30	—	1.50 Ni

A = Air hardening O = Oil hardening D = High carbon, high chromium S = Shock resisting
H = Hot working T = Tungsten base, high speed L = Special purpose, low alloy W = Water hardening
M = Molybdenum base, high speed

Table 24-2. Hardening and tempering treatments for common tool and die steels

AISI Steel Type	Preheat Temperature ° F (° C)	Rate of Heating for Hardening	Hardening Temperature ° F (° C)	Minimum Time at Temperature (Minutes)	Quenching Medium
W1	1,200–1,250 (649–677)	Slow	1,425–1,500 (774–816)	15	Brine or water
W2	1,200–1,250 (649–677)	Slow	1,425–1,500 (774–816)	15	Brine or water
O1	1,200–1,250 (649–677)	Slow	1,450–1,500 (788–816)	15	Warm oil
O2	1,200–1,250 (649–677)	Slow	1,400–1,475 (760–802)	15	Warm oil
O6	1,200–1,250 (649–677)	Slow	1,450–1,500 (788–816)	30	Warm oil
A2	1,450–1,500 (788–816)	Slow	1,725–1,800 (941–982)	30	Air, salt, oil
A8	1,450–1,500 (788–816)	Slow	1,800–1,850 (982–1,010)	30	Air, salt, oil
A9	1,450–1,500 (788–816)	Slow	1,800–1,850 (982–1,010)	30–60	Salt, air
D2	1,450–1,500 (788–816)	Slow	1,800–1,875 (982–1,024)	30	Salt, air
D3	1,450 (788)	Slow	1,750 (954)	30	Warm oil
D4	1,450–1,500 (788–816)	Slow	1,800–1,850 (982–1,010)	30	Air, salt, oil
D7	1,450 (788)	Slow	1,875–2,000 (1,024–1,093)	45	Air, oil

Table 24-2. *(continued)*

AISI Steel Type	Tempering Temperature ° F (° C)	Depth of Hardening	Resistance to Decarburization	As Quenched Hardness R_C	Maximum Tempered Hardness R_C
W1	350–550 (177–288)	Shallow	Best	65–67	62–64
W2	350–550 (177–288)	Shallow	Best	65–67	62–64
O1	350–600 (177–316)	Medium	Good	63–65	62–64
O2	350–600 (177–316)	Medium	Good	63–65	62–64
O6	350–600 (177–316)	Medium	Good	63–65	61–63
A2	350–950 (177–510)	Deep	Fair–good	63–64	62–63
A8	350–950 (177–510)	Deep	Fair	62	60–62
A9	350–1,000 (177–538)	Deep	Poor–fair	58	56–58
D2	350–1,000 (177–538)	Deep	Fair	62–64	61–63
D3	350–950 (177–510)	Deep	Fair	63–65	61–63
D4	400–1,000 (204–538)	Deep	Fair	63–65	61–63
D7	300–1,000 (149–538)	Deep	Fair	65–67	65

Table 24-2. *(continued)*

AISI Steel Type	Preheat Temperature ° F (° C)	Rate of Heating for Hardening	Hardening Temperature ° F (° C)	Minimum Time at Temperature (Minutes)	Quenching Medium
S1	1,400 (760)	Slow	1,700–1,750 (927–954)	15–20	Salt, warm oil
S5	1,250 (677)	Slow	1,550–1,650 (843–899)	15–20	Salt, oil, water
S7	1,200–1,300 (649–704)	Slow	1,725–1,750 (941–954)	20	Oil, salt, air
T1	1,500–1,600 (816–871)	Quickly from preheat	2,100–2,350 (1,149–1,288)	3–7	Salt, warm oil
T15	1,450–1,550 (788–843)	Quickly from preheat	2,175–2,275 (1,191–1,246)	3–5	Salt, warm oil
M1	1,500–1,550 (816–843)	Quickly from preheat	2,150–2,225 (1,177–1,218)	2–5	Salt, warm oil
M2	1,500–1,550 (816–843)	Quickly from preheat	2,175–2,275 (1,191–1,246)	3–7	Salt, warm oil
M4	1,500–1,550 (816–843)	Quickly from preheat	2,150–2,250 (1,177–1,232)	3–7	Salt, warm oil
M7	1,500–1,550 (816–843)	Quickly from preheat	2,175–2,225 (1,191–1,218)	3–7	Salt, warm oil
M42	1,500–1,550 (816–843)	Quickly from preheat	2,150–2,200 (1,177–1,204)	3–7	Salt, warm oil
H12	1,400–1,500 (760–816)	Slow	1,825–1,875 (996–1,024)	30–60	Salt, air
H13	1,400 (760)	Slow	1,825–1,875 (996–1,024)	30–60	Salt, air
H21	1,500 (816)	Slow	2,000–2,250 (1,093–1,232)	10–30	Oil, air
H26	1,500 (816)	Slow	2,100–2,250 (1,149–1,232)	10–30	Oil, air
L6	1,200–1,250 (649–677)	Slow	1,450–1,550 (788–843)	30	Warm oil, water

Table 24-2. (*continued*)

AISI Steel Type	Tempering Temperature ° F (° C)	Depth of Hardening	Resistance to Decarburization	As Quenched Hardness R_C	Maximum Tempered Hardness R_C
S1	300–800 (149–427)	Medium	Fair	57–59	56–58
S5	300–800 (149–427)	Medium	Fair	60–62	59–61
S7	400–1,000 (204–538)	Medium	Poor–fair	58–60	57–59
T1	1,000–1,100 (538–593)	Deep	Good	64	64–65
T15	1,000–1,100 (538–593)	Deep	Fair–good	64–66	66–68
M1	1,000–1,150 (538–621)	Deep	Fair	64–66	65–67
M2	1,000–1,100 (538–593)	Deep	Fair	65–66	64–65
M4	1,000–1,100 (538–593)	Deep	Fair	65–66	65–66
M7	1,000–1,100 (538–593)	Deep	Fair–good	66–67	65–67
M42	975–1,075 (524–579)	Deep	Fair–good	64–66	67–69
H12	1,000–1,100 (538–593)	Deep	Fair	51–53	51–53
H13	1,000–1,150 (538–621)	Deep	Fair	54–56	53–56
H21	950–1,250 (510–677)	Medium–deep	Fair	53–55	54–55.5
H26	1,000–1,200 (538–649)	Deep	Fair–good	57–59	56–58
L6	350–600 (177–316)	Medium	Good	62–63	62

A = Air hardening O = Oil hardening D = High carbon, high chromium S = Shock resisting
H = Hot working T = Tungsten base, high speed L = Special purpose, low alloy W = Water hardening
M = Molybdenum base, high speed

Table 24-3. Comparison of the basic characteristics of tool and die steels

AISI Steel Type	Non-deforming Properties	Safety in Hardening	Toughness	Resistance to Softening Effect of Heat	Wear Resistance	Decarburization Risk During Heat Treatment	Brittleness	Machinability
W1	L	F	G	L	G	L	L/M	B
W2	L	F	G	L	G	L	L/M	B
O1	G	G	G	L	F	L	L	G
O2	G	G	G	L	F	L	L	G
O6	G	G	G	L	F	L	L	B
A2	B	B	F/G	F	G	L	L	F/G
A8	G/B	B	B	G	G	M	L	F/G
A9	B	B	B	B	G	M	L	F/G
D2	B	B	L/F	F	G	L	L	L
D3	G	B	L	F	G/B	L	L	L
D4	B	B	L	F	G/B	L	L	L
D7	B	B	L/F	F	B	L	L	L
S1	F	F/G	B	F	F/G	M	L	G
S5	F	G	B	F	G	L	L	G
S7	F	G	B	F	G	M	L	G
T1	G	G	G	B	G/B	L	L	F
T15	G	G	L	B	B	L	M	L/F
M1	G	G	G	B	G/B	L	L	F

Table 24-3. (continued)

AISI Steel Types	Non-deforming Properties	Safety in Hardening	Toughness	Resistance to Softening Effect of Heat	Wear Resistance	Decarburization Risk During Heat Treatment	Brittleness	Machinability
M2	G	G	G	B	G/B	L	L	F
M4	G	G	G	B	G/B	L	L	F
M7	G	G	G	G/B	B	L	L	F
M42	G	G	G	G/B	G/B	L	L	F
H12	G	G/B	G/B	G/B	F/G	L/M	L	F/G
H13	G	G/B	G/B	G/B	F/G	L/M	L	F/G
H21	G	G	G	G/B	F/G	L/M	L	F/G
H26	G	F	F	B	B	L/M	L	F/G
L6	F/G	G	G	L	F	L/M	L	G

Steel Types

W = Water hardening
O = Oil hardening
A = Air hardening
D = High carbon, high chromium
S = Shock resisting
T = Tungsten base, high speed
M = Molybdenum base, high speed
H = Hot working
L = Special purpose, low alloy

Comparison Scale

L = Low
F = Fair
M = Medium
G = Good
B = Best

Table 24-4. Examples of applications of tool steels for cold-work operations

Application	AISI Steel Type/Group	Hardness R_C
Blanking dies and punches (short run)	O, A, W	58–60
Blanking dies and punches (long run)	A, D	60–62
	M	61–63
Bending dies	O, A, D	58–60
Coining dies	S	52–54
	W	57–59
	D	58–60
Drawing dies	H	52–54
	W, O	58–60
	D	50–62
Dies—cold extrusion	W	59–61
	D	60–62
	M	62–64
Dies—embossing	W, L, O, A, D	58–62
Dies—lamination	D, A, M	60–62
	M, D	58–60
Dies—sizing and ironing	W	59–61
Punches—embossing	S	58–60
Punches—trimming	W, D, O, A	58–60
Punches—notching	M	60–62

A = Air hardening
D = High carbon, high chromium
H = Hot working
L = Special purpose, low alloy
M = Molybdenum base, high speed
O = Oil hardening
S = Shock resisting
T = Tungsten base, high speed
W = Water hardening

The die-repair activity should carefully track die-repair costs as an aid to selecting the most cost-effective tooling materials and processing methods. This information should be used to update the die standards for each type of tooling. In this way, tooling dependability can be continuously improved and costs minimized.

W—Water-hardening Tool Steels

Types W1 and W2 tool steels are both readily available and low in cost. Containing vanadium, W2 provides a finer grain size attended by lower hardenability and higher toughness. Both are shallow hardening. In large sections, this results in a hard case and a softer internal core with high toughness. Closely controlling the vanadium content helps assure a better level of hardenability. They are quenched in water or brine and are subject to substantial size changes when heat treated.

In the past, types W1 and W2 steels were more popular because of their low cost and because they were widely used for short-to-medium-run tooling. They are still a good choice for a variety of applications where cost is a prime factor, such as short-run automotive applications. For example, automotive body-panel dies seldom exceed one-million pieces per model year.

O—Oil-hardening Tool Steels

Steels O1 and O2 were once the workhorses of the industry. These are known as manganese oil-hardening steels. They are readily available and low in cost. These steels, which are normally quenched in oil, have less size change than the water-hardening steels. Compared to the water-hardening steels, they are easier to harden throughout the workpiece and are equally tough. Type O6, which contains free graphite, has excellent machinability.

A—Air-hardening Die Steels

The most popular air-hardening die steel is A2. This steel has small size change when hardened, and has higher toughness than the oil-hardening die steels. Compared to oil-hardening steels, the wear resistance is equal or superior. The availability of the popular A2 steel is excellent. Type A8 is the toughest steel in this group, but its low carbon content makes it less wear-resistant than A2.

D—High-carbon, High-chromium Die Steels

The principal steels of wide application for long-run dies are high-carbon, high-chromium steels. Grade D2, containing 1.50% carbon, is of moderate toughness and intermediate wear resistance. Grades D3, D4, and D7 contain additional carbon, which increases wear resistance; however, the toughness is somewhat lower. Selection between the grades is based on the length of run desired, machining, and grinding problems. Containing molybdenum, D2 and D4 are quenched in air and have small size change when hardened. Type D2 tends to be specified more than the S and M series.

S—Shock-resisting Tool Steels

Shock-resisting steels contain less carbon and higher toughness than the high carbon types that are more wear resistant. They are used where heavy cutting or forming operations are required and chipping or breakage of high-carbon, wear-resistant steels is a problem. Choice among the grades is usually determined by finding which type provides the best results in service. All S-type steels are readily available, with S5 and S7 being widely used. Grade S5 is an oil-hardening type of silico-manganese steel and is more economical than S1. Compared to S1, S5 has equivalent toughness properties with greater wear resistance.

Type S7 can be quenched in air in sections up to 3-in. (76-mm) thick. For thicker sections, interrupted oil or complete oil quenching may be used. The S-series tool steels are superior to most other steels for severe work where chipping is a problem. As a class, the S-series tool steels should be specified more often, especially where extreme toughness and a reduction of chipping problems are required.

T and M—Tungsten and Molybdenum High-speed Steels

High-speed steels T1 and M2 are approximately equal in performance and have excellent properties for cold-work dies. They have higher toughness than most other tool steels, combined with excellent wear resistance.

High-speed steels are often less costly and give better performance than the popular D2 tool steels. They are also readily available. Types T15 and M4 are hardened in a neutral rather than a carburizing atmosphere because they already have very high carbon combined with high vanadium content.

Type M1 may occasionally be used in place of T1 and M2, but it is more susceptible to decarburization. Steel T15 is the most wear-resistant of all steels in the list. Steel M4 has slightly better wear resistance than tool steels such as D4. These steels are more difficult to machine and grind than the other high-speed steels, but the improved performance often justifies the extra machining expense.

H—Hot-working Steels

Die-casting dies, extrusion dies, hot-forming dies, and hot-drawing mandrels are typical applications for hot-working tool steels. Type H13 is widely used for plastic molding dies. Modern cavity sinking is nearly all done with high-speed computer numerical control (CNC) machining or die-sinking electrical discharge machining (EDM). These processes permit rapid material removal and good surface finishes. To take advantage of the high metal-removal rates of modern CNC profile milling machines, coated ceramic and tungsten carbide cutters capable of high metal-removal rates are used. Mold steels such as H13 are available as prehardened material.

L—Low-alloy Tool Steels

Steel L6, a chromium-nickel steel, is one of the many low-alloy steels effective as die materials. In large sizes, it is water-quenched and has a hard case and a soft core with high overall toughness. In small sizes, it may be oil-quenched.

Other Steels

Many alloy steels used in machine and aircraft construction find application as tool steels. Of these, AISI-SAE 4140 is very popular. It is readily available in a wide variety of sizes including material prehardened for toughness. Alloy 4140 is weldable and can be coated for wear resistance. It is a good choice for drawing and forming tooling that must be built quickly in an emergency. While it is not recommended for long-run tooling, dies built of 4140 to meet emergency production requirements have held up for long production runs over several years without the need to build higher-quality tooling.

HEAT TREATMENT OF DIE STEELS

Iron has two distinct and different atomic arrangements—one existing at room temperature (and again near the melting point) and one above the transformation temperature. Without this phenomenon, it would be impossible to harden iron-based alloys by heat treatment.

Simplified Theory

Figure 24-1 represents what happens in the heat treatment of die steels. Starting in the annealed machinable condition *A*, the steel is soft, consisting internally of an aggregate of ferrite and carbide. Upon heating above the iron-carbon eutectoid-transformation temperature of 1,333–1,700° F (723–927° C), the crystal structure of ferrite changes. The ferrite becomes austenite *B*, and dissolves a large portion of the carbide. This new structure, known as austenite, is always a prerequisite for hardening. By quenching or cooling rapidly to room temperature, the carbon is retained in solution, and the structure known as martensite *C* results. This is the hard matrix structure in steels.

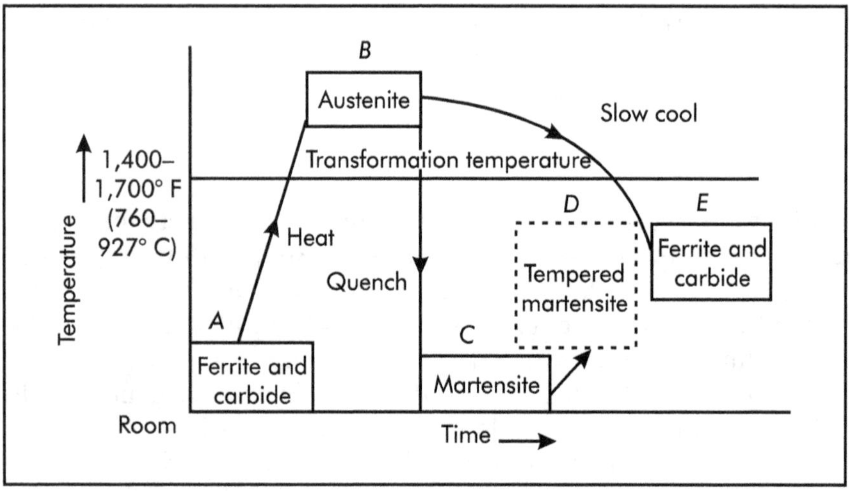

Figure 24-1. A simplified chart illustrates the sequential process for hardening steel.

Quenching and Tempering Tool Steels

Rapid cooling results in high internal stresses (Fletcher 1952). The transformation from austenite to martensite involves some volumetric expansion. This adds further stresses, particularly in parts of varying cross section. These stresses together with the hard, brittle nature of martensite can be sufficient to cause cracking. To avoid this, the steel is reheated to an intermediate temperature (*D* in Figure 24-1) to soften the part to the desired hardness level. This operation, known as *tempering* or *drawing*, also serves to relieve those residual stresses that otherwise would cause brittleness in the steel.

If quenching is not rapid enough, the austenite reverts to ferrite and carbide (*E* in Figure 24-1), and high hardness is not obtained. The rate of quenching required to produce martensite depends primarily on the alloy content. Low-alloy steels require rapid cooling in water or oil, while highly alloyed steel usually can be air-quenched at a much slower rate.

Throughout the heat-treating reactions, most die steels retain excess or undissolved carbides, which take no direct part in the hardening. The high-carbon high-chromium steels, for example, have large quantities of excess iron-chromium carbide, which give them the high degree of abrasion-resistance characteristic of this class of steel.

Influence of Heat Treatment on Die Life

Each type of die steel must be handled slightly differently from any other for optimum results. Different temperatures, different heating and cooling rates, and variable tempering procedures must be used as recommended.

In general, it may be said that the harder a given die, the longer it will wear, while the softer a die is, the tougher it becomes. Assuming the proper die steel is used, dies that are wearing out too quickly should be made harder for improved life, and dies that are breaking or cracking should be made softer.

Within limits, heat treatment can be used to adjust these variables to the best advantage. Oil-hardening steel may work best on one application at R_C 62, and on another involving higher stresses and shock at R_C 58. Adjustments of the drawing temperature easily produce the hardness desired.

Double drawing and, in some instances, triple drawing is desirable for tools in severe applications. This is because steels retain austenite when quenched. The first temper affects the martensite formed during quenching and conditions the austenite so that it transforms upon air cooling from the draw. Double drawing is necessary to affect the martensite, which forms after the first draw. Triple drawing eliminates nearly all retained austenite, further increasing toughness.

Cold and Cryogenic Treatment

Cooling to very low temperatures as part of the drawing process will maintain hardness while actually improving the toughness and fatigue strength of hardened tool steel. This improvement occurs due to the continued transformation of retained austenite into the more desirable martensite at temperatures from –120° to –300° F (–84° to –184° C).

In some tool steels, the martensite finish temperature, which is the temperature at which the austenite-to-martensite transformation is completed, is below room temperature. Here, cold treatment at –120° F (–84° C) extends the range of heat-treating temperatures and enhances the transformation of austenite. Cold treatment is often done between the last two draws.

Cryogenic treatment occurs at the temperature of liquid nitrogen, which is approximately –300° F (–184° C). Size change is negligible but can occur if the material is not properly heat treated. Cryogenic treatment should not be used to attempt to correct poor heat-treatment practices. Multiple draws or tempers should not be eliminated knowing that the cold process will convert austenite to martensite. The value of cryogenic treatment is not fully agreed upon within the tooling industry.

Control of Surface Chemistry

Furnaces for heat-treating tool steels are often equipped with gas generators that provide an atmosphere containing a controllable amount of carbon. Without a protective atmosphere, surface decarburization to a considerable depth can occur because of the oxidizing effects of free oxygen, water vapor, or carbon monoxide. Typical decarburization rates are 0.010–0.030 in. (0.25–0.76 mm) per hour. The loss of surface carbon can produce poor wear resistance. It is customary to make tools sufficiently oversized so that they can be ground to remove the surface affected by chemical change during heat treatment.

If the atmosphere used to heat dies for hardening is strongly reducing, it is possible to carburize the surface, although the carbon content is already quite high. A carburized case may be beneficial to tool life if not carried to excess.

Shops doing only a small amount of heat treating often wrap the die details to be heat treated in a stainless steel or titanium foil especially produced for this purpose. This procedure effectively avoids decarburization without the expense of maintaining an atmosphere furnace. The most popular tool steel heat treated in this way is A2.

Molten salt baths provide decarburization protection for die steels. Careful maintenance involving desludging and rectification is necessary to maintain a neutral condition. Various molten salts are used for heating steels to quenching temperature and as quenching mediums. In some cases, the recommended quenching procedure might involve a high-temperature initial quench in molten salt followed by cooling to room temperature in oil or air.

Control of Dimensional Change

Dimensional changes occur during the hardening operation on dies because hardened steel occupies a greater volume than the annealed steel from which it came. Unfortunately, the dimensional changes, which result from the volume of change, are usually not the same in all three directions, making it impossible to predict the changes accurately. The size and shape of the individual workpiece are the most important factors influencing this variable.

The dimensional changes resulting from hardening may vary from nil up to approximately three parts per thousand. Dimensional changes of this amount are of no concern, provided enough grind stock is allowed for complete removal of scale and decarburization.

If the furnace atmosphere or salt bath is carefully controlled, there may be little or no scale or decarburization. If grind stock is not required, the avoidance of dimensional changes is of great importance.

Air-hardening tool steels usually produce dimensional changes, which are less than one part per thousand, and therefore they are widely used where minimal size change in heat treatment is needed. It is also possible to control the size changes in tools made of type D2 high-carbon high-chromium steel to so-called "zero size change" by an austenite-martensite balance control obtained by multiple tempering.

Keeping data on the size changes that actually occur when identical die details are heat-treated is a good practice. From this information, dimensional allowances for expected size changes can be made.

TOOL STEEL PRODUCTION METHODS

Tool steels are produced by a variety of methods. At one time, essentially all tool steels were made by the cast-ingot process. These products are known as cast and wrought tool-steel alloys. The cast-ingot process is still a high-volume method of producing tool steels. Great care is required to produce uniform tool steel by any process.

The problem of alloying element segregation tends to limit the ingot size of highly alloyed tool steels made by the cast-ingot process. Vacuum degassing may follow the melting process. Producing tool steels in a vacuum is superior to melting in the presence of atmospheric gases.

There is still a considerable volume of tool steel, including A, D, and H series, produced by the cast-ingot process. Cast ingots are rolled or press forged into the finished product. Finally, the material is annealed and finished as required. This process is used for the production of most of the tool-steel grades. For higher carbon, higher alloy steels, or for applications requiring steels with exceptionally good shock or wear resistance, alternative production methods are used. Using careful procedures to insure clean homogeneous steel, the cast-ingot process remains popular for the majority of commercial tool steels. The main limiting factors are the extent to which the steel is alloyed, the alloying elements used, and the size of the ingot required.

Electroslag Remelting

Electroslag remelting is a process used to produce a variety of quality steel alloys including tool steels. A highly useful application early in the development of this process was for the production of critical alloyed steel parts such as turbine shafts. Very clean homogeneous tool steels are produced by use of the electroslag-remelting process.

The tool-steel alloy is first cast into a conventional ingot using melt and cast technology. Next, the ingot is then refined in a vertical cylindrical furnace by means of electroslag remelting. The conventionally cast ingot to be purified is suspended at the top of the furnace opening. A special molten slag separates the ingot from a metallic starter plate that forms the bottom of the furnace. The current melts the slag and the end of the old ingot being fed into the furnace. Electrical current passes through the ingot, which is remelted and purified as it seeps through the slag and forms a new ingot on the bottom of the water-cooled furnace. Since the slag serves to purify the steel alloy and protect the new ingot from atmospheric gases, the ingot is quite pure and very homogenous.

Vacuum Arc Remelting

Vacuum arc remelting (VAR) of tool steel is used for secondary or premium melting of tool steels. The results are much the same as the electroslag-remelting process. The VAR process produces steels having excellent micro-cleanliness.

Powder Metallurgy Tool Steels

Tool steel produced by the powdered metallurgy process is first atomized in a nonreactive gas atmosphere into very fine particles with uniform properties. This material is placed in large steel canisters, which are evacuated and sealed shut. Placing the canisters in a hot isostatic compacting furnace and heating under pressure fuses the canister's contents into a solid, inclusion-free ingot. A combination of heat and high-pressure gas fuses the steel powder into a homogeneous mass.

The chief advantages of the powder metallurgy process are uniformity and freedom from imperfections due to segregation of constituents associated with the cast-ingot process. Tool steels such as AISI grades D2, M2, and M4 produced by the powder-metallurgy process have better uniformity than the corresponding wrought product.

Higher alloy constituencies than can be obtained by conventional melt-and-cast technology are possible. For example, some powdered-metallurgy tool steels contain 10–15% vanadium and have a substantial amount of vanadium carbide in the heat-treated tool-steel matrix, which exhibit extreme resistance to abrasion.

Grade 3V™ has toughness approaching AISI grade S7 and substantially higher wear resistance. Grades V3™, V10™ and V15™ may outperform most of the tool steels listed in Tables 24-1 through 24-4, depending on the application.

DIE SURFACE COATINGS AND TREATMENTS

The uses of die surface coatings, treatments, and plating is increasingly important to increase tool life and reduce lubricant requirements. The use of coatings or other surface modifications is often an effective means to eliminate the need for extreme pressure additives. Lubricants considered to contain hazardous materials often require disposal as hazardous waste.

Chromium Plating

A thin layer of chromium may be applied to forming and drawing dies to increase wear resistance and reduce galling. This chromium-plated surface has a very low coefficient of friction with excellent non-galling characteristics.

The usual practice is to apply a layer of chromium 0.0005–0.001-in. (0.013–0.025-mm) thick to a very finely ground or polished surface. The time taken to polish the surface is well spent. Any defects or irregularities present on the surface prior to plating will show through and actually be more objectionable after the plating process.

Chromium plating is also used to repair worn dies. It is possible to build up a layer of as much as 0.010 in. (0.25 mm) or more of chromium on a worn surface and thus increase the total production life of the die.

Gas Nitriding

The use of gas nitriding to produce a hard, wear-resistant case on steels has been commercially practiced for many years. This procedure also can be used on some tool steels to improve wear resistance.

Gas nitriding can be used on tool steels that do not temper back excessively at the nitriding temperature, typically 975° F (524° C). This limits gas nitriding largely to the hot-work steels and the high-carbon, high-chromium grades. High-speed steels form an exceptionally brittle nitride case; therefore, gas nitriding should not be used on these steels.

Gas nitriding of tool steels requires from 10–72 hours. Typical case depths range from 0.002–0.018 in. (0.05–0.46 mm).

Ion Nitriding

Applications for ion nitriding range from improving the wear resistance of small tool-steel die sections to large iron-alloy sheet-metal drawing punches weighing 10 tons or more. Unlike the older gas-nitriding process, a *glow discharge* or *ion process* takes place when direct current (DC) voltage is applied between the furnace (as the anode) and the workpiece (as the cathode). The furnace atmosphere consists of a nitrogen gas level that is much less than the atmospheric pressure.

The nitrogen gas in the furnace becomes ionized and emits electrons with a negative charge, and ions of nitrogen, which are positively charged. These are attracted toward the cathode, which is the workpiece. The ions are accelerated to a high velocity by the sharp electrical potential drop just in front of the cathode surface. Here, the ions energetically bombard the workpiece (Denton 1989).

Ion-nitriding furnaces large enough to process the largest dies commonly used in the automotive, appliance, and aerospace industry are available. The ion-nitriding process is often preferred to chromium plating for large die applications.

Titanium Nitride and Titanium Carbide

Titanium nitride (TiN) and titanium carbide (TiC) coatings improve the life of tools by acting as a chemical and thermal barrier to diffusion and fusion. The coatings are very thin, typically 0.0001–0.0003 in. (0.003–0.008 mm), and quite hard. Although the thin coatings are very brittle, they tend to assume the ductility and deformation characteristics of the substrate material. The coatings are also quite lubricious, serving to lower the coefficient of friction between the tool and the workpiece. By depositing TiN or TiC onto a steel or carbide tool, the improvement in lubricity causes the tool to resist galling.

Physical Vapor Deposition of Titanium Nitride

Physical vapor deposition (PVD) of titanium nitride is a coating process carried out in a high vacuum at temperatures between 400–900° F (204–482° C). This range of temperatures does not exceed those used to draw hardened high-speed tool steel. Because there is very little distortion or size change of the workpiece, this coating process is frequently used to coat finished punches and buttons whenever rapid wear or galling is found a problem.

The plasma source coats the workpiece in a straight line-of-sight process. Special rotating fixtures with water cooling may be required to insure that all surfaces are evenly coated and small sections are not overheated. A TiN coating deposited by the PVD process is easily recognized by its gold color.

Chemical Vapor Deposition of Titanium Carbide and Nitride

Chemical vapor deposition (CVD) of titanium carbide and nitride is a coating process done at much higher temperatures than the PVD process, generally 1,740–1,920° F (949–1,049° C). For this reason, it is normal practice to follow the coating procedure with conventional heat treatment of the tool-steel substrate.

Titanium carbide coating is limited to use on tool steel and solid-carbide die materials because the substrate surface must act as a catalyst. It is superior to PVD coatings when extreme abrasive wear is a problem. The coating is deposited from a vapor, so uniform coating including blind slots and blind holes is possible.

A CVD coating is dull gray in color. When a CVD-coated tool is polished, the resultant tool is silver, indistinguishable from the base metal.

Thermal Diffusion

The thermal diffusion (TD) process is performed by immersing parts in a fused salt bath at temperatures of 1,600–1,900° F (871–1,038° C) for one to eight hours. Carbide constituents dispersed in the salt bath combine with carbon atoms contained in the tooling substrate, which must contain at least 0.3% carbon or greater. The carbide layer most commonly produced is vanadium carbide; although, depending on the composition of the salt bath, other carbides can be deposited. These include niobium carbide, chromium carbide and, in some newer processes, a niobium-vanadium combination. There is a die part-size limitation due to salt-bath-size availability, which is a restriction on the application of this process.

Choice of Methods

The CVD-coating method can deposit both TiN and TiC. Coatings can be applied to all tool steels, as well as to solid-carbide tooling. When very high-wearability qualities are required, and the distortion caused by the post-coating heat treatment that is usually needed is not a problem, CVD may be the best choice.

PVD is a low-temperature process that can be applied to all tool steels, but is generally used to increase the wearability of finished high-speed steel parts, solid carbide, and brazed-carbide tooling.

CVD coating requires post-heat-treatment to restore the hardness to the steel substrate. In the TD process, popular tool steels such as D2 and A2 are usually quenched upon removal from the fused salt bath. The diffused layer, typically of vanadium carbide, is quite thin, but is exceedingly hard, with higher hardness than tungsten carbide. In situations requiring high-volume production runs, tungsten carbide is also treatable.

WROUGHT LOW-CARBON STEELS AND CAST IRONS

Wrought-steel plate, rounds, and shapes are often used in the fabrication of brackets, frames, feeding, ejecting and transfer mechanisms, and other die-auxiliary devices where structural strength and weldability, rather than wear resistance, are the primary requirements. Short-run steel dies are sometimes made with carburized hot-rolled-steel wear surfaces. Where the properties of AISI-SAE 1018 or similar steel (boilerplate) suffice, wrought low-carbon steels and cast irons have the advantage of being readily available and economical.

Cast Carbon and Low-Alloy Steels

Cast-steel shoes and other die components are often used for large drawing, forming, or trimming dies where a combination of high toughness and strength are required. These steel castings are usually annealed or normalized to provide a homogeneous structure, free from casting stresses. Heat treating or flame hardening is often employed to obtain the desired strength, wear resistance, and toughness.

Cast and Ductile Irons

The high-compressive strength, manufacturing economy, and ease of casting gray-cast iron make this material useful, especially in large forming and drawing dies. Soft, unalloyed gray irons are widely used for plates, jigs, spacers, and other die parts.

Fully pearlitic irons with random uniform flake-graphite structures are excellent for wear resistance. Resistance to wear is significantly improved by flame hardening draw radii or other wear areas. Alloy additions of chromium, molybdenum, and nickel are commonly used to produce uniform pearlite structures and improve the iron's response to flame hardening.

The ductile (nodular) irons retain the casting advantages of cast iron, but because of the free graphite present in spheroidal shape rather than in flake form, they develop toughness and strength levels approaching those of steel. This combination of properties is especially useful in large forming and drawing dies where heavy-impact loads or high-transverse stresses are encountered. It should be specified for dies where breakage has occurred with gray-iron castings.

NONFERROUS AND NONMETALLIC DIE MATERIALS

Dies made of nonferrous materials are used for a variety of reasons. They are economical for limited production runs and producing experimental models. Often they have superior functioning, such as preservation of part finish, are relatively lightweight, and portable for extremely large tools. Other advantages may include corrosion resistance, ease of fabrication, and low lead-time requirements. Often, they permit fast, easy corrections when design changes are necessary. Nonferrous die materials include aluminum alloys, zinc-based alloys, lead-based alloys, bismuth alloys, cast-beryllium alloys, copper-based alloys, plastics, elastomers, and tungsten carbides.

Nonferrous Cast Materials

Cast aluminum bronzes are used for forming and drawing stainless steel and other difficult-to-work materials. They resist metal pickup, which often causes scratching or galling. Proprietary bronzes (for example, Ampco® metal) that are cast to the die shape are used for such applications. These alloys are also used for die-wear plates and die guide-pin bushings where resistance to wear under the condition of high load is required.

Zinc-based Alloys

Zinc-alloy die materials have higher tensile strength and impact resistance than pure zinc. The alloys can be cast into dies for blanking and drawing a variety of aluminum and steel parts, especially complicated shapes and deeper draws than are possible with plastic or wooden dies. The working surface is dense and smooth, and requires only surface machining and polishing. Dies made of this material are frequently mounted in die sets and used for blanking light gages of aluminum.

Frequently, one member of the die set is composed of a zinc alloy and the other member of a softer material such as lead. Drop-hammer operations are an example. Harder punches are required for forming steel sheets where sharp definition is necessary. Worn and obsolete dies made of zinc alloy and lead are remelted to achieve nearly 100% material reuse. Lead punches, composed of 6–7% antimony, 0.04% impurities, and the remainder lead, have been used with zinc-alloy dies. Health concerns over the hazards of lead toxicity must be considered when using this material.

Various proprietary zinc alloys, such as Kirksite®, are available. A typical composition consists of 3.5–4.5% aluminum, 2.5–3.5% copper, and 0.02–0.10% magnesium. The remainder is 99.99% pure zinc. To minimize health hazards from grinding dusts generated when working with these materials, the impurity levels of lead and cadmium must be very low.

Cast Beryllium, Copper, and Bismuth Alloys

Cast alloys composed of beryllium, cobalt, and copper have characteristics comparable to the Ampco® alloys. Beryllium is a very toxic substance, and proper precautions regarding ventilation and industrial hygiene must be taken when working with it.

Some alloys of bismuth are used as a cast-in-place material to secure punch and die parts in the die assembly. These low-melting-point alloys are cast as punches and dies for short-run forming and drawing operations. The melting temperature of some of these alloys occurs below the boiling point of water.

Carbide Die Materials

Cemented carbides consist of finely divided hard particles of the carbide of a refractory metal. Carbide sintered with one or more metals such as iron, nickel, or cobalt as a binder, form a body of high hardness and compressive strength. Cemented tungsten carbide is the most common form of tooling carbide. However, carbides of titanium and tantalum are also used.

Years ago, Dow Chemical Company developed a pure tungsten carbide made by hot isostatic pressing. Since it contains no binder, it is a ceramic material. Because the material has lower tensile strength than cemented carbides, it has not found application in pressworking tooling. However, in highly abrasive applications, such as waterjet cutting nozzles, the ceramic material has been found to outwear cemented tungsten carbide by 20:1.

Thermal expansion is an important physical characteristic of carbide. For most carbide grades, it ranges from one-third to one-half the rate for steel. This rate must be considered when carbide is attached to a steel support or body.

Application of Cemented Tungsten Carbide

Cemented tungsten carbide is used widely for dies intended for high-volume production of difficult-to-stamp materials. For example, motor- and transformer-lamination cutting dies used for high-volume production are often made of cemented tungsten carbide. Such dies have produced millions of parts before the need for resharpening. Other widespread applications include cutting, drawing, forming, and ironing dies used in high-volume production of parts ranging from razor blades and stainless-steel drawn shells to beverage containers.

Restrictions exist on the use of lubricants on some carbide die materials, particularly those containing sulfur. While tungsten carbide is essentially an inert material, some lubricants attack the cobalt binder material. Electrolytic corrosion can result from stray currents from electrical part-sensing equipment.

Machinable Carbides

Machinable carbides are ferrous alloys sold under the trade name Ferro-Tic®, and are made in 10 standard grades. These range from 25–45% (by volume) titanium carbide, tungsten carbide, titanium-tungsten double carbides, or other refractory carbides as the hard phase. They are contained in a heat-treatable matrix or binder of mainly iron.

The 10 grades are suitable for most applications requiring machinable carbide. The standard grades have been modified by Ferro-Tic to meet specific operating requirements. The heat-treatment method required to harden the matrix is similar to that for conventional tool steels. The manufacturer's recommendations should be followed for proper product application, machining, and heat treatment.

Nonmetallic Die Materials

Wood and Hardboard Laminates

Hardwood can be used for form blocks, but laminated impregnated wood, hardboard, and plastics have largely replaced it. Hard maple and beech are good woods for die applications if they are carefully selected for close grain structure.

High-density panels composed of compressed wood fiber and lignin are used for jigs, dies, fixtures, templates, patterns, and molds. Masonite™ is a cellulose semi-plastic material available in various thicknesses, and finds application as a jig and tooling material. It can be readily laminated with cold-setting adhesives. Such wood-based materials are suitable for short-run dies for prototype work. Such dies are also widely used in the aircraft industry.

Rubber

Molded-rubber female dies and rubber-covered punches are used in difficult forming operations, such as the production of deeply fluted lighting reflectors. Many types of rubbers and rubber compounds are used. Natural rubber, neoprene, and polyurethane all find application in pressworking.

Specifications for rubbers used in the conventional Guerin, Marform, and hydroform processes are determined by the performance needed for the process. Some rubber compounds, especially polyurethane, can be cast in place and cured to form the needed shape. This permits the economical production of forming die components, part strippers, pressure pads, and non-marring automation fingers in the toolroom.

Cork

Soft, medium, and hard cork layers compressed into sheet form are sometimes used with, or in place of, rubber pads. Cork deforms only slightly in any direction other than that of the applied load, while rubber flows in all directions.

Plastics

Like wood, the pressures involved in pressworking limit the use of plastic materials for dies. Selection of plastics is based on economy relative to die life expectancy. Draw radii are a primary source of concern because maximum loads and abrasion occur in these areas.

Draw dies with a metal core of either ferrous or zinc-alloy material, capped with a working face of epoxy, are used in the aircraft, appliance, and automotive industries. Rubber forming dies are made of combinations of cast and laminated epoxy applied to a heavy steel base.

Polyester resins are used for low-volume tooling. The chief advantage this material has over the stronger and more stable epoxy resins is cost.

Polyurethanes

Polyurethanes combine many of the good properties of both elastomers and plastics. They have demonstrated a unique combination of abrasion resistance, tensile strength, and high load-bearing capacity not available in conventional elastomers, plus impact resistance and resilience not available in plastics. Because of their liquid, uncured form, and their excellent cured properties, these polymers are useful in draw dies, drop-hammer dies, forming and stamping pads, press-brake forming dies, mandrels, expanding punches, and other tool design applications.

A major use of polyurethanes is for die-automation components such as kickers, lifter heads, and rollers where the excellent wear resistance of the material, together with its non-marking characteristics, are very useful. Polyurethane is available as a two-part liquid formulation that can be mixed and cast in place to form custom-made die pads and to transfer automation jaws.

REFERENCES

Fletcher, S. G. 1952. "The Selection and Treatment of Die Steels." *The Tool Engineer*, April.

Denton, R. 1989. "Application of Ion Nitriding." Selecting Tooling Materials and Tooling Treatments for Increased Tool Performance Clinic, November. Dearborn, MI: Society of Manufacturing Engineers.

Jarvis, D. 2000. Metallurgy of Tool Steels workshop, May. Dearborn, MI: Society of Manufacturing Engineers.

Smith, David. 1990. *Die Design Handbook*. Section 28, Ferrous Die Materials. Dearborn, MI: Society of Manufacturing Engineers.

Appendix A

Generic Die Standards

The following generic examples do not represent the standards of any actual manufacturer, except by coincidence. These examples are provided solely for study and discussion purposes. Anyone wishing to include these in actual practice is advised to check with their engineering, legal, and safety departments to make an independent determination of appropriateness for a company's manufacturing processes. These examples should not be assumed to be appropriate or safe for any specific application. Anyone using these standards should do so strictly at their own risk.

Some of the examples are written in terms that are mainly applicable to progressive-die design-and-build standards. The actual standards for given classes of work vary. Many of these generic standards may be directly applicable to any specific type of die or pressworking operation.

To aid the reader in understanding the examine standards, notes follow many of them. These notes are intended to give the background information that is the basis for the standard as it is written. Any bias on the writer's part should be considered an opinion based on fact and not expert legal advice.

DESIGN-AND-BUILD PROCESS

A customer's verbal approval does not relieve (name of vendor company) of the responsibilities of designing and manufacturing dies capable of stamping parts that meet or exceed part print specifications. Time is of the essence. Sample part submission and meeting the mutually approved binding and final delivery schedules are all mandatory. Verbal approvals are not binding for (name of vendor company). The vendor is responsible for meeting all (name of customer company) specifications.

A (cite number of pieces) -piece minimum sample run should be made with statistical process control (SPC) sampling to insure that each die produces SPC quality that is equal or exceeding ($C_p \geqq 2.0$, $C_{pk} \geqq 1.33$, or whatever other standard is required). The required sample parts are to be submitted to the (name of customer company) layout department for final approval.

Note: It is essential to place notations on the part print. The notations should include dimensional deviations for noncritical part features. Any dimension should take into consideration ease of assembly and the proper functionality of the assembly during its service. For example, clearance holes, stiffening flanges, formed ribs, and similar part features should have dimensions based on the simplification of tool design for low-cost manufacturing. A part designer must understand and make appropriate use of geometric dimensions and tolerances (GD&T) to define the part. The GD&T standard to be followed must be stated. For example, ASME Y14.5M-1994 is perhaps the most common current standard.

Some dimensions are not critical. For example, stiffening flanges and formed ribs can vary within wide limits and still accomplish the intended function. Applying close SPC data to the variation of such noncritical features will increase the cost of the product. Using GD&T avoids confusion by reducing part-drawing interpretation errors. It is important, however, that the datum points are properly selected for ease of assembly. This will result in enhanced interchangeability and minimized part cost.

PRODUCTION AND SAFETY

The initial and final part approval cannot be validated until the die has properly demonstrated production of stamping parts made to print specifications. Production equipment for the applicable processes, meeting all production rates and safety requirements, including those of (customer company), must be used.

Note: It is generally advised that all deviations from the build schedule and required design be made in writing by a designated authority specified in the purchase order or contract. It is the duty of a representative of the company to be present during the die tryout or sample runs. This tryout determines the tooling capabilities and is a way to check for any pressworking safety violations. It is strongly advised that such runs be allowed only if OSHA and the company's safety standards are being followed. Otherwise, depending on applicable case law, the die maker could end up assuming third-party liability for any injuries that occur.

QUOTES

A quote should be submitted on (name of vendor company) letterhead. It should include any part-print issues such as tolerances, radii, concepts, specifications, and any others where a consideration for a deviation may be requested by (name of customer company).

The vendor is responsible for following (name of customer company) die standards and requirements, the design and build outline, making adjustments and corrections to production operations, and incurring shipping costs and transportation to and from the designated production facility.

Note: It is advisable to leave nothing to the imagination. Requested deviations should be made clear during the design process. The requirement of the vendor to follow specified standards and to pay shipping costs should be spelled out. This will allow shipping costs to be included in the initial quotation.

COMPLETION

A full-size, strip layout for each die must be submitted for approval by the engineering department at (name of customer company). All stations must be shown and labeled. In addition, sectional views for each die should be shown in stages to clarify design concepts. The vendor is responsible for determining the number of stages and the most efficient strip layout.

The engineering department at (name of customer company) will determine if extra perishable punches, buttons, and other details are needed. If so, these details should be included in the die builder's quotation to be shipped with the tooling.

Two sets of prints must be sent to the stamping engineering department of (name of customer company) for approval. One set should be returned with changes before die construction begins. Updates and revisions of die prints should be maintained as well. Final prints are then delivered with the completed tooling.

CANCELLATIONS AND DAMAGES

When tool delivery schedules are greater than (specify one of the following, weekly or monthly), progress reports will need to be submitted to (customer-company individual) on (specify how often and the date). [Tryout dates, approval dates, delivery dates, and delays due to engineering or part revision changes should be included. Monetary damages for failure to deliver the tooling in good working order and on time also may be spelled out in a contract.] For shorter delivery dates or upon the discretion of (name of customer company), Gantt charts and weekly review meetings may be required.

After acceptance of the contract or purchase order of (name of customer company), the failure to remain on or ahead of schedule is a reason to cancel a vendor's contract or purchase order at the discretion of (name of customer company).

Note: It is essential to have a general counsel approve the actual language used in a company's die standards. For example, depending on case law, to have a binding contract that constitutes consideration and timely completion, terms such as "time is of the essence" may have legal meaning. Nothing in this document, however, is intended to serve as legal advice.

DESIGNING STRONG DIE SECTIONS

Die design must result in strong die sections and punches, even if extra stations or operations are required. Thin section walls, sharp edges, and sharp corners must be avoided wherever possible. Weak die conditions also should be avoided whenever feasible. If the design unavoidably results in weak die sections or punches, these results should be designated as perishable details that can be easily and safely replaced.

Note: Die conditions involving weak sections are to be avoided as a matter of good practice. Likewise, sharp corners, especially if in square or rectangular holes, result in increased tool wear and maintenance costs (Miedema 1988).

[Often, large, wire-cut electrical-discharge-machined sections are more economical than several smaller sections. All sections should be designed to permit cutting them apart if necessary. However, engineering changes and emergency welded repairs may require splitting. If splitting is done, additional screw and dowel holes will be required to retain the hardened split sections.]

Die sections are to have 0.050–0.062-in. (1.27–1.57-mm) die land (specify die-land size, based on the customer company's type of work). The remaining surface should be relieved at a 1°-draft angle to avoid slug buildup, unless a different angle is specified by the engineering department at (name of customer company), which is dependent on the material conditions and expected die life.

Note: The draft angle typically varies from 0.25–2°, depending on the type of work and the desired die life.

[A design must allow for a location and/or position of a customer's stamp retainer logo and/or the plant identification number before the strip layout is finalized and the design is completed. It may be necessary to add an extra stage to accommodate the standard stamp retainer, depending on the design of the die and on the general shape and size of the stamping.] All costs to build a fully functional die to specifications by (name of vendor company) are included in the cost quotation for the die.

Note: This again serves to remind the die builder of the basic requirements to build a completely functional die to the standards of the customer company.

All perishable details that are not commonly available catalog items with approved suppliers of (name of vendor company) such as cutting sections, form steels, punches, shedders, and pressure pads, must be detailed fully to allow for exact replacement. Math data should be supplied to (name of customer company) in a mutually agreeable CAD design and die manufacturing (name a choice of formats) format. Math data for wire-burn EDM or computer numerical control (CNC) -cutter paths must be submitted with finalized tool tracings. Completed die tracings should be submitted at the time of shipment. Complete digital data should be provided on a computer disk or on tape in AutoCAD (name of version) or another mutually agreed upon software format.

Note: It is important to be able to reproduce perishable details in-house or at any vendor of choice. When a customer company does not supply math data, a die builder can hold the company responsible for producing exact replacement perishable details.

Pitch and progression strip width must be calculated (list tolerance here) and shown on the strip layout without using any dimensions from the raw stock supplier and/or material specification sheet. The (name of vendor company) is responsible for anticipating and compensating for pitch-change errors.

Note: Pitch change can be a function of deformation of the stock and, to a lesser degree, a deflection and elongation of the die shoes and details under load. It is important to make sure the vendor understands this factor.

Final design approval must be received from (name of plant) before the manufacture of a die is started. Approval includes keeping a signature and date on file from (designate someone, for example, the tool engineer).

Note: Again, always avoid any hint of an understanding that a verbal agreement is valid and/or binding. Specify in any contract the name of the management authority permitted to sign any agreement or requested deviation.

Prototype samples and prototype dies are guidelines only. When required, samples for material-flow analysis and determination of thinning conditions should be made on prototype tooling that closely approximates the production-forming rate and surface-friction conditions. Prototype samples should be submitted to the (name of customer company's) tool-engineering department prior to the completion of production die details. All development costs are the responsibility of the (name of vendor company) and must be figured into the pricing of the die. Existing stampings and dies can be used for reference. Since new design processes are expected to continuously improve, a new part print and specifications will govern the final part approval.

Note: In the writer's opinion, soft tool development is overutilized as a "security blanket" in the industry. Due to forming rate and frictional differences between soft and hard tools, successful soft tool parts may not be reproducible on hard tooling. The above statement puts the responsibility for functional tooling on the die builder. You do not want to leave room for the excuse that it was proven with soft or prototype tooling.

Specified dimensional tolerances are to be maintained between mating die details. The tool tracing should reflect complete dimensions, including tolerances for making replacements, without assuming arbitrary shop-practice tolerances.

Note: This statement may not seem necessary, but it is intended to help avoid confusion when specifying locational fits. These specifications should be as precise as possible, while clearances for screw holes should be dimensioned to assure proper mechanical function that is free from interference. Overspecifiying, however, only adds to the cost without adding function or value to the tool.

SCREWS, DOWELS, AND KEYS

Dowel or key all die and punch steels to assure positive location. If blind dowel holes are required, dowels must have threaded pullout holes. The (name of customer company's) tool-engineering department must approve blind dowel holes. Square keys and keyways milled in the die shoes are to be used, in addition to dowels where large lateral forces are developed. Die and punch sections should have slip-fit dowel holes in removable details and a press fit in the die shoe.

It is the policy of (name of customer company) to require certain specification-grade fasteners that meet the requirements of the Fastener Product Assurance Act (FPAA) for die applications. This includes fasteners used in the attachment of the die buildup such as parallels and subplates. The heads of bolts specified for die-setting and die-buildup applications have distinctive head markings. SAE-ASTM-grade eight-hex-head fasteners are identified by six markings. The metric fasteners having similar properties are ISO 10.9- or ISO 12.9-property-class bolts. All SAE-ASTM and ISO fasteners and socket-head screws meeting these specifications must be identified with the manufacturer's logo or identification. Markings for socket-head screws need not be as prominent as those on hex-head fasteners.

Die design and die-setting fastening require an adequate size and number of high-grade fasteners to insure against failure. Mixing or interchanging fasteners for noncritical operations with those that have required specifications meeting requirements of the FPAA is not permitted. Only approved specification grades are to be used to avoid the danger of fasteners for safety-critical applications being replaced with ones intended for noncritical applications.

> *Note: It is important that the bolted joint is designed correctly. Provided the fastener is made of the correct materials and manufactured with total process control (TPC), it should meet specified strength and fatigue requirements. However, many die fasteners are safety critical. Therefore, the FPAA should be followed as required for specification-grade fasteners.*

Jackscrew holes are a requirement for all punches, die sections, inserts, and other details retained by dowels or pockets. All detail prints must show jackscrew-hole locations. Jackscrew holes in hardened sections should be tapped at 0.005 in. (0.127 mm) over size to allow for shrinkage in the heat-treatment process.

Counter-bore depths for screw heads should equal the maximum die life. To insure correct assembly, symmetrical sections are to have asymmetrical dowel and screw holes.

Form steels and wiping-flange steels are to be heeled and keyed, unless they are securely retained in milled pockets, or unless they are backed up with die shoes with cast or welded features.

Note: Most of the above is elementary die making. However, the person designing a die on a CAD system may find it easier not to offset symmetrical holes unless told to do so. The exact amount or class of slip and interference fit for dowels may need to be specified. Use of the last paragraph on pocketing and heeling of form steels is optional, depending on the severity of the operation.

DIE SHOES, SUBPLATES, AND PARALLELS

Precision die shoes are to be ground within 0.002 in. (0.05 mm) of the total indicated run-out. Shoe stock must be normalized after flame cutting and prior to machining. In addition, four-post die sets are to have one pin offset by 0.25 in. (6.4 mm) to avoid backward assembly.

Punch shoes are to be no less than 2-in. (50.8-mm) thick, unless otherwise mutually agreed upon. The lower die shoes are to be no less than 3-in. (76.2-mm) thick, unless otherwise mutually agreed upon.

Note: These are examples that will vary with individual circumstances. In general, the lower shoe must be made thicker than the upper one because it is weakened by cutouts from the discharge of slugs, scrap, and parts. Specifying normalizing is important and costs very little. For some applications, aluminum die sets might be specified.

Sufficient parallels are required to minimize shoe deflection, unless otherwise mutually agreed on. If parallels are used to clamp the die in the press, sufficient grade specification of eight or better of screws should be used to exceed the strength of the die-fastening system.

Accurate die locating is essential for fast and accurate setup of dies. The die must be accurately located in the press. In addition, die-fastening slots need to line up accurately. Therefore, either dowels or keys are to be used to achieve an accurate assembly of parallels, subplates, and other die-buildup components.

Both upper and lower die shoes must be drilled and tapped with hoist swivel hooks or rings. Four tapped holes should be provided per shoe, with two per side located so the hoisting and any required turning of shoes does not interfere with attached parallels and subplates. The tapped hole size must be correct for the hoist rings or swivel hooks recommended by the hook or hoist ring manufacturer for safe handling.

Parallels, attached subplates, and/or die shoes must have U-shaped slots on 6-in. (152.4-mm) Joint Industry Council (JIC) centers (or the company's standard; for example, some metric machines have 160-mm spacing) and be slotted for T-slot bolts. Lower die parallels must have access for lift truck forks. Minimum access width is (whatever specified, such as 6 in. [152.4 mm], to suit the size of dies and die-handling equipment).

> *Note: Die parallel standards may be specified to spell out the amount of deflection permitted. Cutting and forming loads can be determined accurately from standard formulas. There are also engineering standard formulas for determining shoe deflection (Smith 1990).*

DIE SPRINGS

Die pads must contain all necessary springs (name of preferred brand), self-contained nitrogen pressure cylinders, devices, systems, and other required details. Correct nitrogen pressure is to be determined during production and then stamped next to the charging console. In no case should the manufacturer's rated pressure be exceeded. Stamped figures are to be 0.375-in. (9.53-mm) high.

If a hydraulic die-pressure unit is required to obtain a higher force-per-unit area than is available with nitrogen or pad-return delay, the system is to be designed in conjunction with the hydraulic system's engineering department. This design is essential to insure that the die will function correctly at the maximum stroking speed and length.

When springs are used for pad pressure and to actuate stock lifters, all spring pockets must be counter-bored into the die shoe. Spring cans or cages must be provided to retain broken pieces under pressure pads.

Cam-return springs must have guards to retain the springs and a cam-return rod in case of failure. Sheet-metal or angle-iron guards must be located around all pad openings. Exposed openings around moving and stationary tooling must be closed off. Angle iron painted yellow with a minimum guard thickness of 0.125 in. (3.18 mm) must be used. Unless approved in writing by (name of customer company's) tool engineering department, spring deflection must not exceed the spring manufacturer's recommendations for long spring life. The maximum permissible deflection rating must never be exceeded. Springs must conform to the International Standards Organization (ISO) color codes for die-spring duty classes as adopted by the North American Automotive Metric Standards Group (see Table 9-1).

Note: The preceding paragraphs are an important part of die design standards for essentially all sizes and classes of work. These requirements should be incorporated in all die standards where there is a danger of broken springs or other parts exiting the die violently.

OSHA clearly states that during operation, guide posts that constitute a pinch or nip point are to be guarded. Common sense extends this to other moving die parts handled by pressroom personnel.

Broken parts can travel at high velocities and endanger pressroom personnel. Die pads and draw rings must be guarded if there is any danger of pressroom personnel touching a nip or pinch point that is not in constant engagement, such as the heel block or guide post. Injury can occur if there is a failure to guard the opening around lower-die draw rings and/or moving pads.

The correct functioning of die pressure systems cannot be left to chance. In the writer's experience, many die-process capability problems are directly traceable to an incorrect nitrogen system specification. High nitrogen and spring pressures can damage the press due to the back-loading of the gear reduction train as the press rolls through bottom. In such cases, a hydraulic system is the best system choice because the cylinder-refilling rate is controllable. In addition, hydraulic systems are able to concentrate greater force in a given area than nitrogen systems.

The issue of allowable spring deflections must be watched carefully in the design stage. Fortunately, there are ISO standards for die spring color-coding identification. Some spring manufacturers have copyrighted spring color-coding schemes that lead to an unfortunate amount of confusion. The use of the ISO standard is highly recommended to permit standardization of spring identification.

The containment of broken springs is a basic safety requirement. Likewise, not exceeding the spring manufacturer's deflection rating is important from a safety standpoint.

STRIP STARTING STOPS

Details such as lifter pins and stock-guiding rails that are subject to wear from strip feeding are to be made of hardened- (specify the grade) or higher-grade steel. For low-production requirements, an inlay of air-hardening weld at the wear point is acceptable. (Name of customer company) will determine low-production requirements.

The pitch or feed progression per press stroke must be stamped on the lower die shoe where it is easily visible from the starting stop. Stamped figures should be 0.375 in. (9.53 mm) high.

All progressive dies must feed from the right-hand side of the press and die. Unless otherwise agreed upon in writing, all dies must have a positive starting stop to insure starting the strip at the correct location. A positive spring-loaded first-hit stop must be provided. A minimum alternative to this is a sight stop if mutually agreed upon in writing by (name of person or title) in the tooling engineering departments of (name of customer company) and the (name of vendor company).

Starting a new strip and removing the old progression strip must be easily accomplished without the necessity of an operator reaching into a point of operation. If the die produces partial parts at the marked starting point of the strip, instructions for removal must be stamped on the die shoe near the starting stop. All progressive dies must have hardened stock guides with the front side adjustable at the beginning of the strip.

Note: Most progressive die damage occurs at the running-out end of the old strip and at the starting of the new one. It is critical to provide instructions to the operator for accomplishing these operations safely and without damage to the die. The die should be built with solid-stock lifters whenever possible. This helps avoid the problem of the operator reaching into a point of operation to change the strip.

COMMERCIAL PERISHABLE DETAILS

Purchased punches or die buttons should be standard replacement items. Recommended sources for these materials are (list primary source) or (list secondary source).

Die buttons must be made of grade (specify grade; D2 is often specified, M2 and M4 are specified less often) tool steel or better.

Punches must be specified to minimize deflection when used in asymmetrical cutting operations.

Use (name brand name) or equivalent spring-loaded shedder pins in punches where slug-retention problems are expected. Retrofitting spring-loaded shedder pins must be done at the die builder's expense if slug pulling is determined by (name of customer company) to be a problem during production tryout and acceptance.

A means for positive slug retention or slug removal is required in all die sections. This cost is (name of vendor company's) responsibility, including damages to assure slugs do not pull back into the die during the tryout and production approval stages. The use of a reverse draft angle and a less-than-standard recommended punch-and-die clearance requires a written deviation from the (name of customer company's) tool-engineering department.

When ball-lock punches are specified (specify the type and duty class), conventionally headed punches must be designed to avoid punch-head breakage. Punch retainers and backing plates must be made of tool steel hardened to 48–55 R_C (or other hardness specified).

Note: It is the writer's opinion that ball-lock punches tend to be specified most often. An exception is made for punches and/or buttons that must be removed or changed to make a different style of product. In most cases, punches do not break in service. An expectation that punches will break and must be changed frequently is indicative of poor shop practice. This can lead to an incorrect belief that misfeeds, which break punches, are acceptable occurrences.

PADS AND STRIPPERS

Removable windows are to be provided in all pads and strippers around cutting and forming details for access to perishable details. Dies that are designed and built for a family of parts must have removable access plates to accommodate punch removal and model changes within the stamping press.

An acceptable alternative method for model changes is the use of gag bars to engage and disengage punches by a pneumatic or manually operated gag-bar system. Quote both alternatives.

Lancing or cutting on the spring pad must be avoided. If necessary to do so, four standard guide pins and bushings can be used to guide the pad.

Punch retainers and heel blocks are not to be used as leveling blocks for timing and/or tool plates. If the pressure per unit area on pad-bottoming blocks is calculated to be over 15,000 psi (103,419 kPa), use hardened tool-steel blocks and pad-striking inserts.

Pads and stripper plates are to be no less than (state thickness) thick unless mutually agreed between the (name of vendor company's) and (name of customer company's) tool-engineering departments.

Note: Pad or stripper-plate windows should only be provided if style changes are required in the press. A second exception is for severe pressworking conditions that require replacement of perishable details during the normal press run. Secure pad-window retention is required to avoid the possibility of the window falling out and causing extensive damage.

Gag bars that can engage and disengage punches manually or by automatic programmable means are an under-utilized technology. The thickness of the stripper plates can be critical from a strength standpoint for slower, heavy press-working tools. For high-speed operations, stripper and pad weights should be minimized to avoid inertial shock.

GENERAL TOOL STEEL DETAILS

Shedder pins are required in all blanking and piercing sections. Shedder pins shall be provided in trimming sections only if required for foolproof scrap shedding.

Cutoff, notching, and forming steels are to be designed and built to minimize deflection. If necessary, use heel blocks with keys to minimize deflection.

Unless mutually agreed in writing, do not cut stock on opposite sides of the same insert or against form steel. Use separate trim-and-pierce or form steels even if an idle stage is required.

Welded modifications of tool steel sections and the use of shims in new dies requires written approval from (name of customer company's) tool-engineering department.

Shear angles and staggering of punch entry are required to reduce cutting tonnage and snap-through energy release. A record of shear angles and timing is to be supplied on the final die prints.

All forming steels are to be made of (name grades) hardened tool steel as a minimum requirement. Where severe forming and wiping conditions are involved, (name steel, such as D2, M2, etc.) hardened tool steel or better is to be used if needed to facilitate the application of wear-resistant coatings.

Form steels are to be inserted and provisions made wherever possible for inserting shims to compensate for material thickness variations. No other use of shims is permitted unless approved in advance by the (name of customer company's) tool-engineering department.

Die sections must be stamped or etched with the die and detail number. The detail number is to be stamped on the die shoe.

Note: Care must be exercised not to overspecify the tool grade if the cost is a consideration. The goal when designing a good die is to have a tool that will produce good parts without resulting in scrap or unexpected downtime. Provisions must be made to shim either cutting or form steels so the die will be kept in time when sharpening is required. Stamping of die sections generally does not cause cracking during heat treatment because a stress riser is not created. The softening during the heat-treating preheat process will relieve any stress caused by stamping the detail.

GUIDE PIN, BUSHING, HEEL BLOCK, KEEPER, AND WEAR PLATES

Progressive dies use 360°-rotational ball-bearing sleeves with hardened guide pins and bushings. All four post-die sets must have a pin offset by 0.250 in. (6.35 mm) to prevent accidental backward assembly of the die.

All dies must have grease fittings on keepers, cams, and slide gibs. In addition, they must be capable of being greased in the open position. Graphite plug lubrication can be supplied as an option to grease fittings if there is mutual agreement between the (name of vendor company's) and the (name of customer company's) tool-engineering departments. Graphite-plugged wear plates and (name of tool steel and heat-treating process) hardened steel or higher-grade wear plates ground flat are recommended for cam slides. The use of spool-type keepers requires written approval.

[For high-lateral-thrust die operations, graphite-plug bronze wear plates are recommended for corner-post heel blocks on four-post die sets. Graphite wear plates are located on guide-pin bushing blocks.] Type A2 hardened steel or higher-grade wear plates ground flat must be used on guide-pin blocks. If the size of the die shoe does not allow wear plates on corner blocks, then locating heel blocks in the middle of the die shoe is acceptable by mutual agreement.

If heel blocks are removable, they must be keyed in all directions of lateral thrust. Dowel pins should be used for location purposes only. Sufficient high-strength screws must be used to assure secure heel-block retention and to avoid movement under normal working loads.

Heel blocks are not required for short-stroke dies loaded symmetrically. For short stroke operations at low-to-medium speeds (specify speed), bronze or plated-steel bushings may be used upon written approval of the (name of customer company's) tool-engineering department.

Note: Here, experience and good engineering judgment are required to have a die alignment system that avoids over-specification, which will result in excessive cost. It is important to maintain presses so that they impart no lateral loads to the die and build enough lateral load-carrying capability into the die to avoid transferring lateral loads to the press. One recommended way of avoiding lateral loads is to design the die for double-attached parts.

SCRAP DISCHARGE

Dies must have scrap cutters that will chop the carrier strip into pieces that can be easily discharged into scrap hoppers or conveyor systems. Scrap, slugs, etc., must fall freely through the die shoe and clear parallels without air assistance.

Parts should exit freely from the die without air assistance wherever possible. If the best design procedure does not allow the part to exit freely from the end of the die, dropping it through the lower shoe is preferred. The design must avoid mixing parts with scrap. If air assistance is needed, it must be a timed burst with a muffled nozzle.

Provisions must be made for low die-shoe deflection if there are holes in the lower die shoe for discharge of parts and scrap.

If small slugs are produced, mutual consideration by (name of vendor company's) and the (name of customer company's) tool-engineering departments should be given to containing the slugs in slug boxes, rather than discharging them into a scrap conveying system. [Such small slugs can damage conveyors with hinged joints.]

The (name of vendor company) is responsible for designing the die so that the stock strip will feed smoothly and accurately locate in all die stations. Pitch changes due to the forming of the stock must be anticipated and provided for as part of the (name of vendor company's) design and construction responsibility. It is the (name of vendor company's) responsibility to make any needed corrections. It is permissible to lance between each part within the strip, as long as the die feeds smoothly and locates the parts accurately in each station.

Note: Positive part and scrap discharge is an essential requirement for progressive die operation. Likewise, smooth precise feeding is essential for accurate and consistent parts. The issue of small slugs damaging hinged slat-type conveyor flights is very valid and must not be overlooked. A failure of a plant's main scrap conveyor is a serious problem.

STOCK SENSORS

Photoelectric controls, and proximity, sonic, and/or probe sensors must be utilized for part and die protection. An approved electrical-wiring schematic for correct sensor wiring will be provided by the customer.

Single- or double-pitch notch (French stop) stations must be provided with pivoting stops to actuate sensor(s) specified by the (name of customer company's) tool-engineering department. In addition, a sensor is necessary to detect that the part is in the correct position to be severed and exits the die correctly.

Note: These sensor requirements are the minimum that will permit semi-unattended die operations. A sensor is also needed to detect the end of coil at the decoiler.

GENERAL GOOD PRACTICE

[It is important to define die-pass height in the maximum-open position with the stock guides raised.] All setup or stop blocks must be located over parallels or risers to avoid deforming the die shoe in the event of a shut-height error. Screws, pipe fittings, grease fittings, punches, die buttons, etc., must be in American-inch or metric measurements, as specified by the (name of customer company's) tool-engineering department.

Design and build for practical die component access and removal within the stamping press for maintenance, style changes, and adjustment. Sufficient pilots and pilot buttons for precise control of the progression strip must be used. Slug clearance through the die shoe and parallels in the event of a misfeed also must be provided.

Punch-and-die trimming must be used for uniform-flat and parallel-trim bottoms or for the flange on stamped parts with wall or flange designs. Pinch-trim designs must be approved by the (name of customer company's) tool-engineering department to assure that an acceptable smooth edge condition and good die life are obtainable.

OSHA AND ADDITIONAL DIE IDENTIFICATION

The front face of die shoes must be stamped with feeder progression, stock width, part number, the (name of shop) job number, and nitrogen pressure (if used).

Also, to facilitate safe die handling and setup, press tonnage, press stroke, total-die weight, lower-die weight, and punch-shoe weight are required to be stamped on the die.

Note: Making the press tonnage, die weights, and punch-shoe weight available to the die setter is an OSHA requirement. The intention is to aid the die setter in choosing the correct press. To avoid confusion, the recommended procedure is to stamp the information on the die. This is a more certain way of meeting the OSHA requirement in case written instructions are not available to the die setter. Please note that the law requires "stamped" rather than painted or marked.

The stamping of the lower-die weight may not be a requirement of power-press safety law. However, there are occasions when the lower die may need to be worked on when the upper die with the OSHA-required total weight stamped is not available. Therefore, it is highly recommended that the lower-die weight be stamped on the lower shoe to avoid equipment overloading.

REFERENCES

Miedema, A. 1988. "Progressive Dies." Videotape training course and reference guide. Dearborn, MI: Society of Manufacturing Engineers.

Smith, David. 1990. *Die Design Handbook*. Section 3, Die Engineering—Planning and Design. Dearborn, MI: Society of Manufacturing Engineers.

Appendix B

Case Study— Phoenix Specialty Manufacturing Company

Quick-response manufacturing of custom stampings requires every aspect of a company to operate efficiently. This chapter examines one precision contact stamper, Phoenix Specialty Manufacturing Company (Bamberg, South Carolina), as it handles the die buildup for stamping small lots of many customized parts. Every action, from entering client orders properly to shipping on time, must work flawlessly.

Computer integration of the entire manufacturing process is essential, from receipt of a customer's request through die buildup and shipping the order. The key to making money is shipping what the customer wants on time. This requires an efficient die room. An efficient die room is neat and well organized.

ORDER ENTRY

When an order arrives at Phoenix, it may be shipped from existing stock or built from components in stock. Quick response to special orders may require design and computer numerical control (CNC) machining of new die details, which may be produced by wire electrical discharge machining (EDM). Figure B-1 illustrates work order entry, which generates an instruction sheet that follows the job.

Figure B-1. Work orders for standard and custom stampings are entered into the system. The computer-generated work order is also an instruction sheet that specifies the die components required. The work order follows the job. (Courtesy Phoenix Specialty Manufacturing Company)

To produce a custom stamping, most work orders are filled by first assembling the die in a standardized aluminum die set from tool steel or carbide sections neatly stored in indexed cabinets. A work order specifies the parts needed and where they are obtained. A diversified machine shop produces replacement and new die parts as needed.

MACHINE SHOP CAPABILITIES

At Phoenix, a well-equipped machine shop with a variety of milling machines, grinders, lathes, and EDM machines is essential to building and maintaining good dies. Machining centers and wire-EDM machines operating under CNC control are also essential to die making (see Figure B-2).

Accuracy and speed are advantages of CNC machine tools. Achieving efficient use of equipment requires good die designs and a workplace commitment to produce the best possible fitting and functioning dies. Continuous improvement—a plant culture that

Figure B-2. A die maker enters barcode data into a networked computer system at a CNC machine workstation. This simple task assures accurate tracking of job status. (Courtesy Phoenix Specialty Manufacturing Company)

is never satisfied with "good enough"—is the company's commitment to achieve true excellence. A die that just makes good parts is not the goal. Making the best possible parts at the lowest per-part cost is the goal at Phoenix.

Continuous improvement is also applied to machine tools by continuously refining computer-integrated manufacturing systems. Machine tools need software and hardware improvements to do the best possible work at the lowest total cost. For example, job tracking and die maintenance software is upgraded as needed.

Die-making skills overlap with CNC-machining skills. Operating a CNC machine to make die components requires a skilled craftsperson. This is not a job for a "button pusher." New die makers are trained in many skills of the trade during their academic education. Training in die-making theories and practices is provided through on-the-job training from skilled supervisors and journeymen.

PLANT CULTURE

Efficient production and zero defects go hand-in-hand with cleanliness and organization of the workplace. A good plant frequently is a QS-9000, AS-9000, or ISO 9002 -registered facility that practices continuous improvement. A good shop is also a happy shop. A good company considers training important because it encourages everyone to think of ways to improve plant teamwork and operations. A variety of training topics is important and several days per year should be devoted to in-plant updating of skills.

A knowledgeable observer can make a quick audit of a plant's efficiency just by walking around and observing work practices. Simple observation will tell a lot about the plant culture—how people work together and accomplish goals. For example, Figure B-3 shows a die maker at Phoenix sweeping the floor in his work area. All employees must work together to keep the workplace clean. There is no janitorial staff for this task. Some positive outcomes of this attitude are:

- Everyone is intent upon his or her task and working without obvious supervision.
- The ratio of supervisors to craftspersons and machine operators is low.
- Supervisors mainly instruct and provide hands-on help—they do not spend a lot of time in their offices.
- Everything is clean and organized—for example, the machine shop, die storage racks, pressroom, and even the scrap recyclable metal are kept separated to maximize value.
- Workers involved with support activities, including engineering and scheduling, are actively involved on the plant floor.
- The shop is a "happy shop." Everyone is treated as "family" and there is no hesitating to lend a hand to others when needed.

SPARE PARTS STORAGE

At Phoenix, for quick-response manufacturing of custom stampings, many die sections are required to be kept on hand for quick retrieval and die buildup. Figure B-4 shows a journeyman picking the required parts from a drawer in a spare detail cabinet

Figure B-3. A clean workplace requires everyone's cooperation. This tool and die maker is keeping his CNC-machine work cell clean. (Courtesy Phoenix Specialty Manufacturing Company)

for a die buildup work order. The parts are retrieved by referring to work-order routing lists, which tell the location of parts needed. Computer integration ties in with work-order generation, tracking exactly where each section is located, the condition of the sections, and the correct return place for storage or any required maintenance.

ERGONOMICS

Avoiding excessive lifting, bending, and other types of strenuous motions can prevent injury. A properly designed workplace is

Figure B-4. Standard die components are stored in cabinet drawers. The work-order routing sheet lists the location of the parts needed to build up the die. (Courtesy Phoenix Specialty Manufacturing Company)

important in the die and pressroom. The use of ball-and-roller conveyors for movement of die sets and completed dies contributes to efficient work flow and avoids awkward handling. Figure B-5 illustrates the staging station at Phoenix where parts are placed along with an aluminum die set onto a plastic cafeteria tray.

ALUMINUM DIE SETS

Aluminum die sets have been in use for several decades. Fortel™ aluminum alloy is a specially processed material optimized for die shoe applications. Some shops build their own aluminum die sets of Fortel, 6061-T6, and 7075-T651 plate. These alloys are physically hard, easy to machine, and suitable for applications where a lightweight aluminum die set is needed.

When substituting steel for aluminum in die set construction, the differences in the two materials should be considered first. To

Figure B-5. The work-order routing sheet is checked to insure that all die parts and the correct empty aluminum die set are placed in the tray for bench assembly. The use of aluminum die sets has ergonomic advantages because they weigh half as much as an equally strong steel die set. (Courtesy Phoenix Specialty Manufacturing Company)

obtain equal deflection, an aluminum die set must be made two-fifths thicker than steel. This is because the modulus of elasticity of aluminum is only one-third that of steel. Even with the additional thickness, the aluminum plate weighs approximately half that of steel with equal deflection.

The coefficient of expansion of aluminum is approximately twice as great as that of steel. Warping and misalignment with temperature changes can occur if large steel subplates are bolted to aluminum die shoes. In other words, the aluminum-to-steel bolted composite plate acts much like a bimetallic thermometer spring. Good designs consider this factor.

To guide posts and ball bushings, sleeves should not be pressed in. The shoe material may be displaced or sheared and result in pins and bushings being installed crooked. Pins and bushings in should be shrunk in liquid nitrogen or dry ice before inserting them. This procedure is also excellent for steel or cast-iron die shoes.

BENCH DIE BUILD UP

At Phoenix, die sets and parts are moved from the conveyor to the bench where a die maker builds up the tool and checks it manually for correct operation. Work orders with detailed instructions are on top of the die sets on the roller conveyor. Workstations are placed next to the roller conveyors at bench level, permitting the light aluminum die sets to be moved easily without lifting.

Figure B-6 shows a die maker carefully checking a compound die section as a precision shim die is built up from interchangeable component parts. These dies are configured for cutting materials varying from cold-rolled steel strip to beryllium copper, phosphor bronze, and difficult materials such as the Hastelloy series and elgiloy, a watch hairspring alloy. Stampings made from

Figure B-6. This die maker is checking die block hole sizes with precision gages. Note the neat yet busy bench appearance with a calculator and other tools nearby. (Courtesy Phoenix Specialty Manufacturing Company)

these materials are used in products such as automobiles and defense and aerospace applications. Thin and hard materials require very close punch-to-die clearances. The die maker shown in Figure B-7 has completed making a tissue paper cut to verify proper cutting clearances and is double checking the work order to insure correct build up of the die.

QUICK DIE CHANGE CAPABILITIES

In Figure B-7, the die set has U-shaped tie-down slots on the lower die and grooves along the side of the upper die shoe. Many dies use a detachable shank or quill that attaches in the center of the upper die. This shank helps locate the die and incorporates a knockout pin, which is needed for part ejection in compound dies. Most of these dies are run in inclined high-speed precision presses. Timed air blowoff enables near-perfect part ejection.

Figure B-7. A die maker has assembled a complete compound die and made a tissue paper cut to check for exact detail alignment. The final step is double-checking the work order in the plastic protective sleeve. (Courtesy Phoenix Specialty Manufacturing Company)

To meet or exceed requirements for safe die fastening, two clamps are used to hold the upper die shoe, in addition to shanks in some instances. There are expressed concerns about the use of die shanks under OSHA power-press safety law. Some die makers are unfamiliar with why shanks are used with compound and combination press tooling. There is no known OSHA prohibition, however, against using punch shanks with additional bolting to secure the upper die.

AN EYE TO THE FUTURE

At Phoenix, management has a strong civic responsibility as a major employer in the community. The company has expanded its business by adding the capability to produce high-speed progressive dies in-house (see Figure B-8).

CONCLUSION

The metal-stamping industry is essential to national and world economy. There are many successful metal-stamping companies who serve their communities as good employers and trainers of skilled workers.

Well-organized die rooms and machine shops are essential to supporting a successful company. Just as a die consists of many parts that work together to make a product, a wisely managed and partnered work force consists of many individuals who contribute their best efforts to achieve excellence in manufacturing.

Figure B-8. Phoenix president Robert Hurst, Jr. checks a new progressive die. (Courtesy Phoenix Specialty Manufacturing Company)

Index

D

E

F

G

T

www.ingramcontent.com/pod-product-compliance
Lightning Source LLC
LaVergne TN
LVHW010751261025
823898LV00001B/2